METHODS IN MOLECULAR BIOLOGY™

Series Editor
John M. Walker
School of Life Sciences
University of Hertfordshire
Hatfield, Hertfordshire, AL10 9AB, UK

For further volumes:
http://www.springer.com/series/7651

Virus-Host Interactions

Methods and Protocols

Edited by

Susanne M. Bailer

Institute for Interfacial Engineering and Plasma Technology, University of Stuttgart, Stuttgart, Germany

Diana Lieber

Institute of Virology, Ulm University Medical Center, Ulm, Germany

Humana Press

Editors
Susanne M. Bailer
Institute for Interfacial Engineering
and Plasma Technology
University of Stuttgart
Stuttgart, Germany

Diana Lieber
Institute of Virology
Ulm University Medical Center
Ulm, Germany

ISSN 1064-3745 ISSN 1940-6029 (electronic)
ISBN 978-1-4939-6308-9 ISBN 978-1-62703-601-6 (eBook)
DOI 10.1007/978-1-62703-601-6
Springer New York Heidelberg Dordrecht London

Printed on acid-free paper

Humana Press is a brand of Springer
Springer is part of Springer Science+Business Media (www.springer.com)

Preface

Viruses represent a reduced form of life that depends on host cells for propagation. To this end, viruses approach and penetrate cells and usurp cellular machineries for their own benefit. In the course of infection, cellular processes may be attacked, hijacked, or silenced by the invader. Naturally, interactions formed between a virus and its host are very complex. Many viruses exploit the DNA replication machinery for their own genomic amplification. Synthesis of viral proteins takes advantage of transcriptional and translational functions available in the host. Moreover, enveloped viruses remodel cellular membranes to wrap their capsids. Thus, during formation of infectious progeny, all components of a virus including its envelope, structural, and regulatory proteins as well as the genomic material contact host structures.

Usually, a virus aims at maximally exploiting the host functions while maintaining its integrity and vitality. Hosts in turn are equipped with mechanisms for defence involving immunological as well as cell-based means. While the intricate interplay between viral replication and immunological defense has been subject of intensive research for a long time, more recently, virus- and host-encoded small RNAs have emerged as tug-of-war for coexistence of virus and host.

Classical virology in particular of viruses with complex genomes focused on the characterization of individual genes and their function during viral replication with the aim to define their disease-causing potential and to develop antiviral strategies. At the same time, viruses have been used as models to understand cellular processes, e.g., the development of cancer. In addition, viruses allowed insights into basic principles of cell and molecular biology, demonstrated by the identification of protein-embedded signals for nucleocytoplasmic transport. Recent technological improvements have enabled the systematic analysis of the virus–host interplay be it on the genomic, the transcriptomic, or proteomic level. In parallel, bioinformatic tools have emerged in support of the large datasets generated by these high-throughput approaches.

This comprehensive volume consists of 25 chapters and covers various aspects of virological research. Current protocols are provided in great detail where small-scale analyses are equally considered as state-of-the-art systematic approaches. The main intention is to equip researchers with protocols of widely used techniques and to further broaden the repertoire of experienced virologists by adding most recently developed systematic technologies.

Chapters 1–8 on biochemical approaches include molecular interactions and regulatory mechanisms on the protein as well as the RNA level with a strong focus on the manifold possibilities to study protein–protein interactions. As small regulatory RNAs have progressively drawn attention of the field, there are chapters on RNAi screens and functional analyses of miRNAs. Since large-scale analysis has lately become more and more important, bioinformatic methods have been given special consideration in a separate review.

Chapters 9–18 detail cell biological methodologies. These include a number of visualizing and imaging techniques ranging from traditional electron microscopy to advanced high-resolution and live-imaging microscopy, allowing elucidation of protein and infection dynamics. Moreover, tools and methods to monitor characteristic parameters of infectivity and to trace viral genomes or live viruses *in vitro* and *in vivo* such as in 3D tissue or animal models have been included. Finally, this volume concludes with immunological aspects in Chapters 19–25, highlighting viral interactions with immune cells and the usability of immunological assays and mouse genetic models as powerful tools in virology.

Stuttgart, Germany ***Susanne M. Bailer***
Ulm, Germany ***Diana Lieber***

Contents

Contributors

REBECCA J. AICHELER • *Institute of Infection and Immunity, School of Medicine, Cardiff University, Cardiff, UK*
SUSANNE M. BAILER • *Institute for Interfacial Engineering and Plasma Technology IGVP, University of Stuttgart, Stuttgart, Germany*
STEFAN BECKER • *Institute of Virology, Ulm University Medical Center, Ulm, Germany*
SONJA BLASCHE • *Genomics and Proteomics Core Facility, German Cancer Research Center, Heidelberg, Germany*
JENS BERNHARD BOSSE • *Department of Molecular Biology, Princeton University, Princeton, NJ, USA*
HICHAM BOUABE • *Laboratory of Lymphocyte Signalling and Development, Babraham Institute, Cambridge, UK*
MELANIE M. BRINKMANN • *Helmholtz Centre for Infection Research, Braunschweig, Germany*
ANKE BURGER-KENTISCHER • *Molecular Biotechnology, Fraunhofer Institute for Interfacial Engineering and Biotechnology IGB, Stuttgart, Germany*
ANDREW M. BURRAGE • *Department of Microbiology and Immunology, Loyola University Medical Center, Maywood, IL, USA*
ILEANA M. CRISTEA • *Department of Molecular Biology, Princeton University, Princeton, NJ, USA*
LARS DÖLKEN • *Department of Medicine, Addenbrooke's Hospital, University of Cambridge, Cambridge, UK*
LUISE FLORIN • *Department of Medical Microbiology and Hygiene, Johannes Gutenberg University Mainz, Mainz, Germany*
GIADA FRASCAROLI • *Institute of Virology, Ulm University Medical Center, Ulm, Germany*
CAROLINE C. FRIEDEL • *Institute for Informatics, Ludwig-Maximilians-Universität, München, Germany*
SAMANTHA J. GRIFFITHS • *Division of Pathway Medicine, University of Edinburgh, Edinburgh, UK*
WALI HAFEZI • *Institute of Medical Microbiology Clinical Virology, University Hospital Münster, Münster, Germany*
VERENA HOERR • *Institute of Clinical Radiology, University Hospital Münster, Münster, Germany*
INA HOGK • *Institute for Interfacial Engineering and Plasma Technology IGVP, University of Stuttgart, Stuttgart, Germany*
GERHARD JAHN • *Institute of Medical Virology and Epidemiology of Virus Diseases, University Hospital Tuebingen, Tuebingen, Germany*
BENEDIKT B. KAUFER • *Institute of Virology, Freie Universität Berlin, Berlin, Germany*
MANFRED KOEGL • *Boehringer Ingelheim Oncology, Vienna, Austria*
KERSTIN LAIB SAMPAIO • *Institute of Medical Virology and Epidemiology of Virus Diseases, University Hospital Tuebingen, Tuebingen, Germany*
DIANA LIEBER • *Institute of Virology, Ulm University Medical Center, Ulm, Germany*
STEFAN LIENENKLAUS • *Helmholtz Centre for Infection Research, Braunschweig, Germany*

Vladimir Gonçalves Magalhães • *Helmholtz Centre for Infection Research, Braunschweig, Germany*
Ruben Martinez • *Microbiologie Fondamental et Pathogénicité, MFP CNRS UMR 5234, University of Bordeaux SEGALEN, Bordeaux, France*
Tobias May • *Helmholtz Centre for Infection Research, Braunschweig, Germany*
Thomas Mertens • *Institute of Virology, Ulm University Medical Center, Ulm, Germany*
Klaus Okkenhaug • *Laboratory of Lymphocyte Signalling and Development, Babraham Institute, Cambridge, UK*
Ruth Popa-Wagner • *Division of Tumorvirology, German Cancer Research Center, Heidelberg, Germany*
Jessica Ragues • *Microbiologie Fondamental et Pathogénicité, MFP CNRS UMR 5234, University of Bordeaux SEGALEN, Bordeaux, France*
Daniell L. Rowles • *Department of Molecular Biology, Princeton University, Princeton, NJ, USA*
Steffen Rupp • *Molecular Biotechnology, Fraunhofer Institute for Interfacial Engineering and Biotechnology IGB, Stuttgart, Germany*
Martin Schauflinger • *Department of Cell Biology, University of Texas Southwestern Medical Center, Dallas, TX, USA*
Konstanze D. Scheffer • *Department of Medical Microbiology and Hygiene, Johannes Gutenberg University Mainz, Mainz, Germany*
Evgenia Scheibe • *Helmholtz Centre for Infection Research, Braunschweig, Germany*
Tina Schmidt • *Department of Transplant and Infection Immunology, Saarland University, Homburg(Saar), Germany*
Martina Sester • *Department of Transplant and Infection Immunology, Saarland University, Homburg(Saar), Germany*
Christian Sinzger • *Institute of Virology, Ulm University Medical Center, Ulm, Germany*
Richard J. Stanton • *Institute of Infection and Immunity, School of Medicine, Cardiff University, Cardiff, UK*
Hannah Striebinger • *Max von Pettenkofer-Institute, Ludwig-Maximilians-Universität, München, Germany*
Scott S. Terhune • *Department of Microbiology and Molecular Genetics, Biotechnology and Bioengineering Center, Medical College of Wisconsin, Milwaukee, WI, USA*
Stefania Varani • *Unit of Microbiology, Department of Experimental, Diagnostic and Specialty Medicine, University of Bologna, Bologna, Italy*
Clarissa Villinger • *Institute of Virology, Ulm University Medical Center, Ulm, Germany*
Jens von Einem • *Institute of Virology, Ulm University Medical Center, Ulm, Germany*
Paul Walther • *Central Facility for Electron Microscopy, Ulm University, Ulm, Germany*
Kezheng Wang • *Institute of Biology and Medical Sciences, Soochow University, Jiangsu Suzhou, P.R. China*
Shuai Wang • *Institute of Biology and Medical Sciences, Soochow University, Jiangsu Suzhou, P.R. China*
Siegfried Weiss • *Helmholtz Centre for Infection Research, Braunschweig, Germany*
Christopher M. Wiethoff • *Department of Microbiology and Immunology, Loyola University Medical Center, Maywood, IL, USA*
Harald Wodrich • *Microbiologie Fondamental et Pathogénicité, MFP CNRS UMR 5234, University of Bordeaux SEGALEN, Bordeaux, France*
Zeguang Wu • *Institute of Virology, Ulm University Medical Center, Ulm, Germany*
Chunfu Zheng • *Institute of Biology and Medical Sciences, Soochow University, Jiangsu Suzhou, P.R. China*

Chapter 1

A High-Throughput Yeast Two-Hybrid Protocol to Determine Virus-Host Protein Interactions

Hannah Striebinger, Manfred Koegl, and Susanne M. Bailer

Abstract

The yeast two-hybrid (Y2H) system is a powerful method to identify and analyze binary protein interactions. In the field of virology, the Y2H system has significantly increased our knowledge of structure and function of viral proteins by systematically assessing intraviral protein interactions. Several comprehensive approaches to determine virus-host interactions have provided insight into viral strategies to manipulate the host for efficient replication and to escape host-derived countermeasures. To expand our knowledge of intraviral and virus-host protein interactions, we here present a Y2H protocol that is well suited for high-throughput screening. Yeast mating followed by liquid handling in a 96-well format as well as fluorescent readout of the reporter system provides a highly standardized and fully automated screening situation. The protocol can either be applied to screen complex host cDNA libraries or protein pairs arrayed for cross-testing. The ease of use, the cost-effectiveness as well as the robotic handling allows for extensive and multiple rounds of screening providing high coverage of protein-protein interactions. Thus, this protocol represents an improved "deep" screening method for high-throughput Y2H assays.

Key words Yeast two-hybrid, Y2H, High-throughput, Liquid handling, Automation, Yeast mating protocol, Virus-host interactions, Protein-protein interactions, Herpesvirus

1 Introduction

Soon after its development in 1989 [1], the yeast two-hybrid (Y2H) system turned into the most valuable high-throughput method to determine binary protein interactions. In the Gal4-based Y2H system, two proteins to be tested for interaction are fused to either the Gal4 DNA binding domain (bait) or the Gal4 activation domain (prey), and co-expressed in budding yeast cells. In case of interaction of the candidate proteins, the Gal4 transcriptional activator is reconstituted to form a functional entity able to activate one or several reporter genes present in the screening strain.

Over the years, the Gal4-Y2H system has been widely used to determine protein-protein interactions (PPI) in screening situations of varying complexity (for review *see* ref. 2). Initially, individual proteins were tested for interaction by transforming yeast

Susanne M. Bailer and Diana Lieber (eds.), *Virus-Host Interactions: Methods and Protocols*, Methods in Molecular Biology, vol. 1064, DOI 10.1007/978-1-62703-601-6_1, © Springer Science+Business Media, LLC 2013

reporter cells with both one bait and prey plasmid. A more open approach followed the same principle though by exposing a defined bait protein to a whole collection of potential preys which were provided as a cDNA prey library co-expressed with the bait protein. More recently, the yeast mating protocol that takes advantage of the natural ability of haploid yeast cells to mate in order to form diploids provided a much more efficient way to bring bait and prey plasmids together. In contrast to the *yeast transformation protocol* which is difficult to control and unlikely to result in a screening system covering all potential combinations, the *yeast mating protocol* allows for efficient cross-combination of a large number of pre-transformed bait and prey pairs providing an important prerequisite for high-throughput Y2H screening. Depending on the preys available, Y2H screening can be performed against a complex mixture of preys provided as cDNA libraries [3], alternatively, individual bait and prey pairs can be cross-tested for interaction using array-based mating systems.

Naturally—like all assays—the Y2H system is challenged by limitations. The use of the yeast organism as such could potentially disable the interaction of proteins that require particular species-specific modifications. But in general the environment seems to be sufficiently natural for the analysis of other species as has been demonstrated [4]. Similarly, the nuclear reporter system disables modifications provided by membrane compartments and integral membrane proteins need to be trimmed of their membrane anchors to access the nucleus. And finally, transcriptional activators are not suited to be tested in the Y2H system due to the use of a transcriptional reporter system.

Sceptics perceive the Y2H system as being prone to false-positive interactions. Large-scale analysis, however, revealed that the Y2H system is comparable in specificity to other assays detecting binary protein interactions [4–6]. Clearly however, the screening protocol is instrumental in providing sufficient sensitivity: (1) To ensure a "deep" screening situation, all proteins potentially interacting have to be exposed to each other in pairwise combinations. High-quality complex cDNA libraries enhance the screening success [3]. Alternatively, large ORF collections like the Mammalian Gene Collection (http://mgc.nci.nih.gov/) have been generated based on whole genome sequencing and subsequent recombinational cloning of all genes of a particular organism. Collections of this size and complexity can be screened either in batch procedures or by cross-testing of all potential protein interactions using arrays. (2) Classically, bait and prey proteins are tagged at their amino-terminal end with the bait fused to the Gal4 DNA binding domain and the prey fused to the Gal4 transcriptional activator. Fusion of tags to the carboxy-terminal end of a bait or a prey as well as testing of a given protein both as bait and prey fusion, allows for more refined screening [7, 8].

This reveals otherwise unrecognized interactions—*false negatives*—due to misfolding of tagged proteins or steric constraints that a particular tag may impose on the Gal4 reporter system. (3) Technical limitations significantly influence the screening output. Therefore, multiple rounds of screening of the same set of proteins will provide large and more reliable data sets. A Y2H mating protocol that allows for automated and standardized robotic screening is thus of utmost importance in replicate screening of large sets of protein pairs.

Post-screening analysis is required to identify interaction pairs of biological relevance. Prey and bait counts reveal the frequency a given prey or bait has been identified in one or several replicate screen(s). Interactions that were identified repeatedly with one particular bait or prey in screens of varying setups are likely to be specific, while preys isolated only once should be considered as uncertain interactors. Preys that are found very often while screening different baits, are probably "sticky," that is to say they may interact with other proteins in a nonphysiological manner and should be excluded from the evaluation as likely-false positives [3], while frequently identified bait proteins may partially self-activate the reporter system. Next, high-confidence data sets are subjected to computational analysis (*see* Friedel, Chapter 8 of this series). Database searches are performed to generate virus-host protein interactomes which are then connected to other high-throughput screens such as RNAi screens (*see* Griffiths, Chapter 5 of this series). Altogether these data are likely to provide insights into so far unknown viral strategies and to reveal viral or host factors that may serve as targets for antiviral therapy.

Meanwhile several methods have been developed that are suited to validate high confidence PPIs identified by Y2H analysis (for review *see* ref. 2). These include *methods for affinity-isolation* of co-expressed bait and prey proteins, e.g., the luminescence-based mammalian interactome (LUMIER) pull-down assay [9], and *protein fragment complementation assays*, e.g., the bimolecular fluorescence complementation assay (BiFC; *see* Becker and von Einem, Chapter 3 of this series). While most of these methods are of limited use in high-throughput applications either because the method is rather time-consuming, costly or requires extensive optimization on the level of a single protein pair, the LUMIER assay [9] has been adapted for high-throughput analysis (*see* Blasche and Koegl, Chapter 2 of this series). In brief, this assay is based on co-expression of two proteins in mammalian cells: while the bait protein is tagged by the protein A- or Flag-tag and used to immobilize the complex, the co-isolated interacting prey protein is detected via the enzymatic activity of its luciferase fusion partner. Most importantly, like the Y2H system, the LUMIER assay primarily detects binary interactions, provides high confirmation rates and thus represents the best method of choice to validate Y2H interactions.

Whole species Y2H interactomes have been generated for yeast [10, 11], humans [12, 13], *D. melanogaster* [14], *C. elegans* [15], and various pathogens (for review *see* ref. 2, 16; Friedel, Chapter 8 of this series). The analyses of intraviral interactions are of particular importance in complex viruses like Vaccinia virus [17], herpesviruses [7, 18–23], papillomaviruses [24], and SARS coronavirus [25]. Moreover, the identification of virus-host protein interactions is instrumental in understanding viral strategies to manipulate the host for efficient replication. Several very recent insights into the virus-host interplay come from high-throughput Y2H studies involving HCV [21], Influenza virus [26], SARS coronavirus [27], the gammaherpesviruses EBV [22], and MHV68 [23] as well as a collection of tumor viruses [24].

Current efforts combine Y2H screening with affinity-purification of larger protein complexes [24]. Since these two strategies are rather complementary than confirmatory, they are well suited to increase the coverage of interactomes. Individual proteins are fused to ProtA-, FLAG-, or GFP-tag to serve as baits for the isolation of protein complexes the composition of which is subsequently analyzed by mass spectrometry (*see* Rowles et al., Chapter 4 of this series). While functional tagging in the chromosomal context is still not feasible for higher eukaryotic proteins, the BAC technology which is available for numerous virus systems allows for genomic tagging of viral genes and thus their use in affinity purification of virus-host protein complexes. This way, PPIs of a particular viral protein can be investigated at different time-points of viral infection and upon infection of various cell systems. Most importantly, in case of viral genes with relevance for the infection cycle, the functional integrity of the viral fusion product can easily be assessed. A powerful combination of the Y2H system and affinity-isolation of protein complexes has recently been presented for four tumor virus families (the papilloma viruses, the polyoma viruses, the herpesviruses, and the adenoviruses) providing high stringency virus-host interactomes [24].

In order to generate a PPI data set of high coverage and high confidence, a high-throughput Y2H screening system is necessary. Here we present a Gal4-Y2H protocol that is well suited for high-throughput screening of intraviral as well as virus-host protein interactions. This method not only allows for simple and cost-effective screening. In particular, it can be fully automated and performed using a liquid handling robot and thus enables screening of extensive bait and prey combinations as well as multiple rounds of library screenings, thereby overcoming major drawbacks of the Y2H system. Thus, this protocol represents an improved "deep" screening method for high-throughput robotic Y2H assays.

2 Materials

If not indicated otherwise, all media and stock solutions are prepared using deionized water, sterilized by sterile filtration (pore size 0.2 µm), and stored at 4 °C. To avoid bacterial contamination, all media are supplemented with penicillin/streptomycin prior to use. Yeast cells are cultivated in a shaking incubator at 30 °C and 200 rpm unless otherwise stated.

2.1 Yeast Strains and Vectors

1. *Saccharomyces cerevisiae* strains Y187 (MATα, *ura3-52, his3-200, ade2-101, trp1-901, leu2-3, 112, gal4Δ, met-, gal80Δ, MEL1, URA3::GAL1*$_{UAS}$-*GAL1*$_{TATA}$-*lacZ*) and AH109 (MATa, *trp1-901, leu2-3, 112, ura3-52, his3-200, gal4Δ, gal80Δ, LYS2::GAL1*$_{UAS}$-*GAL1*$_{TATA}$-*HIS3, GAL2UAS-GAL2TATA-ADE2, URA3::MEL1*$_{UAS}$-*MEL1*$_{TATA}$-*lacZ, MEL1*) of opposite mating type are used throughout all yeast two-hybrid analyses.
2. Y2H vectors pGADT7 (prey) and pGBKT7 (bait) containing the genes of interest are applied.
3. *S. cerevisiae* strain Y187 is transformed with the prey plasmid pGADT7, while *S. cerevisiae* strain AH109 is holding the bait vector pGBKT7 (*see* **Note 1**).
4. For library-based Y2H screening pretransformed cDNA libraries can be purchased, usually consisting of a tissue-specific cDNA library cloned into prey vector pACT2 or pGADT7-Rec and transformed into the yeast strain Y187.

2.2 Media and Stock Solutions

1. Prepare 10× concentrated stock solution of SD medium (a synthetic minimal medium) according to the manufacturer's protocol (*see* **Note 2**).
2. Prepare 10× dropout supplements according to the manufacturer's protocol. Dropout supplements (DO) needed are DO-Leucine (-L), DO-Tryptophane (-W), DO-Leucine/Tryptophane (-L/W), and DO-Leucine/Tryptophane/Histidine (-L/W/H).
3. Purchase or prepare 100× concentrated sterile penicillin/streptomycin (Pen/Strep) solution.
4. SD-L: Dilute 100 mL 10× SD medium, 100 mL 10× DO-L, and 5 mL 100× Pen/Strep in 795 mL autoclaved water under aseptic conditions.
5. SD-W: Dilute 100 mL 10× SD medium, 100 mL 10× DO-W, and 5 mL 100× Pen/Strep in 795 mL autoclaved water under aseptic conditions.
6. SD-L/W: Dilute 100 mL 10× SD medium, 100 mL 10× DO-L/W, and 5 mL 100× Pen/Strep in 795 mL autoclaved water under aseptic conditions.

7. SD-L/W/H: Dilute 100 mL 10× SD medium, 100 mL 10× DO-L/W/H, and 5 mL 100× Pen/Strep in 795 mL autoclaved water under aseptic conditions.
8. 2× YPD medium: Weigh 40 g peptone, 20 g yeast extract and 40 g glucose into a beaker and add water to a volume of 1 L (*see* **Note 3**).
9. Prepare 100× Adenine solution by weighing 0.2 g Adenine into 100 mL water.
10. YPDA-medium: Dilute 500 mL 2× YPD medium, 5 mL 100× Adenine solution, and 5 mL 100× Pen/Strep in 490 mL autoclaved water under aseptic conditions.
11. Prepare a 0.15 M 4-Methylumbelliferyl-α-D-galactopyranoside (4-MUx) solution with sterile DMSO (dimethyl sulfoxide) and store at −20 °C.
12. Prepare a 1 M solution of 3-Amino-1,2,4-triazole (3-AT) and store at 4 °C.
13. Agar plates: Prepare a 2× Agar solution in 500 mL bottles by adding 15 g Agar to 250 mL deionized water and autoclave. Prepare 250 mL of 2× medium from the respective SD medium stock with sterile water. Bring both solutions to a temperature of 55 °C, mix, add 5 mL 100× Pen/Strep solution and spread onto Petri dishes under aseptic conditions.
14. YPD medium (20 % PEG): Weigh 10 g peptone, 5 g yeast extract, 10 g glucose, and 20 g Polyethylenglycol (PEG 6000) into a beaker and add water to a volume of 500 mL.
15. 0.25 % SDS aqueous solution.

2.3 Equipment and Consumables

1. Shaking and non-shaking incubator at 30 °C.
2. For high-throughput analyses the use of a robotic liquid-handling platform is recommended. If neither is at hand, it is also possible to complete the Y2H screening with a manually supported 96-well pipetting system (2–200 μL), e.g., the Liquidator96, Mettler Toledo.
3. A fluorescence microplate reader (excitation spectrum 365 nm, emission spectrum 448 nm) preferably equipped with an automatic stacker.
4. 96-well microtiter plates (MTPs) with both F- and U-bottom shape.
5. A centrifuge for microplates.
6. Polystyrene tubes of 50 mL volume.
7. Appropriate sterile pipet tips and reservoirs.
8. Self-adhesive foil for MTPs.
9. Laboratory plastic foil.

10. Sterile Petri dishes.
11. 96-well PCR plates.
12. Taq DNA polymerase kit of your choice, preferably equipped with loading dye.
13. PCR cycler.
14. PCR DNA primer to amplify prey vector inserts and accordingly a sequencing primer.
15. 0.8 % agarose gel.

3 Methods

3.1 Array-Based Yeast Two-Hybrid Screening

This procedure allows cross-testing of single proteins of interest with a defined set of possible interaction partners that are arranged as array (Fig. 1). Hence, any potential interaction pair is tested individually. The protocol displayed here is performed in 96-well MTPs and liquid medium. For convenience, the assay is described using a single bait protein to test against a prey library, although naturally it can also be carried out in the opposite setting.

1. Inoculate SD-W medium with a single yeast colony transformed with a bait vector (e.g., strain AH109) and incubate in a 50 mL tube over night at 200 rpm and 30 °C (*see* **Note 4**). To control for auto-activation of reporter gene activity potentially induced by the prey collection, also prepare a culture with yeast cells carrying an empty bait vector.
2. In an MTP (F-bottom), inoculate 150 μL SD-L medium with your choice of preys transformed into a yeast strain of opposing mating type (e.g., Y187) arranged in the preferred matrix, cover plate with adhesive foil, and incubate over night at 30 °C (*see* **Note 5**). It is strongly recommended to occupy three wells with controls: (1) blank control containing only medium but no yeast, (2) negative control containing yeast transformed with prey vector lacking an insert, and (3) as positive control a bait-prey pair proven to give a positive result (*see* **Note 6**).
3. On the next day the yeast mating is performed in an MTP equipped with U-bottom. For this purpose, 25 μL of each bait and prey culture is added to 100 μL YPDA and the MTP is gently centrifuged (1 min, 180 × *g*) to collect the yeast cells at the bottom of the wells. Incubate over night.
4. To select for diploid yeasts, transfer 15 μL of mating plate to a fresh MTP (F-bottom) with 150 μL SD-L/W medium and incubate for 2 days. Add the positive control to the designated well at this step.
5. To monitor the reporter gene activity, the diploid yeasts are then divided onto five different MTPs (F-bottom). 10 μL each

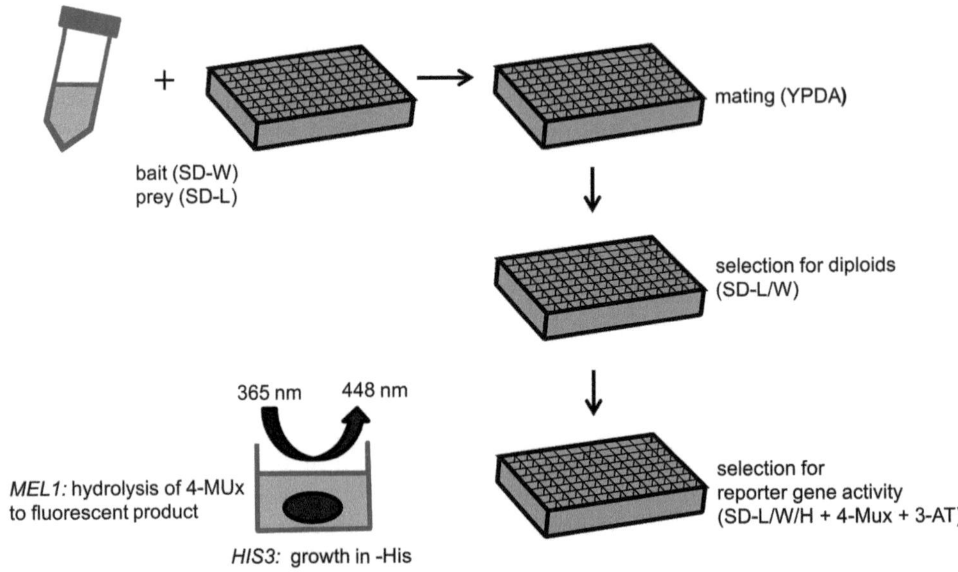

Fig. 1 The array-based yeast two-hybrid (Y2H) screening allows the cross-testing of a single protein against a defined set of possible interaction partners. The haploid yeast strain carrying the bait vector is propagated under selective conditions in synthetic minimal medium lacking Tryptophane (SD-W), while yeast cells of opposing mating type transformed with the various prey vectors are grown arrayed in 96-well microtiter plates (MTP) in SD medium lacking Leucine (SD-L). The combination of the bait yeast strain with the multiple prey strains in complete medium (YPDA) over night leads to mating, resulting in diploid cells carrying both bait and prey vectors, which are selected by growth in SD-L/W medium. The reporter gene activity is examined in SD medium lacking additionally Histidine (SD-L/W/H) and supplemented with 4-Methylumbelliferyl-α-D-galactopyranoside (4-MUx) and 3-Amino-1,2,4-triazole (3-AT) in a fluorescence plate reader (excitation 365 nm, emission 448 nm). The interaction of the bait protein fused to Gal4 DNA binding domain with the prey protein fused to Gal4 activation domain leads to the reconstitution of Gal4 transcription factor and subsequently expression of *HIS3* and *MEL1* reporter genes. Expression of the reporter genes allows growth in a Histidine-free environment as well as the hydrolysis of 4-MUx to fluorescent methylumbelliferone, whose fluorescent activity is a measure of protein-protein interaction. Usually 3-AT is added in increasing concentration as competitive inhibitor of the leaky expression of the *HIS3* reporter gene

are transferred to 150 μL SD-L/W/H medium containing 50 μM 4-Methylumbelliferyl-α-D-galactopyranoside (4-MUx) and increasing concentrations of 3-Amino-1,2,4-triazole (3-AT) (typically 0, 1, 2.5, and 5 mM) as competitive inhibitor of the leaky expression of the *HIS3* reporter gene. Also, 10 μL are used to inoculate 150 μL SD-L/W medium for growth control. All MTPs are incubated for 3 days.

6. The MTP with SD-L/W is controlled by eye for yeast growth. If there is no detection of growth in some wells, the corresponding wells of the SD-L/W/H MTPs cannot be analyzed.
7. The SD-L/W/H MTPs are then analyzed for reporter gene activity in a fluorescence plate reader (excitation 365 nm,

emission 448 nm). While the gene product of *HIS3* permits growth in Histidine-free medium, the α-Galactosidase encoded by *MEL1* catalyzes hydrolysis of 4-MuX to fluorescent methylumbelliferone.

8. The resulting data is analyzed by comparing the relative fluorescence intensity of each bait/prey-pair to the negative controls (empty vectors). For this, the relative fluorescence units (RFU) for all 3-AT concentrations of each protein pair are plotted on a bar diagram and compared to the values produced by expression of each bait or prey with the respective empty vector (*see* **Note 7**). The values of the positive control help to ease the interpretation of results. Cutoff values for positive interactions are usually defined individually for each bait, as RFU values tend to take a broad range in between the individual tested proteins.
9. Inclusion of several fragments of one possible prey to your matrix as well as repeating the procedure with the opposite vector combination (test your protein of interest as prey against a bait matrix) will help to raise the confidence in your resulting data.

3.2 Library-Based Yeast Two-Hybrid Screening

This protocol describes the yeast two-hybrid screening of individual bait proteins against a prey library consisting of cDNA usually derived from a certain tissue type cloned into a prey vector and pretransformed in yeasts (Fig. 2). This allows testing of single proteins of interest for interaction with a broad spectrum of possible cellular interaction partners. The library-based Y2H screening is performed in 96-well MTPs and liquid medium in analogy to the previously described matrix-based assay (*see* Subheading 3.1).

1. As a first step, the optimal 3-AT concentration for each individual bait should be identified in a prescreen.
2. Inoculate 5 mL SD-W medium with a single yeast colony transformed with a bait vector (e.g., strain AH109) and incubate in a 50 mL tube over night at 200 rpm and 30 °C.
3. Measure OD_{600} of the overnight culture and of the prey library (e.g., strain Y187) you want to apply. Inoculate 15 mL SD-W medium to an OD_{600} of 0.1 using the overnight bait culture and accordingly inoculate 15 mL SD-L with the prey library (*see* **Note 8**). Incubate at 30 °C and 160 rpm until OD_{600} of 0.9–1.0 is reached, which is usually about 16 h later.
4. To monitor the mating efficiency, prepare 10^{-1} to 10^{-4} dilutions of both cultures with sterile water and streak out on agar plates with the respective medium for viability count. Incubate at 30 °C for 3 days. Count the colonies on the plate with the appropriate dilution and calculate the colony forming units (cfu) per mL of culture.

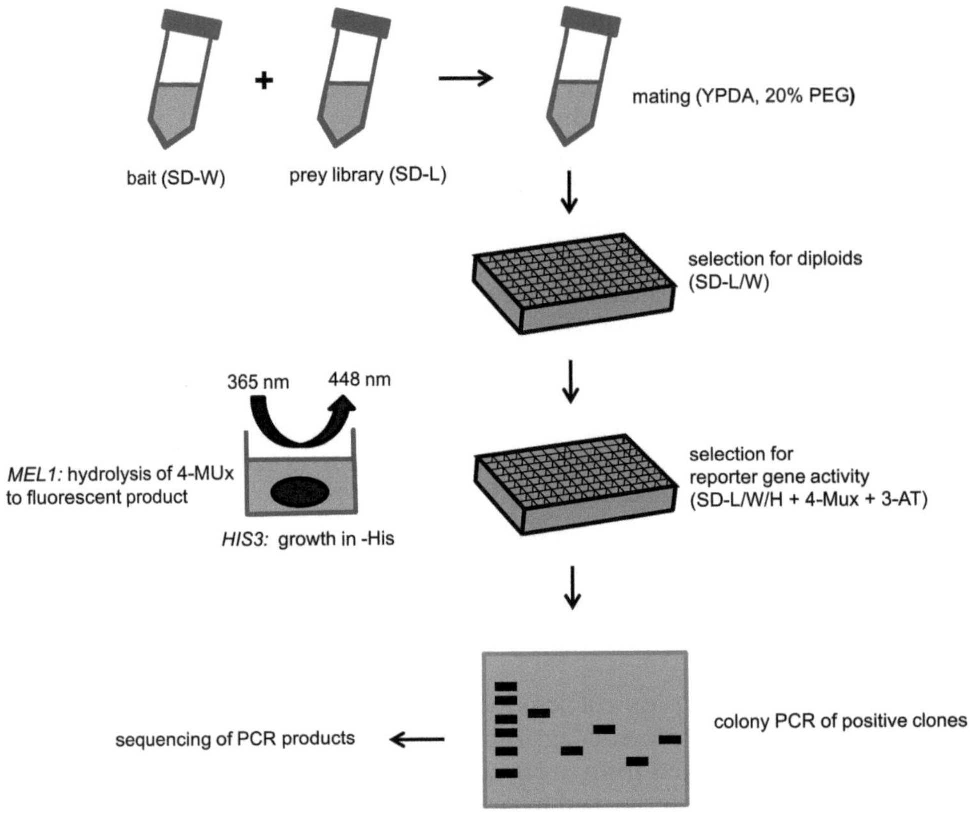

Fig. 2 The library-based yeast two-hybrid (Y2H) screening is applied to test an individual bait protein for interaction against a complex prey library, usually consisting of cDNA of a certain tissue type and pretransformed in yeasts. Haploid yeast strains of opposing mating type carrying either bait or prey vectors are propagated under selective conditions in synthetic minimal medium lacking Tryptophane (SD-W) or Leucine (SD-L), respectively. The mating process to gain diploid cells containing both bait and prey vector is performed in complete medium (YPDA + 20 % Polyethyenglycol) under gentle shaking. Afterwards, the yeast culture is spread on 96-well microtiter plates (MTP) and incubated under selective conditions (SD-L/W). The reporter gene activity is then examined in SD medium lacking additionally Histidine (SD-L/W/H) and supplemented with 4-Methylumbelliferyl-α-D-galactopyranoside (4-MUx) and 3-Amino-1,2,4-triazole (3-AT) in a fluorescence plate reader (excitation 365 nm, emission 448 nm). The interaction of the bait protein fused to the Gal4 DNA binding domain with a prey protein fused to the Gal4 activation domain leads to reconstitution of the Gal4 transcription factor and subsequently expression of *HIS3* and *MEL1* reporter genes. The reporter genes allow growth in a Histidine-free environment as well as the hydrolysis of 4-MUx to fluorescent Methylumbelliferone, whose fluorescent activity is a measure of protein-protein interaction. The optimal 3-AT concentration as a competitive inhibitor to leaky reporter gene *HIS3* expression was defined in a previously performed prescreen. To identify the prey genes that resulted in a positive outcome, a colony PCR of the respective diploid cells is carried out. The agarose gel electrophoresis reveals those hits, where only one prey gene is present. These PCR products then undergo sequencing analysis and their gene ID is identified using an alignment tool like NCBI's BLAST (http://blast.ncbi.nlm.nih.gov/)

5. The mating process is performed in 50 mL tubes. A similar number of yeast cells of both bait and prey culture should be applied. Calculate the culture volume needed to represent an OD_{600} of 12 in 1 mL ($12/OD_{600}$), mix both volumes in a tube and centrifuge for 2 min at 860 × *g*. Discard supernatant and resuspend the cell pellet in 25 mL YPD medium (20 % PEG). Incubate at 30 °C with gentle agitation (100 rpm) for exactly 3 h (*see* **Note 9**).
6. After mating, centrifuge cells at 860 × *g* for 3 min. Resuspend the cell pellet in 19 mL SD-L/W and strictly avoid vortexing (*see* **Note 10**). Prepare 10^{-1} to 10^{-4} dilutions for viability count and streak out on SD-L/W agar plates. Incubate for 3 days, then count the colonies to calculate the colony forming units (cfu) per mL.
7. Centrifuge cells again and resuspend the cell pellet in 1 mL SD-L/W/H.
8. Prepare SD-L/W/H medium containing 50 μM 4-MuX sufficient for seven MTPs with 200 μL medium per well (134.4 mL) plus the dead volume needed for the reservoir. Add the mated yeasts.
9. Now the mated yeasts are divided onto seven MTPs (F-bottom) with increasing 3-AT concentration (0, 0.5, 1, 2.5, 5, 10, 50 mM). First, transfer 200 μL each to an MTP without any 3-AT (0 mM plate). Then add the appropriate volume of 1 M 3-AT to the remaining yeast culture to obtain a 0.5 mM concentration in the remaining volume and fill the next MTP. Repeat this pipetting scheme until all seven plates contain cells in SD-L/W/H medium with 50 μM 4-MuX and the respective 3-AT concentration (*see* **Note 11**).
10. Stack the MTPs, cover the topmost with self-adhesive foil and wrap the stack in plastic foil. Incubate for 6 days at 30 °C.
11. To monitor the mating efficiency, divide the cfu/mL of diploids by cfu/mL of the limiting partner (the strain which yielded the fewest viable cells). Multiply with 100 to obtain the mating efficiency in percent. The mating efficiency should lie between 2 and 5 %.
12. The MTPs are analyzed after 6 days of incubation for reporter gene activity in a fluorescence plate reader (excitation 365 nm, emission 448 nm).
13. Plot the resulting RFU values in a line chart so that each plate is represented by a line. Choose the 3-AT concentration that results in a reasonable number of positive results to perform the Y2H screening (*see* **Note 12**).
14. To perform the actual Y2H screen repeat **steps 1–7**.
15. Prepare SD-L/W/H medium containing 50 μM 4-MuX and the optimal 3-AT concentration for the individual bait evaluated

by prescreen sufficient for ten MTPs with 200 μL medium per well (192 mL) plus the dead volume needed for the reservoir. Add the mated yeasts. Transfer 200 μL of mated yeasts to each well of the ten MTPs.

16. Repeat **steps 10–12**.
17. Plot the resulting RFU values in a line chart so that each plate is represented by a line. Define a reasonable cutoff value and identify the position of positive hits.
18. Unite the positive colonies on one or several MTPs by transferring 10 μL of each well with positive results to a fresh MTP containing 150 μL SD-L/W/H medium (hitpicking plate). Incubate for 3 days.
19. The identity of positive preys is analyzed by colony PCR of positive hits and subsequent sequencing of PCR products. Prior to performing the colony PCR, 5 μL of each well of the hitpicking plate is transferred to 20 μL of 0.25 % SDS solution in 96-well PCR plates and boiled for 5 min at 95 °C in a thermal cycler in order to break down the yeast cell walls. Spin down shortly (180 × *g*, 1 min).
20. Now a colony PCR is performed to amplify the cDNA inserts of positive prey vectors with DNA primers specific for prey vector sequences upstream and downstream of an insert sequence. Apply 5 μL of boiled yeast suspension in SDS solution as template. The primer sequence naturally depends on the prey vector applied. Use any Taq Polymerase Kit, preferably equipped with a reaction buffer already containing gel loading dye. We performed a nested PCR with two pairs of short primers partially overlapping to amplify the insert sequence from a pACT2 prey vector (*see* **Note 13**).
21. Load an aliquot of PCR reaction onto an 0.8 % agarose gel (e.g., 20 μL from a total volume of 50 μL). Those PCR reactions that exhibit a single DNA band in the gel are then sent for DNA sequencing to identify the respective cDNA insert (*see* **Note 14**).
22. Perform a blast search to determine the identity of the positive and sequenced prey cDNAs (http://blast.ncbi.nlm.nih.gov/).
23. To evaluate your results choose the following approach. High-confidence interaction partners are those preys, which lead to specific and reproducible results and encode for naturally occurring proteins. This means that preys should be found more than once with your specific bait. Preys that are found very often while screening different baits, are probably sticky preys and should be excluded from evaluation as likely-false positives, as well as noncoding 3′-UTR (untranslated regions) sequences as true false positives. Preys that were isolated only once should be considered as uncertain interactors.

4 Notes

1. Naturally, it is also possible to employ yeast strain Y187 as bait and AH109 as prey instead of the combination proposed by us. But we find it more convenient to adapt our protocols to the pretransformed libraries available on the market and to keep this constellation consistent throughout all analyses.
2. The minimal SD medium can be assembled from product systems by different supply companies, which may vary in their respective composition. Make sure your SD medium contains yeast nitrogen base (YNB) without amino acids but with ammonium sulfate. If glucose as a source for carbohydrates is not included by the manufacturer, it has to be added to a final concentration of 2 % (w/v).
3. Alternatively to sterile filtration, YPD medium can also be autoclaved. However, exposing the medium to boiling temperatures will result in caramelizing of the contained glucose. Although we find that this has no negative impact on growth rates of *S. cerevisiae*, some might prefer to add the glucose as a sterile solution after the autoclaving process.
4. Consider the dead volume needed to provide sufficient capacity when pipetting from a reservoir. Add dead volume to the 2.5 mL of yeast culture actually required to perform the assay (e.g., inoculate at least 30 mL for usage of a manual 96-well pipet).
5. Always prepare a glycerol stock of your prey matrix and store at −80 °C to use as inoculum for future analyses. For this purpose, it is also possible to keep the MTP with the prey library at 4 °C for a short period of time (e.g., 2 weeks).
6. Evidently, the positive control should be added only after the mating process on MTP. We found it convenient to keep the positive control in the form of diploid yeasts as glycerol stock at −80 °C and on SD-L/W agar plates stored in the fridge for a short time.
7. It is crucial to include the screening of the empty bait versus the prey collection in every run that is accomplished. Although the auto-activating preys will produce similar results every time, RFU intensities may vary. Slightly auto-activating bait proteins are controlled with the empty prey, which allows to evaluate only those results above a certain 3-AT concentration.
8. It is recommended to measure the OD_{600} of the prey library once and produce aliquots that result in an OD_{600} of 0.1 when 15 mL SD-L medium are inoculated to avoid repeated freezing and thawing.
9. Yeast cells require about 3 h for doubling. If incubation of mated diploids proceeds for more than 3 h, the diploids divide

resulting in overrepresentation of expanded clones obscuring the prey count.

10. Vortexing could dissociate the mating yeast cells as well as early diploids.
11. Keep in mind that the volume is reduced by 19.2 mL after every pipetting step and consider the 3-AT concentration already included for the previous step to calculate the respective volumes of 1 M 3-AT that have to be added after every step.
12. Typically, a low 3-AT concentration produces a lot of high RFU values, which will abate with increasing 3-AT concentration. It needs some experience to choose the optimal concentration. As a rule of thumb, pick a concentration that results in less than ten hits/plate.
13. To amplify cDNA inserts from a pACT2 library, the primer pair used for the first PCR is 5′-cta gag gga tgt tta ata cca cta caa tgg-3′ and 5′-ggt tac atg gcc aag att gaa act tag agg-3′ in a total volume of 25 μL for each PCR reaction. You may need to add 2 % Triton X-100 as detergent. The primer pair applied for the second PCR is 5′-tgt tta ata cca cta caa tgg atg atg-3′ and 5′-cat aaa aga agg caa aac gat g-3′ in a total volume of 50 μL for each PCR reaction. Use 1 μL of the first PCR as template for the second PCR. Ensure that the elongation time is sufficient for insert sizes up to 6 kb or the maximum insert size included in the cDNA prey library according to the manufacturer's protocol. Apply 26 cycles for both PCR reactions.
14. For DNA sequencing of pACT2 cDNA inserts, an aliquot of 20 μL from a total volume of 50 μL PCR reaction was analyzed by applying the pACT2 forward primer (5′-gat gat gaa gat acc cca c-3′).

References

1. Fields S, Song O (1989) A novel genetic system to detect protein-protein interactions. Nature 340(6230):245–246. doi:10.1038/340245a0
2. Bailer S, Haas J (2009) Connecting viral with cellular interactomes. Curr Opin Microbiol 12(4):453
3. Mohr K, Koegl M (2012) High-throughput yeast two-hybrid screening of complex cDNA libraries. Methods Mol Biol (Clifton, NJ) 812:89–102
4. Braun P, Tasan M, Dreze M et al (2009) An experimentally derived confidence score for binary protein-protein interactions. Nat Methods 6(1):91–97. doi:10.1038/nmeth.1281
5. Yu H, Braun P, Yildirim MA et al (2008) High-quality binary protein interaction map of the yeast interactome network. Science 322(5898):104–110. doi:10.1126/science.1158684
6. Venkatesan K, Rual JF, Vazquez A et al (2009) An empirical framework for binary interactome mapping. Nat Methods 6(1):83–90. doi:10.1038/nmeth.1280
7. Stellberger T, Häuser R, Baiker A et al (2010) Improving the yeast two-hybrid system with permutated fusions proteins: the Varicella Zoster Virus interactome. Proteome Sci 8:8
8. Rajagopala SV, Uetz P (2011) Analysis of protein-protein interactions using high-throughput yeast two-hybrid screens. Methods Mol Biol (Clifton, NJ) 781:1–29

9. Al-Khoury R, Coulombe B (2009) Defining protein interactions that regulate disease progression. Expert Opin Ther Targerts 13: 13–17
10. Uetz P, Giot L, Cagney G et al (2000) A comprehensive analysis of protein-protein interactions in Saccharomyces cerevisiae. Nature 403(6770):623–627. doi:10.1038/35001009
11. Ito T, Chiba T, Ozawa R et al (2001) A comprehensive two-hybrid analysis to explore the yeast protein interactome. Proc Natl Acad Sci USA 98(8):4569–4574. doi:10.1073/pnas.061034498
12. Stelzl U, Worm U, Lalowski M et al (2005) A human protein-protein interaction network: a resource for annotating the proteome. Cell 122(6):957–968. doi:10.1016/j.cell.2005.08.029
13. Rual JF, Venkatesan K, Hao T et al (2005) Towards a proteome-scale map of the human protein-protein interaction network. Nature 437(7062):1173–1178. doi:10.1038/nature04209
14. Giot L, Bader JS, Brouwer C et al (2003) A protein interaction map of Drosophila melanogaster. Science 302(5651):1727–1736. doi:10.1126/science.1090289
15. Li S, Armstrong CM, Bertin N et al (2004) A map of the interactome network of the metazoan C. elegans. Science 303(5657):540–543. doi:10.1126/science.1091403
16. Mendez-Rios J, Uetz P (2010) Global approaches to study protein-protein interactions among viruses and hosts. Future Microbiol 5(2):289–301
17. McCraith S, Holtzman T, Moss B et al (2000) Genome-wide analysis of vaccinia virus protein-protein interactions. Proc Natl Acad Sci 97(9):4879–4884
18. Uetz P, Dong YA, Zeretzke C et al (2006) Herpesviral protein networks and their interaction with the human proteome. Science 311(5758): 239–242
19. Rozen R, Sathish N, Li Y et al (2008) Virion-wide protein interactions of Kaposi's sarcoma-associated herpesvirus. J Virol 82(10): 4742–4750
20. Fossum E, Friedel CC, Rajagopala SV et al (2009) Evolutionarily conserved herpesviral protein interaction networks. PLoS Pathog 5(9):e1000570
21. De Chassey B, Navratil V, Tafforeau L et al (2008) Hepatitis C virus infection protein network. Mol Syst Biol 4:230
22. Calderwood MA, Venkatesan K, Xing L et al (2007) Epstein–Barr virus and virus human protein interaction maps. Proc Natl Acad Sci 104(18):7606–7611
23. Lee S, Salwinski L, Zhang C et al (2011) An integrated approach to elucidate the intra-viral and viral-cellular protein interaction networks of a gamma-herpesvirus. PLoS Pathog 7(10):e1002297. doi:10.1371/journal.ppat.1002297
24. Rozenblatt-Rosen O, Deo RC, Padi M et al (2012) Interpreting cancer genomes using systematic host network perturbations by tumour virus proteins. Nature 487(7408):491–495. doi:10.1038/nature11288
25. Von Brunn A, Teepe C, Simpson JC et al (2007) Analysis of intraviral protein-protein interactions of the SARS coronavirus ORFeome. PLoS One 2(5):e459
26. Shapira SD, Gat-Viks I, Shum BO et al (2009) A physical and regulatory map of host-influenza interactions reveals pathways in H1N1 infection. Cell 139(7):1255–1267. doi:10.1016/j.cell.2009.12.018
27. Pfefferle S, Schöpf J, Kögl M et al (2011) The SARS-coronavirus-host interactome: identification of cyclophilins as target for pan-coronavirus inhibitors. PLoS Pathog 7(10):e1002331

Chapter 2

Analysis of Protein–Protein Interactions Using LUMIER Assays

Sonja Blasche and Manfred Koegl

Abstract

Co-affinity purification methods can test whether two proteins physically engage in a complex. The assay principle is to enrich cellular extracts for a first protein by a purification step, and then test if a second protein is enriched as well. This principle has been optimized for use at high-throughput in LUMIER (luminescence-based mammalian interactome mapping) assays, which use luciferase-tags for the detection of the second proteins. This protocol is ideal for miniaturization of the assay on microtiter plates and supports a throughput of several hundred interaction tests per week.

Key words Protein–protein interactions, Luciferase, Protein networks, High-throughput screens, Interactome

1 Introduction

Most proteins fulfil their functions as part of a multiprotein complex. In the investigation of virus–host interactions, an analysis of which host proteins are physically contacted by viral proteins paves the way to an understanding of host cell takeover by the virus. While the isolation of protein–protein complexes from natural, unperturbed sources by immunoprecipitation using specific antibodies may constitute the least artificial experimental setting to test for the interaction of proteins, it suffers from several limitations. Not the least of limitations, the lack of appropriate antibodies often demands the exogenous expression of epitope-tagged proteins. In such a setting, tagging of the second protein of an interacting pair with an easily detectable tag offers to significantly simplify the interaction test. This idea has been put to practice in LUMIER assays [1] using proteins fused to luciferase for detection of the co-purified protein. The exquisite sensitivity and high dynamic detection range of luciferases make these enzymes ideal for this purpose.

The workflow of the LUMIER method is depicted in Fig. 1a. In brief, two proteins ("X" and "Y" throughout this paper) are

Susanne M. Bailer and Diana Lieber (eds.), *Virus-Host Interactions: Methods and Protocols*, Methods in Molecular Biology, vol. 1064, DOI 10.1007/978-1-62703-601-6_2, © Springer Science+Business Media, LLC 2013

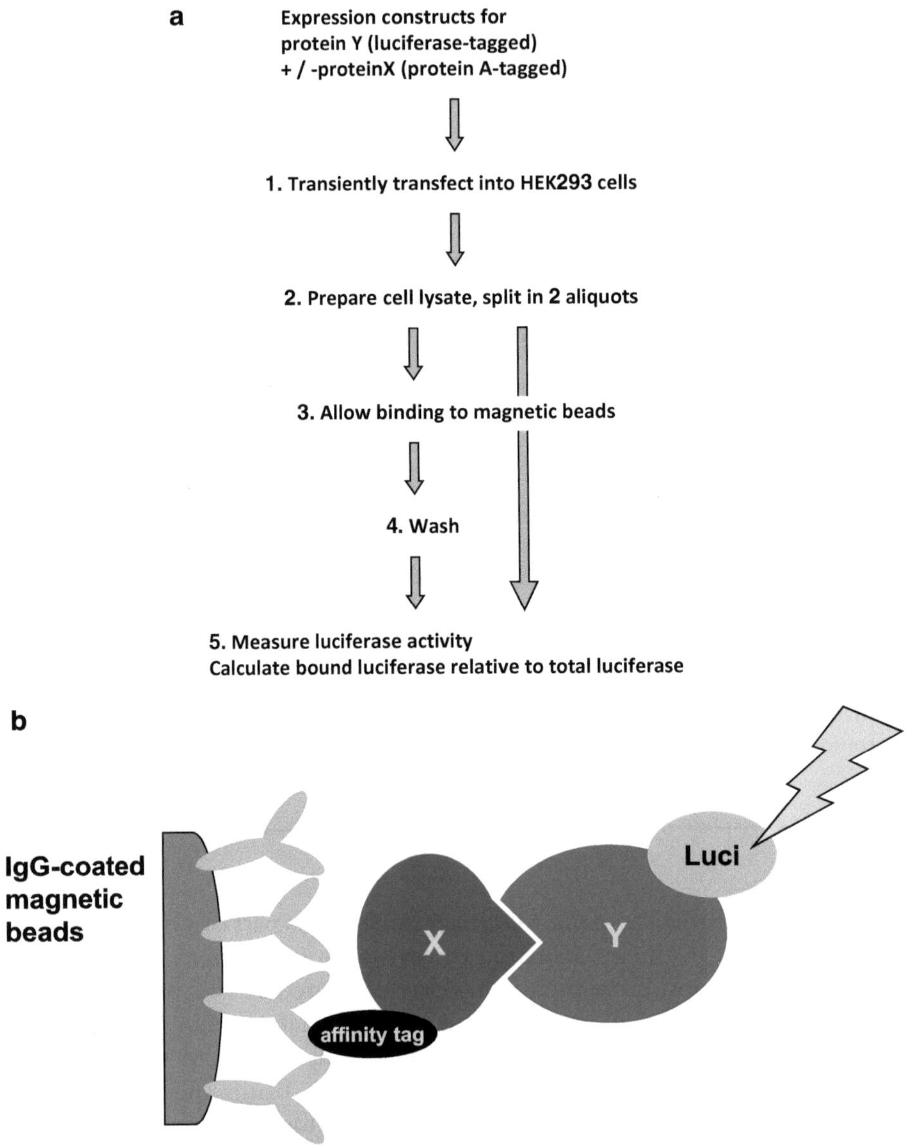

Fig. 1 (**a**) Schematic of the protocol. (**b**) Assay principle

co-expressed by transient transfection in tissue culture cells. X is extended with a tag for affinity purification, Y is tagged with luciferase. After cell lysis, extracts are enriched for X via affinity purification. The presence of luciferase in the enriched extract is compared to a negative control to infer interaction of X and Y (Fig. 1b).

Various negative controls lend themselves to use here, such as the application of non-affinity matrices for purification, or competition of the affinity binding reaction with high amounts of tag without protein fusion partner (e.g., FLAG peptide for FLAG-tagged proteins). In our experience, the most consistent negative control is the expression of luciferase-tagged protein Y with only the affinity tag as a partner. This tests for interactions of Y with the tag or the affinity matrix, which in both cases would lead to elevated luciferase signals in the negative control.

To conclude that two proteins interact, LUMIER data have to be normalized for transfection and expression efficiency, and compared to the noise of the assay, i.e., the signal strength generated by noninteracting pairs of proteins. Normalization is necessary, since the signal generated by the co-purified luciferase is proportional to the amount of luciferase-tagged protein present in the lysate, even for noninteracting proteins (Fig. 2a). If the assay is done at a high-throughput level, a random set of noninteracting proteins can be used to measure the average and standard deviation of the background to transform the interaction signals into z-scores.

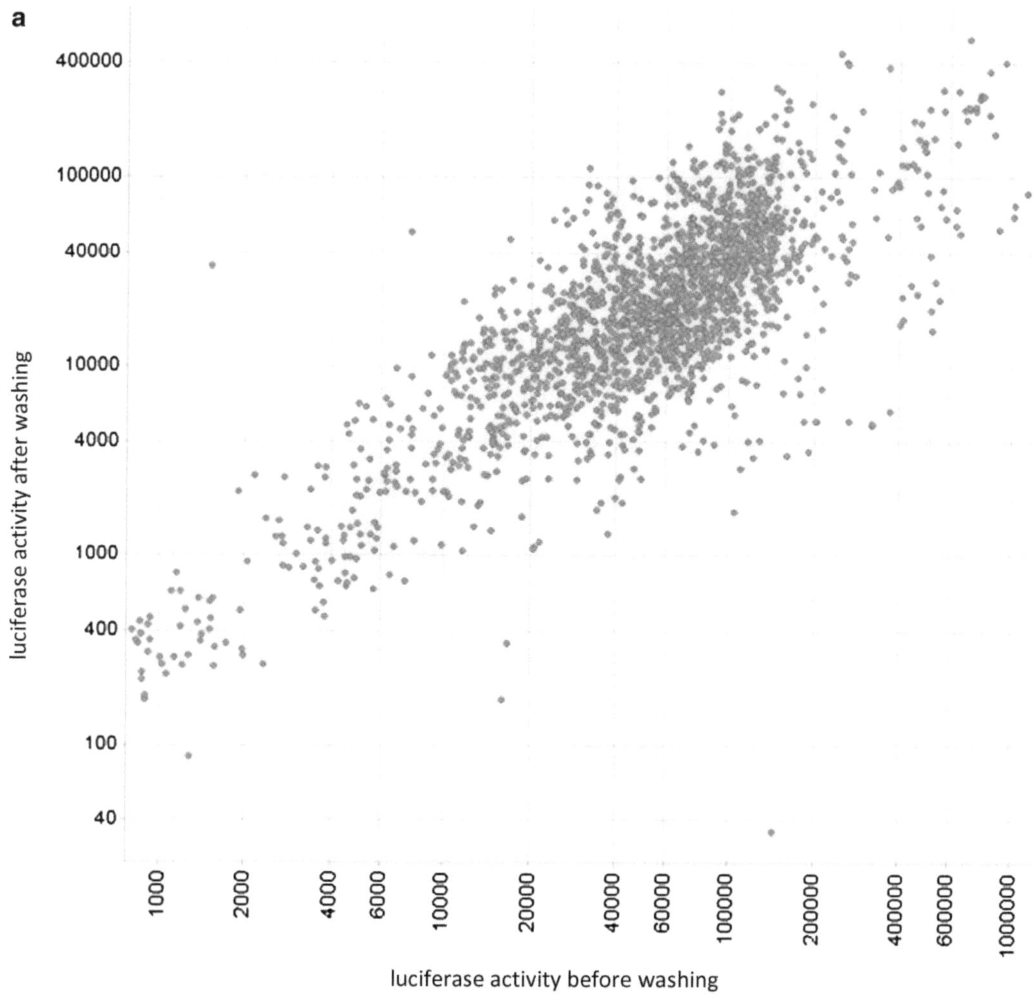

Fig. 2 (**a**) Correlation of the expression level of the luciferase-tagged protein with the raw interaction signal and (**b**) with the normalized signal to background ratio. (**c**) Correlation of z-scores with signal to background ratios. Note that the y-axis in (**c**) is in logarithmic scale. In (**a**) and (**b**), all data are from the interaction of JUN with FOS. *Light gray dots* in (**c**) are positive controls

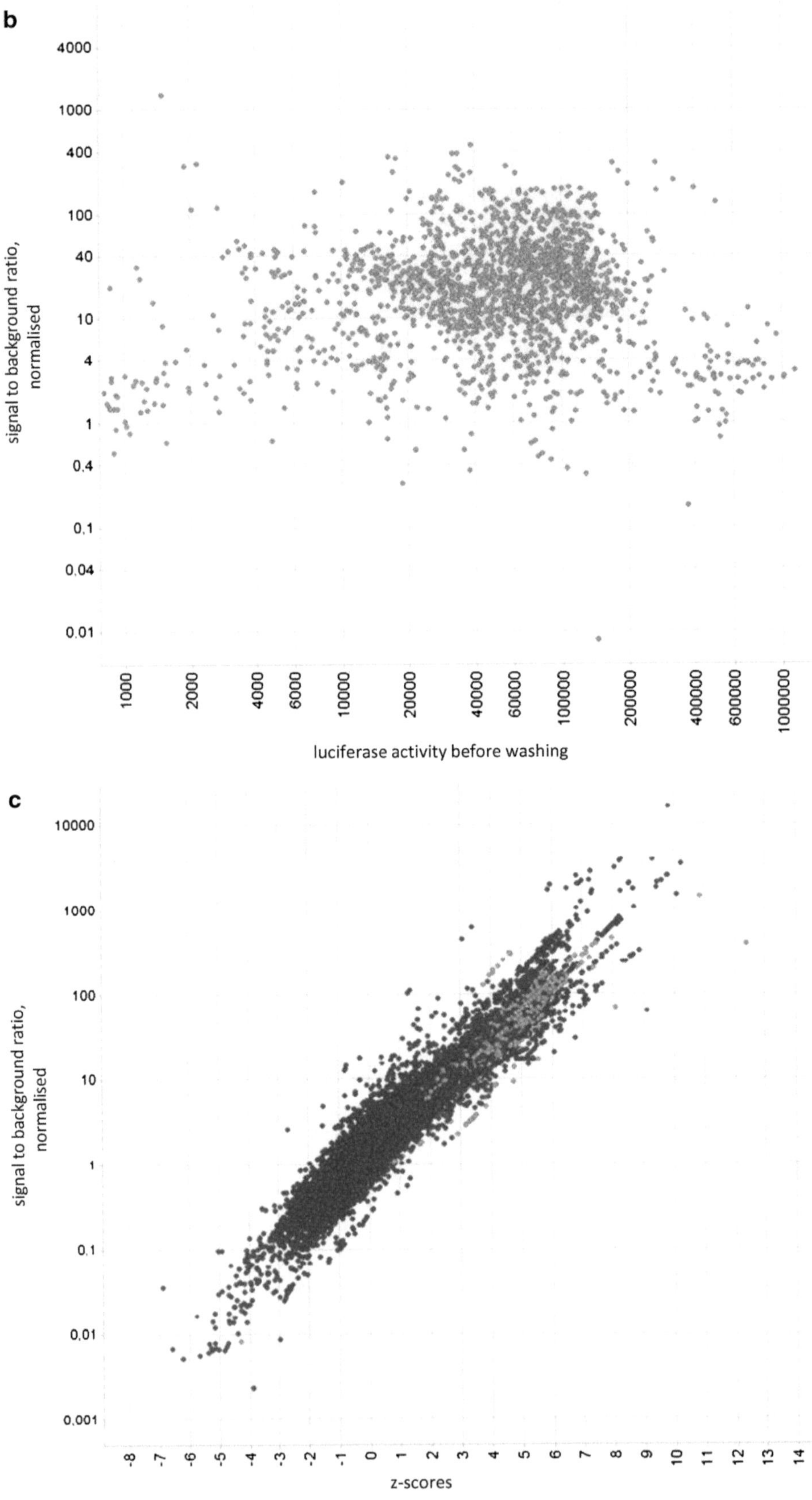

Fig. 2 (continued)

In our lab, proteins of human viruses tested for interactions with bacterial pathogens, which can be reasonably expected to not interact other than by chance, provided such a negative control data set. Alternatively, normalized signal to background ratios can be used to generate an easily comprehensible measure of binding (Fig. 2b, c, *see* Subheading 3).

Trivial to say, the position of the tag at the amino-or carboxy-terminal end of the tagged protein has to fit the posttranslational processing of the protein. For example, proteins entering the secretory route via a signal peptide cannot be tagged at the amino-terminal end. In addition, the luciferase used for secreted proteins has to be stable outside the cytoplasm. While Firefly luciferase works poorly for this purpose, Gaussia luciferase is an excellent choice here.

Regarding the tag used for the affinity purification, the original work by Barrios Rodiles et al. used the hemagglutinin-tag, whereas our lab has made excellent experience using the IgG-binding domain of *Staphylococcus aureus* protein A: the protein A tag binds with high affinity to the Fc part of IgG, which allows to use any IgG-coated surface as an affinity matrix. In our protocol, binding of protein A-tagged proteins to IgG-coated magnetic particles approaches completion in ~15 min (Fig. 3). This short incubation time between lysis and washing helps to keep the dissociation of weakly interacting binding partners at a minimum.

The use of separately detectable luciferases such as those from Firefly and Renilla allow the test of two potential binding partners to an affinity-tagged protein in a single transfection. In such an experiment, proteins can be tested for dependence on a third binding protein for interaction.

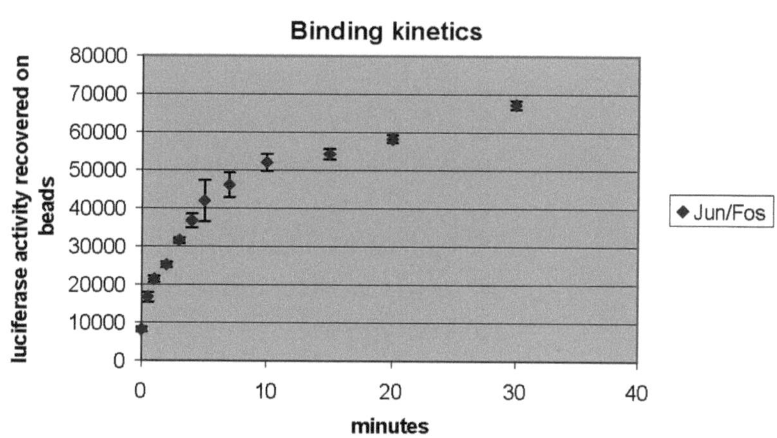

Fig. 3 Binding kinetics for Renilla luciferase-tagged FOS binding to protein A-tagged JUN. Cells were lysed in magnetic bead-containing lysis buffer as described in this paper and incubated on ice for the time indicated before washing. The *y*-axis shows raw luciferase counts

The method detailed below lends itself to use at high-throughput and has successfully been applied in several projects analyzing host–pathogen interactions, both to verify interactions observed using other methods and to systematically look for novel interactions [2, 3]. The availability of large ORFeome collections [4, 5] and the ease of cloning using the GATEWAY system have made it relatively straightforward to generate an array of plasmids for expression of affinity-tagged proteins. To identify novel protein–protein interactions such an array can be interrogated for interaction with luciferase-tagged proteins in one-by-one tests, as exemplified in Fig. 4.

2 Materials

2.1 Human Cell Line

1. Human Embryonic Kidney 293T (HEK-293T) cells.

2.2 Vectors

1. pTREX-dest30-ntPrA Amp^R, gateway cassette, SV40 origin of replication, N-terminal Staphylococcus aureus protein A tag.
2. pcDNA3-Rluc-GW Amp^R, gateway cassette, SV40 origin of replication, N-terminal Renilla reniformis luciferase tag.
3. pTREX-dest30-ntFf Amp^R, gateway cassette, SV40 origin of replication, N-terminal Firefly luciferase tag.

2.3 Stock Solutions

Unless specified otherwise all stock solutions are prepared with deionized water.

1. 1× PBS 1 tablet/200 mL deionized water, autoclaved.
2. 1 M Tris–HCl (pH 7.5 and pH 7.8), autoclaved.
3. Protease Inhibitor Cocktail (Roche 1 836 170), 1 tablet per 2 mL deionized water (25×), store at −20 °C.
4. Phosphatase Inhibitor Cocktail (Roche 04 906 837), 1 tablet per 1 mL deionized water (10×), store at −20 °C.
5. 1 M DTT, store at −20 °C.
6. 5 M NaCl, autoclaved.
7. 0.5 M Na_2EDTA (pH 8.0 with NaOH), autoclaved.
8. 0.25 M EGTA (pH 8.3).
9. 200 mM ATP, store at −20 °C.
10. 1 M $MgSO_4$.
11. 10 mM luciferin, store at −20 °C.
12. 200 mM AMP, store at −20 °C.
13. 1 M phosphate buffer, dissolve 1 M KH_2PO_4 in deionized water, adjust to pH 5.1 with KOH.
14. 10 % (0.1 g/mL) BSA (Serva 11930), store at −20 °C.

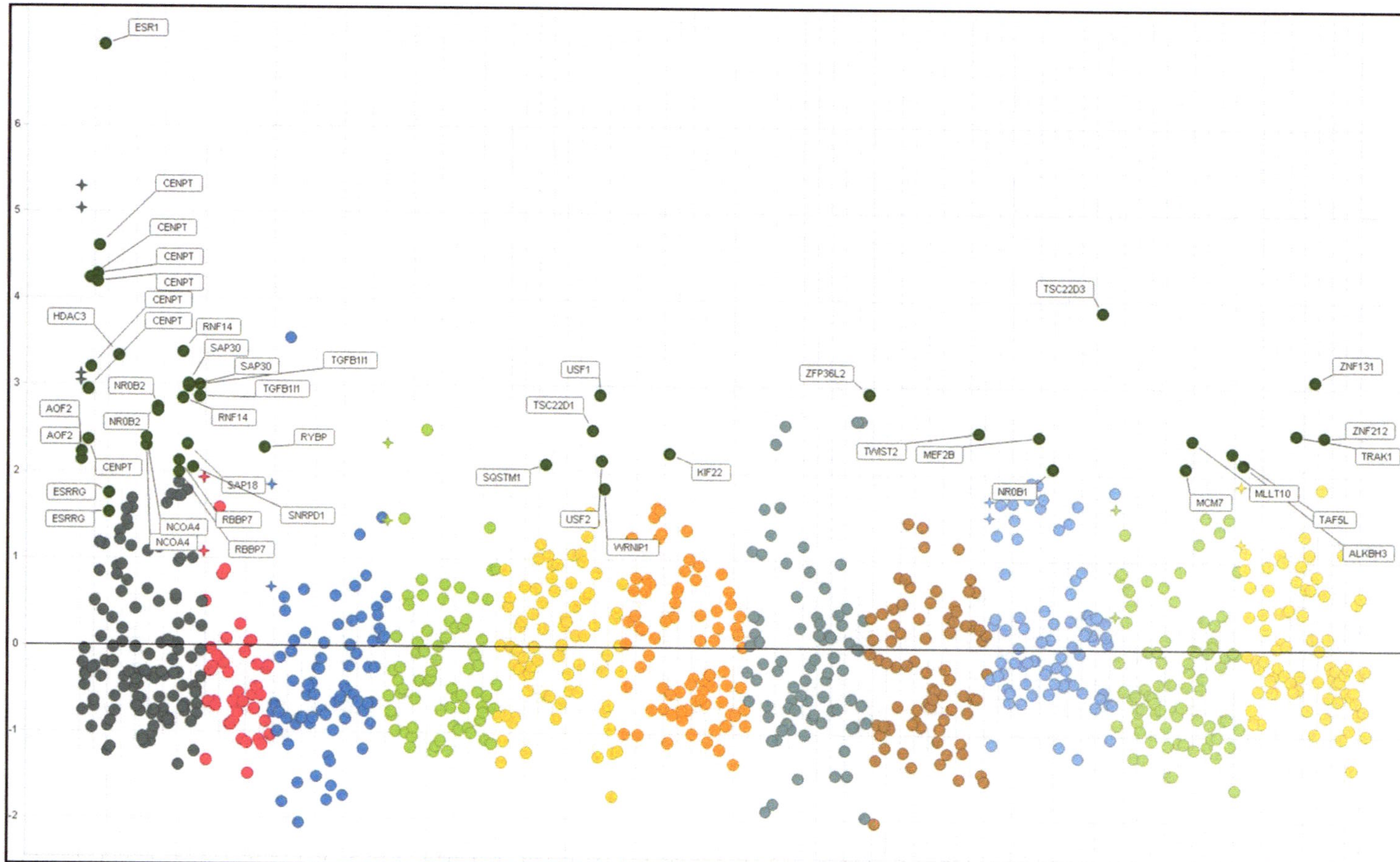

Fig. 4 LUMIER-Screen for CENPT—binding proteins among 855 nuclear proteins. LUMIER assays were done with luciferase-tagged CENPT and an array of 855 different nuclear proteins tagged with protein A as an affinity tag. The 855 expression constructs were arranged on twelve 96-well microtiter plates and transiently transfected into HEK-293T cells along with the CENPT-luciferase fusion. After lysis, extracts were purified on IgG-coated magnetic beads, and co-purifying luciferase activity was measured. *y*-axis represents *z*-scores, *colors* indicate individual 96-well plates. *Stars* represent FOS-JUN binding as the positive control. Selected hits are marked

15. 10 mM coelenterazine, dissolve in 236 μL methanol, store at –20 °C.

2.4 Buffers

1. Lysis buffer Part I, final concentration (1.1×): 22 mM Tris–HCl pH 7.5, 1.1 % Triton X-100, 275 mM NaCl, 11 mM EDTA (pH 8.0), store at 4 °C for up to 1 year.
2. Lysis buffer: For ~1 mL lysis buffer (enough for one 96-well plate) use 900 μL lysis buffer Part I and add 100 μL Phosphatase Inhibitor Cocktail (10×), 40 μL Protease Inhibitor Cocktail (25×), 10 μL DTT (1 M), and 0.5 μL Benzonase (purity >90 %, 25 units/μL), store at 4 °C for 1 week maximum. 200 μL Dynabeads M-280 sheep anti-rabbit IgG, 6–7 × 10^8 beads/mL, approx. 10 mg/mL are used for 1 mL lysis buffer. Prior to use, their storage liquid is removed and the beads are washed three times with 1× PBS using a magnetic stand and finally resuspended in lysis buffer (*see* **Note 1**).
3. Renilla assay buffer: Final concentration: 1.1 M NaCl, 2.2 mM EDTA (pH 8.0), 220 mM phosphate buffer (pH 5.1), 0.44 mg/mL BSA, 2.5 μM coelenterazine. For 50 mL buffer add 27.5 mL deionized water, 11 mL 5 M NaCl, 0.22 mL 0.5 M EDTA (pH 8.0), 11 mL 1 M phosphate buffer, and 0.225 mL 10 % BSA. Store at –20 °C. 12.5 μL coelenterazine are added directly before application (*see* **Note 2**).
4. Firefly luciferase assay buffer: Final concentration: 25 mM Tris–HCl (pH 7.8), 4 mM EGTA, 1 mM ATP, 1 mM DTT, 15 mM $MgSO_4$, 100 μM luciferin (PJK, Bitterfeld, Germany), 0.2 mM AMP. For 50 mL buffer add 46.4 mL deionized water, 1.25 mL 1 M Tris–HCl (pH 7.8), 0.8 mL 0.25 M EGTA (pH 8.3), 0.25 mL 200 mM ATP, 50 μL 1 M DTT, 0.75 mL 1 M $MgSO_4$, and 0.5 mL 10 mM luciferin. Store at –20 °C.

2.5 HEK-293T Cell Culture

1. Minimal essential medium (MEM): MEM (500 mL, 31095 Gibco, Invitrogen), 10 % fetal calf serum, 5 mL nonessential amino acids (100×), 5 mL sodium pyruvate (100 mM, 5 mL penicillin/streptomycin (10,000 units penicillin/mL, 10 mg streptomycin/mL)). Ready to use media are stored at 4 °C.
2. Trypsin–EDTA solution (0.05 % trypsin, 0.02 % EDTA in PBS w/o Ca + Mg).
3. Dulbecco's Phosphate-Buffered Saline (DPBS) 1×.

2.6 HEK-293T Transfection

1. OptiMEM.
2. Lipofectamine™ 2000 Transfection Reagent.

2.7 Plastic Ware

1. 96-well plates, white (LumiNunc), flat bottom.
2. 96-well plates with lid, flat bottom, sterile.

2.8 Equipment

1. Magnetic bead washer, e.g., TECAN HydroFlex.
2. Luminescence plate reader, e.g., TECAN Infinite 200.
3. Magnetic stand for washing of magnetic beads.

3 Methods

3.1 Transient Transfection of HEK-293T Cells

1. For each luciferase-tagged protein (Y) tested for an interaction with a protein A-tagged protein (X), include a negative control which uses the luciferase-tagged protein Y together with the protein A tag only. Work with triplicate transfections for each interaction test.
2. Plate 10,000 HEK-293T cells per well in 100 μL medium in 96-well plate at least 3 h before the transfection.
3. Add 0.075 μL Lipofectamine to 12 μL of OptiMEM (1) and 24 ng of each plasmid DNA to 12 μL of OptiMEM (2), mix.
4. Combine OptiMEM solution 1 and 2.
5. After 30 min of incubation at room temperature, add the mix to the well.
6. Grow cells for 36–48 h.

3.2 Cell Lysis, Washing and Measurement

1. Prepare the magnetic beads: Use 20 μg beads per well (in total 2 mg are required for one 96-well plate). Wash three times with 1× PBS using a magnetic stand (*see* **Note 3**). Resuspend in 10 μL (1 mL) lysis buffer per well (per plate).
2. Pour the medium from the microtiter plates. Knock plates gently onto a tissue to remove most of the medium. Put the microtiter plates on ice.
3. Add 10 μL of ice-cold lysis buffer containing the magnetic beads to the cells. Shake briefly. Incubate for 15 min on ice (*see* **Note 4**).
4. Add 100 μL PBS. Remove 10 μL, corresponding to ~10 % of the lysate to a white F-bottom plate, place on ice (activity plate, no washing steps required).
5. Transfer 90 μL to a white F-bottom plate.
6. Wash beads five times with 200 μL of ice-cold PBS in an automated magnetic bead washer (*see* **Note 5**).
7. Resuspend in 20 μL of PBS (*see* **Note 6**).
8. Measure luciferase activity of the washed beads and of the 10 μL of unwashed beads by injecting 50 μL of luciferase buffer. If both Firefly and Renilla fusions are used, inject 70 μL of Renilla buffer immediately after measuring Firefly activity.

3.3 Data Analysis: Signal to Background Ratio

1. Calculate (signal of washed beads)/(signal of 10 % of the lysate, non-washed) for both the interaction test (X-protein A + Y-luciferase) and the negative control (protein A alone + Y luciferase) to generate "signals normalized for expression levels."
2. Divide the "signals normalized for expression levels" of the interaction test by the "signals normalized for expression levels" of the negative control to derive signal to background ratios (*see* **Note 7**).

3.4 Data Analysis: z-scores

1. Test a large set of randomly chosen protein pairs for interaction signals, including negative controls.
2. Log-transform the luciferase counts to achieve a normal distribution of the values. Calculate (signal of washed beads)/(signal of 10 % of the lysate, non-washed) for both the interaction test and the negative control. Subtract the value of the negative control from the interaction test.
3. From the numbers for the random protein pairs generated in **step 2**, calculate the mean and the standard deviation and generate *z*-scores: Subtract the mean from each value, such that the data scatter around zero. Divide each resulting value by the standard deviation, such that a value of 1 signifies an interaction signal with a distance of 1 standard deviation from the mean. An example of *z*-scores for the test of one luciferase-tagged protein with a large set of protein A-tagged proteins is shown in Fig. 4 (*see* **Note 7**).

4 Notes

1. The Dynabeads are not stable in lysis buffer and therefore have to be prepared directly prior to use.
2. Due to its instability in aqueous solution, coelenterazine is to be stored separately in methanol at −20 °C.
3. Washing of the beads in detergent-containing buffer (e.g., 0.1 % Triton X-100) has in our hands led to significant loss of beads and did not improve signal to background ratios.
4. Binding of protein A-tagged proteins to magnetic beads approaches completion after 10–20 min. Prolonged incubation times do not result in better interaction signals. On the contrary, labile complexes are expected to fall apart upon prolonged incubation due to the dilution of the cytoplasm in the lysis buffer.
5. Do not set the dispense rate higher than 200 μL/s. This is the slowest rate possible for the TECAN HydroFlex™ bead washer.

6. We set the washer, so that its needles remain 8.1 mm above the bottom of the well and thus cannot touch the beads. As a consequence 20 µL of PBS remain in the well after each washing cycle and after the final aspiration.
7. As shown in Fig. 3c, the two methods of normalizing the data yield interaction parameters which correlate well with one another. If a small number of interactions is tested, e.g., comparing the interactions of mutant versions of a protein for binding to a given partner, we recommend to use normalized signal to background ratios, since they are more intuitive. For the analysis of screening campaigns, i.e., interaction tests of a single protein against a large set of potential binding partners, *z*-scores are more practical, since they incorporate the level of noise in the assays in their calculation. In our setup, a normalized signal to background ratio of 10 typically corresponds to a *z*-score around 3 and can be considered a highly significant interaction.

Acknowledgments

We are grateful to Kerstin Mohr and Gabriella Siszler for excellent technical support and to Gabriella Siszler for critical reading of the manuscript and helpful comments.

References

1. Barrios-Rodiles M, Brown KR, Ozdamar B et al (2005) High-throughput mapping of a dynamic signaling network in mammalian cells. Science 307:1621–1625
2. Tahoun A, Siszler G, Spears K et al (2011) Comparative analysis of EspF variants in inhibition of Escherichia coli phagocytosis by macrophages and inhibition of E. coli translocation through human- and bovine-derived M cells. Infect Immun 79:4716–4729
3. Pfefferle S, Schopf J, Kogl M et al (2011) The SARS-coronavirus-host interactome: identification of cyclophilins as target for pan-coronavirus inhibitors. PLoS Pathog 7: e1002331
4. Strausberg RL, Feingold EA, Grouse LH et al (2002) Generation and initial analysis of more than 15,000 full-length human and mouse cDNA sequences. Proc Natl Acad Sci USA 99: 16899–16903
5. Lamesch P, Li N, Milstein S et al (2007) hORFeome v3.1: a resource of human open reading frames representing over 10,000 human genes. Genomics 89:307–315

Chapter 3

Detection of Protein Interactions During Virus Infection by Bimolecular Fluorescence Complementation

Stefan Becker and Jens von Einem

Abstract

The bimolecular fluorescence complementation (BiFC) allows not only the investigation of protein interactions but also the visualization of protein complexes in living cells. This method is based on two nonfluorescent fragments of fluorescent proteins (FPs) which can reassemble into a fluorescent complex. The formation of the fluorescent complex requires association of the nonfluorescent fragments which is facilitated by their fusion to two proteins that interact with each other. It is necessary to confirm the specificity of the BiFC signal, e.g., by using proteins with a mutated interaction site. Here, we describe a BiFC protocol adapted for the investigation of protein–protein interactions during herpesvirus infection.

Key words BiFC, Protein interaction, Virus, Bimolecular fluorescence complementation, Herpesvirus

1 Introduction

Protein–protein interactions between viral proteins or viral and cellular proteins play crucial roles in all viruses and in many steps of their replication cycle [1]. The identification and investigation of these complex networks of interactions is important for a better understanding of the processes during virus infection, leading to the production of infectious virus progeny and in the end may help to identify new targets for the development of antiviral therapy [2, 3].

Bimolecular fluorescence complementation (BiFC) allows to investigate protein–protein interactions or protein complexes including their visualization in living cells under conditions with minimal perturbation of the environment [4–6]. The possibility to perform BiFC in living cells virtually eliminates artifacts caused by fixation methods or cell lysis. BiFC has been successfully utilized to confirm known protein–protein interactions or to determine their intracellular interaction site as well as for the identification of assumed or unknown interactions [7, 8]. This technique makes use of the discovery that a fluorescent protein (FP) can be split into

Susanne M. Bailer and Diana Lieber (eds.), *Virus-Host Interactions: Methods and Protocols*, Methods in Molecular Biology, vol. 1064, DOI 10.1007/978-1-62703-601-6_3, © Springer Science+Business Media, LLC 2013

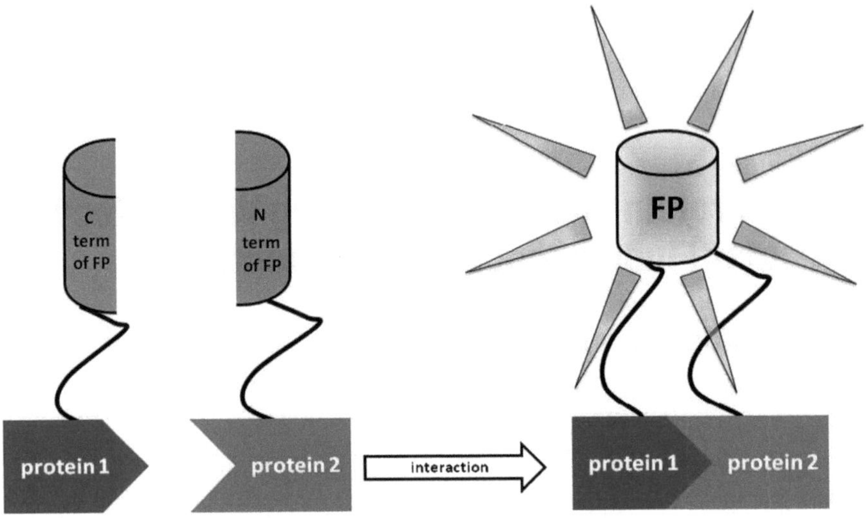

Fig. 1 Principle of the BiFC. Proteins of interest fused with an N-terminal (N-term of FP) and a C-terminal (C-term of FP) fragment of an FP. The co-expression of both constructs within one cell leads to the formation of the fluorescent complex under recovery of fluorescence, in case of an interaction of the two proteins of interest

two nonfluorescent fragments from which a fully functional and energetically preferred fluorescent protein complex can be formed. An interaction of two interacting proteins fused to the two nonfluorescent FP fragments brings the nonfluorescent fragments in close proximity to each other and thus dramatically increases the likelihood of their association for the length of time needed to form the fluorescent complex (Fig. 1) [8]. The spatial resolution of the BiFC lies in the range of approx. 50 nm depending on several circumstances such as linker sequences between the nonfluorescent fragments and the interacting proteins [9].

One prerequisite for the BiFC is that the investigated interaction partners tolerate fusion to fluorescent protein fragments without negatively affecting their function or localization (*see* **Note 1**). Furthermore, it has to be taken into account that association of the nonfluorescent fragments and their refolding into a fluorescent complex stabilizes the interaction complex itself. Therefore, it is necessary to ensure that the functionality of the interaction complex is maintained and that the stabilization is not harmful for the cell.

As other methods, BiFC has limitations and strengths. The above-mentioned stabilization of the interaction complex is likely to affect dissociation of the interaction partners. Consequently, formation of the fluorescent protein complex appears to be irreversible in most of the reported interactions, making BiFC not applicable for studying dynamics of temporal interactions [10]. Moreover, the formation of the fluorophore is a slow process that requires some time [11]. Although improved FPs with faster refolding kinetics are available [12], the BiFC assay is not suitable for real-time detection of rapid changes in interactions. Another critical point of the BiFC is that a spontaneous association of the fluorescent protein fragments independent of an interaction between proteins fused to the fragments cannot completely be eliminated despite the use of noninteracting FP fragments. This is the major source of background signal in the BiFC. It is mainly influenced by the level of expression [13], because high expression levels of the fusion proteins can result in abnormal protein accumulations thus increasing the likelihood for unspecific fluorescence complexes. Due to mostly unavoidable unspecific background signals, the specificity of the BiFC signal must be verified by using controls, e.g., the nonfluorescent FP fragments. In that regard, the preferred control is the use of proteins that are mutated in their interaction site, however, this requires knowledge of the interaction domain (*see* **Note 2**).

The strategy of the BiFC assay in virus infected cells is dependent on the investigated protein interaction during virus infection (e.g., interaction of viral with cellular or viral with viral or cellular with cellular proteins). In case of an interaction between viral proteins a recombinant virus can be generated where the two interacting viral proteins of interest are either fused to the N-terminal or to the C-terminal fragment of the FP. This strategy ensures that the expression of both fusion proteins during virus infection is under control of the natural viral promoters and thus at physiological expression levels. However, negative effects on the viral replication cycle or on other functions of these proteins cannot be excluded, because of an increased stability of the interaction complex due to the formation of an irreversible fluorescent complex (see above). Alternatively, the BiFC components can be provided by expression plasmids [14]. This would also allow the investigation of protein interactions where the insertion of the fluorescent protein fragment into the viral genome is not tolerated.

We have successfully used a strategy where one interaction partner is fused to the N-terminal fragment of the FP and expressed from a recombinant virus, whereas the second interaction partner fused to the C-terminal FP fragment is provided by an inducible expression plasmid (Fig. 2a). This setup is also suitable for the investigation of protein–protein interactions between viral and cellular proteins. One advantage is that at least the interaction partner encoded by the recombinant virus is expressed at physiological levels, while time and level of expression of the other interaction partner can be controlled by using an inducible expression plasmid. This strategy provides sufficient flexibility to optimize the conditions for the BiFC and thus likely minimizes the above-mentioned negative effects. Furthermore, the implementation of negative controls and investigation of more than one interaction partner for the protein of interest is easier and faster to perform,

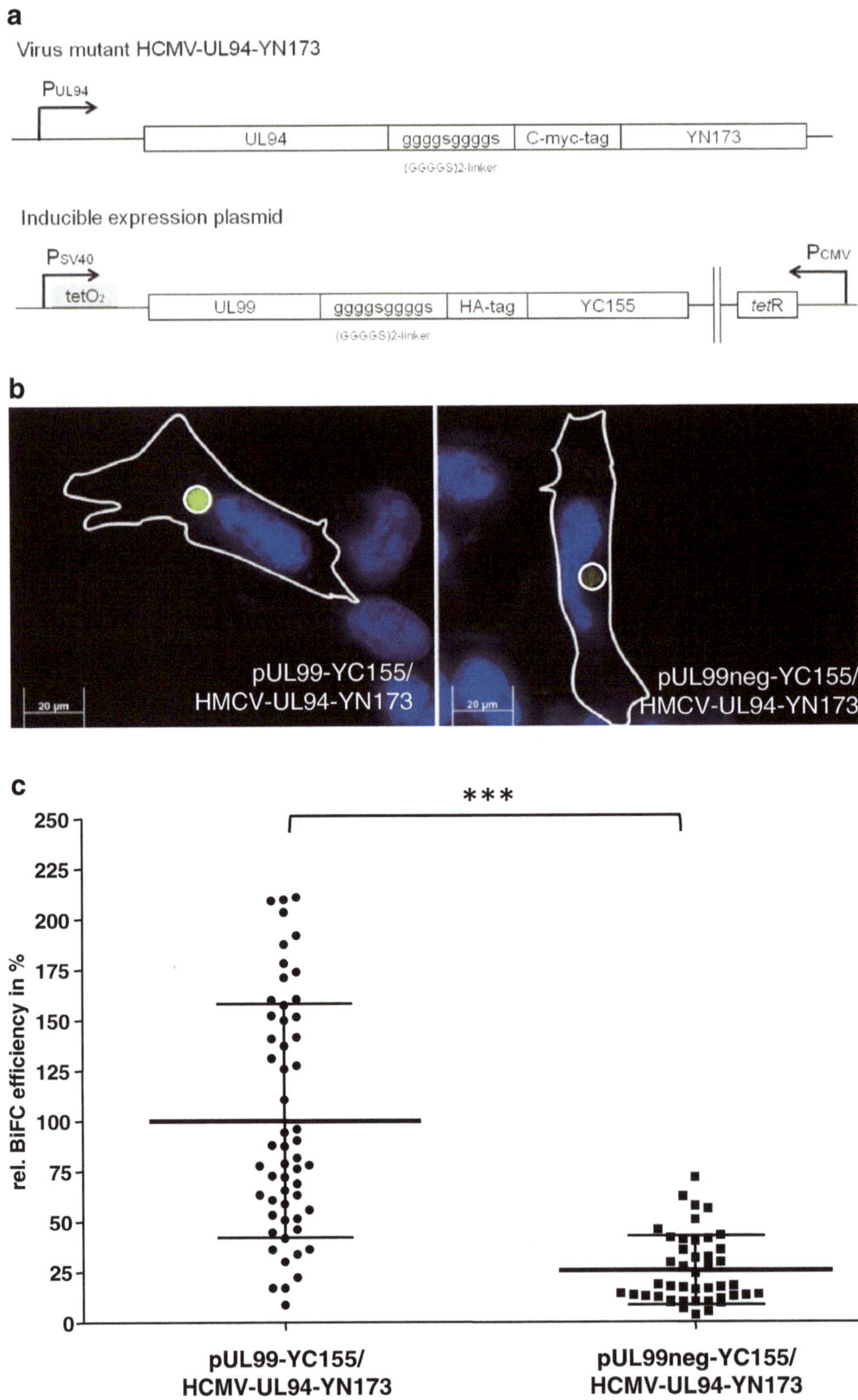

Fig. 2 Visualization and quantification of BiFC signals during HCMV infection in human fibroblasts 5 days postinfection. (**a**) For the investigation of the pUL94/pUL99 interaction during HCMV infection, an UL94 mutant virus (HCMV-UL94-YN173) was generated, expressing a pUL94-YN173 fusion protein. The N-terminal fragment, YN173, of Citrine was fused to the C-terminus of pUL94 separated by the flexible $(GGGGS)_2$-linker

because the time consuming generation of mutant viruses can be avoided by constructing expression plasmids instead. In addition, it allows also the investigation of essential interactions during the viral replication cycle with all necessary controls (*see* **Note 2**).

In the following we describe a protocol for the detection of protein–protein interactions during human cytomegalovirus (HCMV) infection by BiFC. It is conceivable that also protein–protein interactions in other herpesviruses can be studied using the same strategy but with an adaptation of the BiFC protocol [15]. The methods for the generation of recombinant viruses and cloning of expression plasmids are only briefly described, while mainly focusing on the procedure of the BiFC.

2 Materials

For the generation of a recombinant virus expressing one FP fragment fused to a viral protein of interest we used a bacterial artificial chromosome (BAC) of HCMV in combination with a markerless BAC mutagenesis technique (*see* **Note 3**). Below, we describe the procedure of BiFC during virus infection for the known interaction of HCMV proteins pUL94 and pUL99 [16]. In the given example, the N-terminal fragment (YN173) and the C-terminal fragment (YC155) of Citrine [13, 17] are fused to the C-terminus of pUL94 and pUL99, respectively, and are separated by a linker sequence (Fig. 2a) (*see* **Note 4**).

2.1 Generation of Recombinant Viruses

1. To generate the recombinant virus HCMV-UL94-YN173 (Fig. 2a), we used the *en passant* RED-GAM recombination technique [18, 19], thereby inserting the sequences corresponding to a $(GGGGS)_2$-linker sequence, the c-myc-tag and the N-terminal fragment of the yellow fluorescent protein

Fig. 2 (continued) and the myc-tag. In addition, Tet-inducible expression plasmids [21], encoding pUL99-YC155 (shown), their interaction-deficient negative control (pUL99neg-YC155) and mCherry were generated (both not shown). For the BiFC, mCherry together with either pUL99-YC155 or pUL99neg-YC155 were co-expressed in human fibroblasts and infected with HCMV-UL94-YN173 mutant virus. Expression of pUL99-YC155, pUL99neg-YC155, and mCherry was induced by Doxycycline (1 μg/mL) 24 h prior fixation at day 5 postinfection. (**b**) The BiFC signals were determined with identical settings by using an YFP-filter (excitation 500/20, emission 535/30), a 63× objective, and the Zeiss Observer Z1 fluorescence microscope. Cell nuclei were stained using DAPI. Infected cells (*white border*), which were identified by using an anti-HCMV antibody. (**c**) BiFC efficiency was determined by measuring the Citrine and mCherry fluorescence of at least 50 single-infected cells of three independent experiments by using the AxioVision 4.8 software module Colocalization Analysis (Zeiss) and calculation of the ratio of Citrine/mCherry signal intensities. The mean ratio pUL99-YC155/pUL94-YN173 was set to 100 %. Given is the standard deviation and statistical analysis by using the Wilcoxon-signed rank test (***p-value < 0.0001)

(YFP) variant Citrine [20], containing amino acids 1–173 (YN173). The insertion of the fluorescent protein fragment into the viral genome must be tolerated and should neither affect the function of the viral protein nor the expression of neighboring genes. Therefore, it is necessary to ensure that the mutant virus exhibits no replication defect (*see* **Note 2**). The YN173 sequence was first PCR amplified from an YFP-transfer construct retrieved from reference [19] using primers adding a linker sequence and a tag sequence ($(GGGGS)_2$-linker followed by c-myc-tag) (*see* **Note 5**). This product was subsequently amplified with primers adding the regions of sequence homology to the HCMV BAC which determine the insertion site for the RED recombination. The principle and procedure of the *en passant* recombination technique has been described in detail in above-mentioned references.

2.2 Generation of Inducible Expression Vectors

1. The inducible expression vector for expression of the interaction partner fused to the other FP fragment was obtained from reference [21]. This vector contains the simian virus 40 (SV40) early enhancer including two *tet* operators to drive the fusion protein expression and the tetR gene encoding the Tet-Repressor under control of a HCMV immediate-early promoter enhancer (Fig. 2a). The DNA encoding for pUL99, the interaction partner of pUL94, was cloned in frame with a $(GGGGS)_2$-linker sequence, the Hemagglutinin-tag (HA-tag), and the C-terminal half of the YFP variant Citrine (YC155) containing amino acids 156–239 and mutation A206K (*see* **Note 6**).
2. As negative control an interaction-deficient pUL99 (pUL-99neg) was constructed with the same strategy. The YC155 fragment was obtained from [22].
3. An additional inducible plasmid expressing the red fluorescent protein variant mCherry [23] is required as internal reference for protein expression levels and used for the determination of the BiFC efficiency (*see* **Note** 7).

2.3 Cells and Cell Culture

1. Human lung fibroblasts (MRC-5) not older than passage 27 were used for electroporation of the different plasmids and for infection with the HCMV recombinant virus (*see* **Note 8**).
2. Dulbecco's modified eagle medium (DMEM, Gibco-BRL) supplemented with 10 % fetal calf serum (FCS, Gibco-BRL), penicillin/streptomycin (100-fold, PAA Laboratories), 1 % L-glutamine (200 mM, PAA Laboratories) 1 % nonessential amino acids (Biochrom AG).
3. Phosphate-buffered saline (PBS): 137 mM sodium chloride, 10 mM sodium phosphate, 2.7 mM potassium chloride, and 2 mM potassium phosphate.

4. PBS supplemented with 0.02 % EDTA.
5. 0.05 % Trypsin–EDTA.
6. Doxycyclin (Sigma Aldrich).

2.4 Electroporation Components

1. OptiMEM™ (Gibco-BRL) supplemented with 5 % FCS for dilution of the DNA for electroporation (*see* **Note 9**).
2. Electroporation cuvettes with 0.4 mm gap.
3. Electroporator EasyjecT optima (EquiBio) for transfection of MRC-5.

2.5 Visualization of BiFC Signals and Quantification of Signals

1. Fluorescent microscope (Axio Observer.Z1, Zeiss) with high resolution objectives (×63) equipped for live cell imaging. The microscope should be provided with a sensitive digital camera for the acquisition of fluorescence signal intensities and software for the analysis of fluorescence signals.
2. Appropriate sets of filters for excitation/emission of Citrine at 515/535 nm and of mCherry at 587/600 nm and for the dye Alexa Fluor 647 at 640/690 nm.
3. μ-slide IBIDI 8-well format (IBIDI) (*see* **Note 10**).
4. Fixation solution: 4 % paraformaldehyde (Sigma Aldrich) in PBS.
5. Permeabilization solution: 0.1 % Triton-X100 (Carl Roth) in PBS.
6. Blocking solution: PBS supplemented with 1 % BSA (bovine serum albumin, Sigma Aldrich), and 10 % human serum of HCMV sero-negative individuals (*see* **Note 11**).
7. Washing solution: PBS supplemented with 1 % BSA and 0.1 % Triton-X100.
8. DAPI (4′-6-diamidino-2-phenylindole, Roche): 1 μg/mL in PBS.
9. Primary and secondary antibodies for detection of preferably viral proteins localizing to the region of interest (ROI).

3 Methods

For the BiFC assay, MRC-5 were first transfected by electroporation with the inducible expression plasmids. The transfection is performed 1 day before infection with a mutant virus expressing the corresponding putative interaction partner fused to YN173. This is followed by inducing the expression of the protein coupled to YC155 and mCherry (*see* **Note 12**).

3.1 Transfection of MRC-5 with Inducible Expression Plasmids (Day 1)

1. For electroporation, MRC-5 cells with a confluency of 80–90 % were used.
2. Cells were washed three times with 5 mL PBS–EDTA prior incubation with Trypsin–EDTA at 37 °C for 5–10 min.

3. Trypsinized cells were transferred to a 50 mL Falcon tube and filled up with PBS for washing of the cells.
4. The cells were pelleted by centrifugation at approximately 100 rcf for 5 min at room temperature.
5. After centrifugation and removal of PBS, the cell pellet was gently resuspended in OptiMEM 5 % FCS. For each electroporation, approximately 2.5×10^5 cells were resuspended in 250 μL OptiMEM 5 % FCS.
6. A total of 3 μg of plasmid DNA was diluted in a total volume of 40 μL OptiMEM 5 % FCS. DNA of the mCherry expression plasmid was mixed in equal quantities with DNA of the expression plasmid either expressing pUL99-YC155 or the negative control pUL99-neg-YC155. mCherry serves as a reference for the level of expression and is required for the quantification of the BiFC efficiency (*see* **Note 13**).
7. 250 μL of cells in OptiMEM 5 % FCS were mixed with 40 μL of the DNA mixture and transferred to an electroporation cuvette.
8. Electroporation was performed with a pulse of 260 V and 1,050 μF.
9. After the pulse, cells were immediately recovered from the electroporation cuvette by rinsing with 1.5 mL prewarmed DMEM.
10. Cells were then seeded into a μ-slide IBIDI 8-well for direct imaging using 200 μL per well (*see* **Note 14**).

3.2 Infection with the Recombinant Virus (Day 2)

1. The electroporated cells were incubated for at least 24 h post electroporation at 37 °C and 5 % CO_2 to let the surviving cells recover and to reach a confluence of about 80 %.
2. Dead cells and cell debris were removed by two washings with PBS and PBS finally replaced with fresh prewarmed DMEM.
3. Subsequently, cells were infected with the mutant virus, HCMV-UL94-YN173, using a multiplicity of infection (MOI) of 2 plaque forming units (pfu) per cell.
4. After 4 h of infection, the inoculum was removed and replaced with fresh prewarmed DMEM.

3.3 Induction of the Inducible Expression Plasmid (Day 6)

The following steps vary depending on the used herpesvirus and have to be adjusted for each investigated interaction. The interaction of the given example (pUL94-YN173 and pUL99-YC155) during HCMV infection was investigated at 5 days postinfection (Fig. 2).

1. Expression of mCherry and either pUL99-YC155 or pUL-99neg-YC155 was induced by adding Doxycycline to the culture medium (up to 1 μg/mL).

2. The induction time for the indicated proteins was between 12 and 36 h. First BiFC signals (YFP fluorescence) can be observed after 12 h of induction in live cell imaging (*see* **Note 15**).

3.4 Fixation and Indirect Immunofluorescence for the Analysis of BiFC (Day 7)

Indirect immunofluorescence was performed to identify infected cells for BiFC analysis. This is especially important in case of poorly infected cells with no clear visible cytopathic effect or in case of no BiFC signal for identification of the ROI.

1. For the quantification of BiFC signal cells were fixed. Therefore the cells were incubated in Fixation solution for 10 min at 4 °C followed by two washings with PBS (*see* **Note 16**).
2. For permeabilization, cells were incubated with Permeabilization solution for 5 min at room temperature.
3. Cells were washed three times with PBS.
4. Then, cells were incubated in Blocking solution for 30 min.
5. Primary antibody was diluted in Blocking solution and incubated with the cells for 45 min.
6. Cells were washed three times with Washing solution.
7. Secondary antibody conjugated with Alexa Fluor 647 was diluted in Blocking solution and incubated with the cells for 45 min.
8. Cells were washed three times with Washing solution.
9. Incubation of the cells followed for 10 min with DAPI.
10. Finally, cells were washed three times with PBS.

3.5 Quantification of BiFC Efficiency

For quantification of the BiFC efficiency, it is necessary to acquire several images of individual cells for further analysis (*see* **Note 17**) by using high resolution objectives (e.g., ×63). Identical settings of the microscope (e.g., fixed exposure times for Citrine and mCherry) must be used for the acquisition in order to compare BiFC signals from different cells. The fluorescence intensity of Citrine and mCherry is determined with an appropriate software such as the AxioVision Software 4.8 (Zeiss). The BiFC efficiency is generally calculated by dividing the fluorescence intensity of Citrine by the fluorescence intensity of mCherry in a selected ROI.

$$\text{BiFC efficiency} = \frac{\text{fluorescence intensity of Citrine(BiFC)}}{\text{fluorescence intensity of mCherry(reference FP)}}$$

In the given example, the ROI of the interaction of pUL94-YN173 with pUL99-YC155 was an HCMV-induced complex, the so-called assembly complex identified by staining of a viral or cellular protein localizing at this complex (Fig. 2).

BiFC specificity is determined by comparing the BiFC efficiency of cells expressing interaction–competent partners with those expressing negative controls. The interaction is specific when the corresponding negative control exhibits significantly reduced BiFC efficiency or no BiFC signals compared to the BiFC efficiencies of interaction–competent proteins (Fig. 2c).

4 Notes

1. Proteins fused to fluorescent protein fragments should be compared to their respective non-fused proteins in indirect immunofluorescence experiments to ensure similar intracellular localization. Furthermore, Western blot analyses can be performed to control for the correct size and for similar expression thus indicating similar protein stability. For these experiments both expression plasmids and recombinant viruses can be used. For the latter, effects on viral growth by the insertion of fluorescent protein fragments can be controlled by comparing viral growth of the recombinant virus with that of parental virus, e.g., by performing replication kinetics. In some cases, it is possible to test for already known protein functions of the interaction partners, given that respective methods are available.
2. The specificity of BiFC signals must be verified by including controls. The best control is the use of interaction partners that are mutated in their interaction site. However, in absence of such a control, alternative proteins can be used that exhibit comparable characteristics to the interaction partners particularly regarding the subcellular localization. Such a control accounts for BiFC signals by spontaneous association of the two FP fragments due to close proximity of their fusion partners. Inadvisable is to choose control proteins with a different intracellular localization that prevents fluorophore formation by physical distance.
3. Several herpesvirus genomes are cloned as BACs in bacteria (summarized in ref. 24). Transfection of permissive eukaryotic cells with herpesvirus BAC DNA reconstitutes viral growth. The main advantage of BACs is the possibility to manipulate the viral genome with several DNA manipulation techniques in bacteria, including recently developed markerless BAC mutagenesis techniques. The latter allow introduction of minor changes such as point mutations, but also the deletion of small or large sequences or the insertion of large DNA fragments such as FPs into herpesvirus genomes. Several markerless BAC mutagenesis techniques exist and all of them can be used for the generation of mutant viruses for BiFC [19, 25–28]. Therefore it is a matter of choice which of these techniques is used.
4. Several FPs, e.g., the yellow fluorescent protein (YFP), the cyan fluorescent protein (CFP), the monomeric red fluorescent

protein (mRFP), and the green fluorescent protein (GFP) with differing nonfluorescent fragments have been successfully used for fluorescence complementation experiments [29, 30]. The FPs differ in their characteristics such as emitted wavelength of fluorescence, their time for maturation, their stability, and in their signal intensity of the complex and thus in their usability for BiFC. With the diversity of the available FPs even a multicolor BiFC is possible which enables either to visualize multiple interactions within the same cell [31] or to investigate competition of two possible interaction partners for one protein [22].

5. Other linker sequences have also been used [8, 31]. It is also recommended to include small protein tag-sequences such as HA- or myc-tag that can be inserted between the linker sequence and the fluorescent protein fragment. These protein tags allow the easy detection of the fusion proteins. Additional optimization steps can include the use of alternative FPs (mentioned in **Note 4**) or the use of different length of fragments or combinations. Other possible optimization may require fusion of the FP fragments to either the C- or the N-terminal end of the protein of interest.
6. The A206K mutation in YC155 reduces unspecific accumulations at increased protein expression levels and by that reduces the background signal.
7. Also other FPs, e.g., CFP is applicable. One requirement is that the fluorescence spectra of these FPs do not overlap with that of the FP used for BiFC (e.g., Citrine).
8. Cells have to be transfectable and permissive to the investigated virus. Electroporation is the most efficient method for transfection of MRC-5. Other cell lines may have higher transfection efficiencies with other methods, e.g., lipofection, which can be used if virus entry is not affected by the transfection reagent.
9. Addition of FCS to OptiMEM increased survival of cells after electroporation. Other cell culture media without antibiotics can be used for electroporation instead of OptiMEM.
10. Cells can be grown either on cell culture dishes from other manufacturers for direct imaging or on normal cell culture dishes and imaged with long-working distance objectives.
11. Blocking with human serum of HCMV sero-negative individuals is specific for indirect immunofluorescence staining of HCMV infected cells.
12. Appropriate negative controls are required in form of an interaction-deficient protein or a protein with similar localization fused to the complementing fragment of the FP or at least the complementing FP fragment alone cloned into the inducible expression plasmid.

13. Low transfection efficiencies may require optimization of the procedure, e.g., by using different DNA amounts.
14. The number of cells in each well can be influenced by altering the volume of medium for the uptake of cells after electroporation and the number of cells for electroporation.
15. Optimal expression conditions (concentration of inductor and time of induction) are crucial for the specificity of the BiFC and thus for the quantification of the BiFC efficiency. The optimal conditions of expression have to be determined for each protein–protein interaction to reduce unspecific background signals. The optimal time of expression can vary between 12 and 36 h. The level of protein expression from the inducible plasmids can be controlled by using different concentrations of inductor. Optimal conditions for each new protein–protein interaction have to be determined. Among those are infection rates, amount of DNA for transfection, time of induction as well as level and length of induction of the expression from the inducible expression plasmids.
16. The fixation of cells prior to imaging ensures equal conditions (e.g., time of expression) for all combinations of interacting proteins and their respective negative controls and minimizes unspecific background signals, due to higher autofluorescence levels in living cells.
17. Signal intensities of mCherry provide an indication for the level of expression from the inducible plasmids. Very high expression levels increase the risk of unspecific signals or localization artifacts which can be avoided by investigating cells exhibiting moderate mCherry fluorescence.

References

1. Fossum E, Friedel CC, Rajagopala SV et al (2009) Evolutionarily conserved herpesviral protein interaction networks. PLoS Pathog 5(9):e1000570. doi:10.1371/journal.ppat.1000570
2. Sackett DL, Sept D (2009) Protein-protein interactions: making drug design second nature. Nat Chem 1(8):596–597
3. Wells JA, Mcclendon CL (2007) Reaching for high-hanging fruit in drug discovery at protein-protein interfaces. Nature 450(7172):1001–1009
4. Blondel M, Bach S, Bamps S et al (2005) Degradation of Hof1 by SCF(Grr1) is important for actomyosin contraction during cytokinesis in yeast. EMBO J 24(7):1440–1452
5. Ohad N, Shichrur K, Yalovsky S (2007) The analysis of protein-protein interactions in plants by bimolecular fluorescence complementation. Plant Physiol 145(4):1090–1099
6. Ventura S (2011) Bimolecular fluorescence complementation: illuminating cellular protein interactions. Curr Mol Med 11(7): 582–598
7. He J, Yu T, Pan J et al (2012) Visualisation and Identification of the Interaction between STIM1s in Resting Cells. PLoS One 7(3):e33377. doi:10.1371/journal.pone.0033377
8. Hu CD, Chinenov Y, Kerppola TK (2002) Visualization of interactions among bZIP and Rel family proteins in living cells using bimolecular fluorescence complementation. Mol Cell 9(4):789–798
9. Zal T (2011) Visualization of protein interactions in living cells. Self Nonself 2(2): 98–107

10. Magliery TJ, Wilson CG, Pan W et al (2005) Detecting protein-protein interactions with a green fluorescent protein fragment reassembly trap: scope and mechanism. J Am Chem Soc 127(1):146–157
11. Demidov VV, Dokholyan NV, Witte-Hoffmann C et al (2006) Fast complementation of split fluorescent protein triggered by DNA hybridization. Proc Natl Acad Sci USA 103(7):2052–2056
12. Kodama Y, Hu CD (2012) Bimolecular fluorescence complementation (BiFC): a 5-year update and future perspectives. Biotechniques 53(5):285–298
13. Shyu YJ, Liu H, Deng X et al (2006) Identification of new fluorescent protein fragments for bimolecular fluorescence complementation analysis under physiological conditions. Biotechniques 40(1):61–66
14. Hernandez FP, Sandri-Goldin RM (2011) Bimolecular fluorescence complementation analysis to reveal protein interactions in herpes virus infected cells. Methods 55(2): 182–187
15. Hernandez FP, Sandri-Goldin RM (2010) Herpes simplex virus 1 regulatory protein ICP27 undergoes a head-to-tail intramolecular interaction. J Virol 84(9):4124–4135
16. Phillips SL, Bresnahan WA (2012) The human cytomegalovirus (HCMV) tegument protein UL94 is essential for secondary envelopment of HCMV virions. J Virol 86(5):2523–2532
17. Walter M, Chaban C, Schutze K et al (2004) Visualization of protein interactions in living plant cells using bimolecular fluorescence complementation. Plant J 40(3):428–438
18. Tischer BK, Smith GA, Osterrieder N (2010) En passant mutagenesis: a two step markerless red recombination system. Methods Mol Biol 634:421–430
19. Tischer BK, Von Einem J, Kaufer B et al (2006) Two-step red-mediated recombination for versatile high-efficiency markerless DNA manipulation in Escherichia coli. Biotechniques 40(2):191–197
20. Griesbeck O, Baird GS, Campbell RE et al (2001) Reducing the environmental sensitivity of yellow fluorescent protein. Mechanism and applications. J Biol Chem 276(31): 29188–29194
21. Rupp B, Ruzsics Z, Sacher T et al (2005) Conditional cytomegalovirus replication in vitro and in vivo. J Virol 79(1):486–494
22. Waadt R, Schmidt LK, Lohse M et al (2008) Multicolor bimolecular fluorescence complementation reveals simultaneous formation of alternative CBL/CIPK complexes in planta. Plant J 56(3):505–516
23. Shaner NC, Campbell RE, Steinbach PA et al (2004) Improved monomeric red, orange and yellow fluorescent proteins derived from Discosoma sp. red fluorescent protein. Nat Biotechnol 22(12):1567–1572
24. Tischer BK, Kaufer BB (2012) Viral bacterial artificial chromosomes: generation, mutagenesis, and removal of mini-F sequences. J Biomed Biotechnol 2012:1. doi:10.1155/2012/472537
25. Warming S, Costantino N, Court DL et al (2005) Simple and highly efficient BAC recombineering using galK selection. Nucleic Acids Res 33(4):e36
26. Adler H, Messerle M, Koszinowski UH (2003) Cloning of herpesviral genomes as bacterial artificial chromosomes. Rev Med Virol 13(2):111–121
27. Borst EM, Hahn G, Koszinowski UH et al (1999) Cloning of the human cytomegalovirus (HCMV) genome as an infectious bacterial artificial chromosome in Escherichia coli: a new approach for construction of HCMV mutants. J Virol 73(10):8320–8329
28. Messerle M, Crnkovic I, Hammerschmidt W et al (1997) Cloning and mutagenesis of a herpesvirus genome as an infectious bacterial artificial chromosome. Proc Natl Acad Sci USA 94(26):14759–14763
29. Jach G, Pesch M, Richter K et al (2006) An improved mRFP1 adds red to bimolecular fluorescence complementation. Nat Methods 3(8):597–600
30. Kerppola TK (2009) Visualization of molecular interactions using bimolecular fluorescence complementation analysis: characteristics of protein fragment complementation. Chem Soc Rev 38(10):2876–2886
31. Hu CD, Kerppola TK (2003) Simultaneous visualization of multiple protein interactions in living cells using multicolor fluorescence complementation analysis. Nat Biotechnol 21(5): 539–545

Chapter 4

Discovery of Host–Viral Protein Complexes During Infection

Daniell L. Rowles, Scott S. Terhune, and Ileana M. Cristea

Abstract

Viruses have co-evolved with their hosts, developing effective approaches for hijacking and manipulating host cellular processes. Therefore, for their efficient replication and spread, viruses depend on dynamic and temporally regulated interactions with host proteins. The rapid identification of host proteins targeted by viral proteins during infection provides significant insights into mechanisms of viral protein function. The resulting discoveries often lead to unique and innovative hypotheses on viral protein function. Here, we describe a robust method for identifying virus–host protein interactions and protein complexes, which we have successfully utilized to characterize spatial–temporal protein interactions during infections with either DNA or RNA viruses, including human cytomegalovirus (HCMV), herpes simplex virus type 1 (HSV-1), pseudorabies virus (PRV), human immunodeficiency virus (HIV-1), Sindbis, and West Nile virus (WNV). This approach involves cryogenic cell lysis, rapid immunoaffinity purification targeting a virus or host protein, followed by identification of associated proteins using mass spectrometry. Like most proteomic approaches, this methodology has evolved over the past few years and continues to evolve. We are presenting here the updated approaches for each step, and discuss alternative strategies allowing for the protocol to be optimized for different biological systems.

Key words Virus–host interactions, Infection, Epitope tag, Immunoaffinity purification, Cryogenic cell lysis, Proteomics, Protein complex, Protein–protein interactions, Mass spectrometry

1 Introduction

The ability to identify host targets of viral proteins during the progression of an infection is invaluable for determining the functions of different viral proteins and identifying targets for therapeutic intervention. With the development of effective approaches to isolate protein complexes and identify and quantify co-isolated proteins using highly sensitive and accurate mass spectrometers, we are now able to quickly determine viral protein targets. Cristea et al. [1] have previously developed a rapid immunoaffinity purification approach of protein complexes that has been used successfully during infections with either DNA or RNA viruses. Furthermore, this approach has been used to isolate viral and cellular proteins as

Susanne M. Bailer and Diana Lieber (eds.), *Virus-Host Interactions: Methods and Protocols*, Methods in Molecular Biology, vol. 1064, DOI 10.1007/978-1-62703-601-6_4, © Springer Science+Business Media, LLC 2013

baits during infection, allowing the study of protein–protein interactions from either the virus or host perspective. Here, we present this strategy that, in our hands, has proven to be effective and robust for isolating and identifying protein complexes (Fig. 1). Numerous approaches are available for isolating protein complexes, and selecting the appropriate approach is dependent upon your biological system and available resources. Ultimately, discovering the targets of viral proteins within infected cells provides significant insights into viral protein function.

Using proteins within the yeast nuclear pore complex as baits, Cristea et al. [1] initially developed the method discussed in this chapter following extensive optimization to integrate information regarding protein localization with the rapid isolation of native protein complexes from eukaryotic cells. Optimizations included determining effective cell-lysis conditions, the composition of lysis buffer for protein complex extraction, the amount of resin used for immunoaffinity purification, and the incubation time of the antibody with the lysate. The method was later successfully applied to study virus–host protein interactions using the Sindbis viral protein nsP3 as bait during virus replication in mammalian cells [2]. Efficient isolation of protein complexes is achieved by combining cryogenic disruption of the infected cells into a frozen powder, *see* Subheading 3.2.3, with optimized lysis buffer conditions for each target protein complex, *see* Subheading 3.4.3 [3]. The method is also optimized using magnetic resin (beads) to which antibodies are covalently attached [4]. These steps reduce the amount of nonspecific proteins and IgG contamination and are discussed in Subheading 3.3. Finally, and central to the success of this method, is the time of incubation of the antibody-conjugated beads with the cell lysate. Incubation times longer than 60 min resulted in increased isolation of nonspecific proteins with no significant increase in specific interactions [1]. We discuss incubation times within Subheading 3.4.1. In addition to optimizing lysis conditions and the efficiency of isolation, the ability to interpret the results and determine the specificity of identified interactions requires the use of appropriate controls. Although the method has been optimized to reduce nonspecific interactions, these interactions still occur, and we discuss the various sources of nonspecific interactions in Subheading 3.5. The use of appropriate controls will help distinguish between specific and nonspecific binding proteins.

Over the last few years, the method presented in this chapter or aspects of it have been applied to isolating several viral proteins using different epitope and antibody combinations. We discuss these in Subheading 3.1. These included antibodies to GFP (green fluorescent protein) [2, 5–7] or FLAG epitope tags [8, 9], protein-specific antibodies [10], as well as single-step isolations using IgG to target the Protein A component of a TAP (tandem affinity purification) tag [6, 11] (Table 1). In all the studies in which we have utilized the approach described in this chapter, identifying viral–

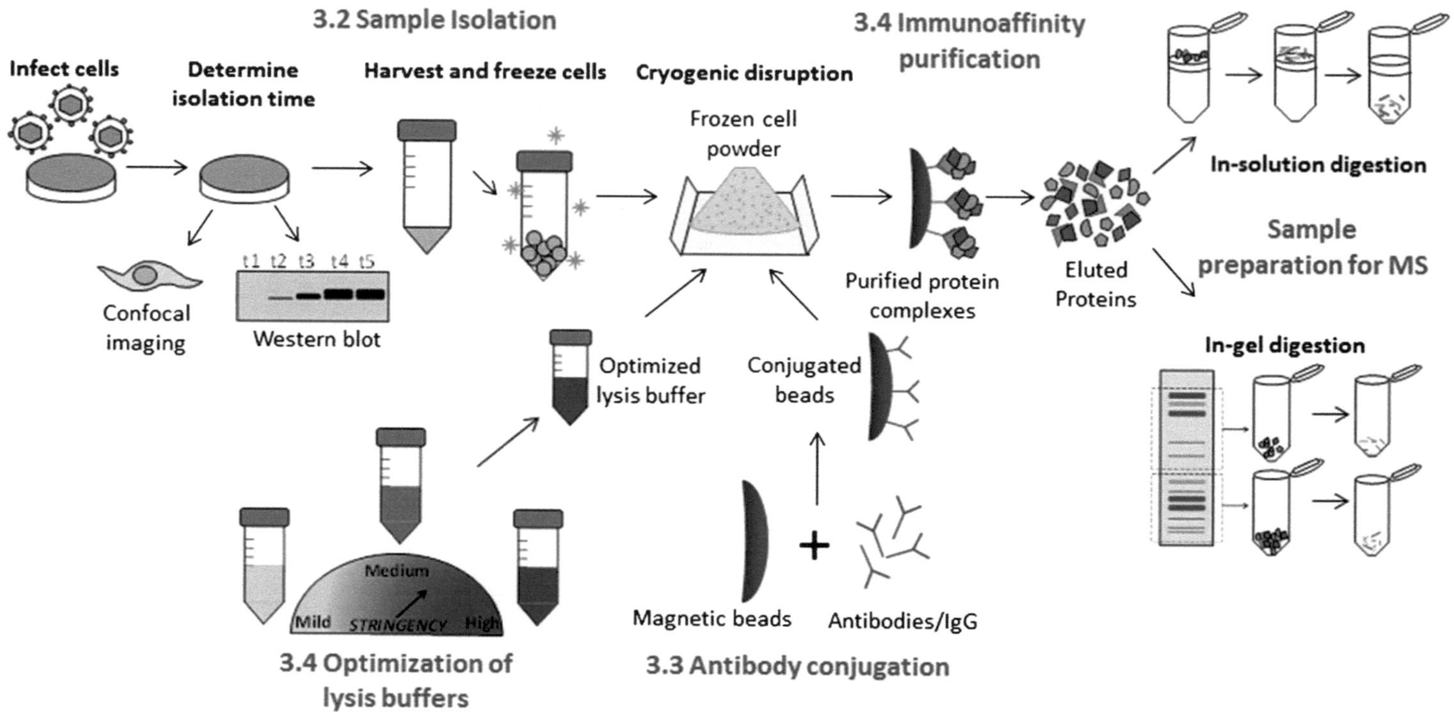

Fig. 1 Workflow for the immunoisolation of protein complexes from viral infected cells in culture

Table 1
Examples of epitope and epitope-specific antibodies that we have utilized for studying virus–host protein interactions

Epitope tag[a]	Mass	Antibody	Reference
TAP 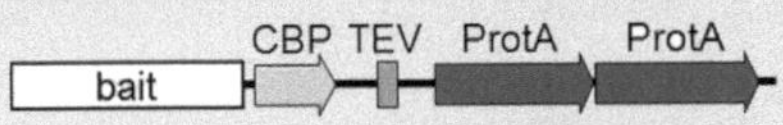	20 kDa	IgG	[6, 11]
GFP 	27 kDa	Anti-GFP	[1, 2, 5–7, 12–14, 24, 25]
FLAG 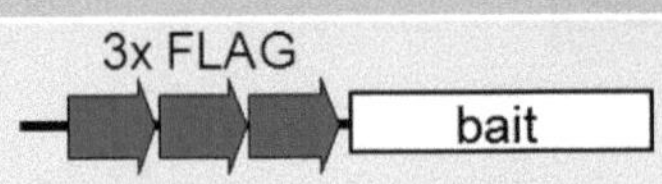	1 kDa/FLAG	Anti-FLAG	[8, 9]

[a]Epitopes can be introduced at either the amino or carboxyl terminus of the bait protein, or on an internal flexible loop of the protein, provided that it does not interfere with protein function

host protein interactions provided significant mechanistic insight into viral protein activity during infection (Table 2). For example, using a full-length replication-competent Sindbis virus (SINV) expressing the viral nonstructural protein nsP3 tagged with GFP, Cristea et al. [2] were able to track viral protein localization during infection using fluorescence microscopy, as well as identify changes in nsP3 interactions at different time points of SINV infection. Using a Protein A tag, Moorman et al. [11] identified tuberin (TSC2), a negative regulator of mTOR (mammalian target of rapamycin), as the target of the human cytomegalovirus (HCMV) protein, pUL38. Expression of pUL38 was demonstrated to be both necessary and sufficient to maintain mTOR *kinase* activity during conditions of nutrient stress [11]. Similarly, building spatial–temporal interaction networks for several GFP-tagged HCMV proteins has provided insights into processes involved in virion assembly, indicating the presence of multiple intermediate virion structures that traffic independently from each other prior to merging into larger vesicles [5]. Using a FLAG epitope, Reitsma et al. [9] determined the target of the HCMV protein, pUL27, to be the cellular acetyltransferase complex, Tip60. Expression of pUL27 was shown to be necessary and sufficient to degrade Tip60 resulting in the expression of the CDK inhibitor, p21CIP1 and a transient cell cycle arrest [9]. And, most recently, Youn et al. [10] used a specific antibody against West Nile virus (WNV) NS1 protein to determine an interaction with WNV NS4B. These studies identified a possible mechanism linking NS1 to regulating WNV RNA replication [10]. Beyond viral proteins, immunoaffinity purification of protein complexes using the approach described in this chapter has also been used for identifying targets of cellular proteins including histone deacetylase 1 [12] and 5 [13], sirtuin 7 [14], metadherin [15], and interferon-inducible protein 16 [16] (Table 3). For these applications, GFP epitope or antibody against

Table 2
Examples of lysis buffers that in our hands have yielded successful isolations of different viral proteins with a range of properties

Bait	Description and localization	Tag	Virus	Cell type	Optimized lysis buffer[a]	Reference
nsP3	Nonstructural protein 3 (cytoplasm)	GFP	Sindbis Virus	Rat2 cells	1 % Triton, 0.5 % deoxycholate, 500 mM NaCl, 25 units/mL DNase	[2]
pUL38	Early phase nonstructural protein (dynamic localization in nucleus and cytoplasm)	ProtA	HCMV	HFF	1 % Triton X-100, 250 mM NaCl, 4 μg/mL DNase, 20 mg/ml phenylmethylsulfony fluoride, 0.4 mg/mL pepstatin A	[11]
pUL99, pUL32	Both involved in assembly (cytoplasm)	GFP	HCMV	HFF	1 μM $ZnCl_2$, 1 μM $CaCl_2$, 1 % Triton, 250 mM NaCl	[5]
pUL83	Tegument protein; (nuclear and cytoplasmic)	GFP	HCMV	HFF	1 % Triton X-100, 250 mM NaCl, 4 μg/mL DNase, 20 mg/ml phenylmethylsulfony fluoride , 0.4 mg/mL pepstatin A	[6]
		ProtA				
pUL27	Early phase protein (nuclear)	3xFLAG	HCMV	HFF	1 % Triton X-100, 250 mM NaCl, 4 μg/mL DNase	[9]
nsP4	Nonstructural RNA-dependent RNA polymerase (cytoplasmic)	3xFLAG	Sindbis Virus	Rat2 cells	1 % Triton X-100, 0.5 % deoxycholate, 500 mM NaCl, 25 U DNase and RNase/mL	[8]
NS1	Nonstructural glycoprotein (cell surface and secreted)	No tag	West Nile Virus	BHK21-15	1 μM $ZnCl_2$, 1 μM $CaCl_2$, 1 % Triton, 0.5 % sodium deoxycholate, 0.3 % sodium N-lauroylsarcosine, 0.1 M NaCl, and phosphatase inhibitor cocktail	[10]

[a]All lysis buffers contained 20 mM K-HEPES pH 7.4, 110 mM KOAc, 2 mM $MgCl_2$, 0.1 % Tween 20, protease inhibitor mixture in addition to optimized lysis buffer components listed

Table 3
Examples of lysis buffer conditions that we previously used to isolate mammalian host proteins from both cultured cells and tissue

Bait	Description and localization	Tag	Cell type	Optimized lysis buffer[a]	Reference
Nup37 Nup43	Member of nuclear pore complex and subunit of Nup107-160 subcomplex (nuclear membrane)	GFP	HeLa	0.5 % Triton, 200 mM NaCl, 20 mg/mL PMSF, 0.4 mg/mL pepstatin A	[1]
GluRδ2	Postsynaptic densities; (cerebellar excitatory synapses)	GFP	Mouse tissue	10 mM HEPES[b], pH 7.4, 2 mM $CaCl_2$, 132 mM NaCl, 3 mM KCl, 2 mM $MgSO_4$, 1.2 mM NaH_2PO_4, 0.5 % Triton X-100, 1/100 (v/v/) protease inhibitor cocktail	[25]
HDAC1	Histone deacetylase 1 (nucleus)	GFP	HFF	1 μM $ZnCl_2$, 1 μM $CaCl_2$, 0.5 % Triton X-100, 250 mM NaCl	[12]
H3	Histone 3 isoforms (nucleus)	YFP	mouse ES cells	0.5 % Triton, 300 mM NaCl	[24]
HDAC5	Histone deacetylase 5 (nucleus and cytoplasm)	GFP	HEK293	1 μM $ZnCl_2$, 1 μM $CaCl_2$, 0.5 % Triton X-100, 250 mM NaCl, 4 μg/mL DNase, and phosphatase inhibitor cocktails	[13]
SIRT7	Sirtuin 7 (nucleoli)	GFP	HEK293	1 μm $ZnCl_2$, 1 μm $CaCl_2$, 0.5 % Triton X-100, 250 mm NaCl, 4 μg/ml DNase, and phosphatase inhibitor cocktails	[14]
IFI16	Interferon inducible protein 16 (nucleus and cytoplasm)	No tag	CEM T	1 μM $ZnCl_2$, 1 μM $CaCl_2$, 0.6 % Triton X-100, 200 mM NaCl, 10 μg/mL DNase I, phosphatase inhibitor cocktail	[16]

[a]All lysis buffers contained 20 mM K-HEPES, pH 7.4, 110 mM KOAc, 2 mM $MgCl_2$, 0.1 % Tween 20, 1/100 protease inhibitor mixture in addition to optimized lysis buffer components listed

[b]This buffer is the complete buffer composition without core components listed above

the endogenous protein was used. This chapter describes a protocol used to immunoaffinity purify protein complexes from infected cells and identify their protein composition (Fig. 1).

2 Materials

2.1 Epitope Selection

1. TAP sequence containing Protein A [17] using IgG.
2. GFP sequence [18] or variants using an anti-GFP antibody.
3. Multiple repeats of the FLAG sequence [19] using anti-FLAG antibody.

2.2 Selection, Preparation, and Lysis of Infected Cells

2.2.1 Preparing Infected Cells

1. Infected mammalian cells in tissue culture dishes or flasks.
2. DPBS pH 7.4 Dulbecco's Phosphate-Buffered Saline 1× (Invitrogen).
3. Rubber policeman or a plastic cell scraper.
4. Centrifuge at 4 °C.
5. 250 mL cone tipped centrifuge bottle or 50 mL conical tubes.
6. Cryo Buffer: 20 mM Na–HEPES, 1.2 % (w/v) polyvinylpyrrolidone, pH 7.4, 1/100 (v/v), protease inhibitor cocktail (Sigma) added immediately before use.

2.2.2 Freezing Cell Pellet

1. Metal tube rack.
2. Ice bucket or liquid nitrogen safe container (e.g., Styrofoam).
3. Liquid nitrogen.
4. 50 mL screw cap tube.
5. 18-gauge needle.
6. 10 mL syringe.

2.2.3 Cryogenic Disruption of the Cell Pellets

1. Retsch MM 301 Mixer Mill.
2. Two 25 mL or two 10 mL stainless steel jars with tungsten carbide grinding balls (Retsch, Newtown, PA).
3. Liquid nitrogen in an open container.
4. Long forceps.
5. Metal spatula.
6. Windex, methanol, and ultrapure H_2O for cleaning.
7. 50 mL conical tubes.
8. Alternative method (*see* Subheading 3.2.5): 2.0 mL adaptor (Retsch), 3 mm grinding balls (Retsch), and 2.0 mL Safelock tubes (Eppendorf).

2.3 Antibody Conjugation

1. Dynabeads M-270 Epoxy (Invitrogen).
2. Magnetic tube rack.
3. Round bottomed microcentrifuge tubes (Eppendorf).
4. Rotating tube rack in 30 °C incubator or equivalent environment.
5. Tube shaker (e.g., Tomy shaker).
6. Immunoglobin G (IgG) or high affinity purified antibodies against the selected tag or bait protein.
7. Sodium Phosphate Buffer: (0.1 M, pH 7.4): 19 mM NaH_2PO_4, 81 mM Na_2HPO_4 in water.
8. Ammonium sulfate (3 M): Prepared in 0.1 M Sodium Phosphate Buffer.
9. Glycine–HCl (100 mM, pH 2.5): Prepared in water and pH adjusted with HCl. Filter sterilize with 0.2 μm filter.
10. Phosphate-Buffered Saline (PBS).
11. 0.02 % sodium azide (NaN_3): dissolved in PBS.
12. 0.5 % Triton X-100: dissolved in PBS.
13. Tris–HCl (10 mM, pH 8.8): Prepared in water and pH adjusted with HCl.
14. Triethylamine (100 mM): Prepared immediately before use in water.

2.4 Immunoaffinity Purification: Strategy, Optimization, and Assessment of Efficiency

2.4.1 Immunoaffinity Purification of Protein Complexes

1. Polytron homogenizer (e.g., PT 10-35 Polytron from Kinematica).
2. Centrifuge and rotor rated to spin 8,000 × *g* at 4 °C (rotor must fit a 50 mL conical tube).
3. Rotating tube rack at 4 °C.
4. Round bottomed microcentrifuge 1.5 mL tubes.
5. Axygen microcentrifuge tube (to reduce polymer contamination).
6. Heat block at 70 °C.
7. Tube shaker (e.g., Tomy shaker).
8. Magnetic tube rack for 1.5 mL tubes and bar magnet for conical tubes.
9. Vacuum evaporator (e.g., speedvac).
10. Frozen cell powder (*see* Subheading 3.2.3).
11. Optimized Lysis Buffer and wash buffer (*see* Subheading 3.4.3).
12. Antibody-conjugated magnetic beads (*see* Subheading 3.3).
13. Ammonium Hydroxide (14.8 M): (Sigma) Stored at 4 °C.
14. Base Elution Buffer: 5 μL of 0.5 M EDTA, pH 8.0, and 169 μL of Ammonium Hydroxide diluted in 4.826 mL of ultrapure H_2O, prepared fresh before use.

15. 4× LDS elution buffer: 0.666 g of Tris–HCl, 0.682 g of Tris-Base, 0.8 g of LDS, and 0.006 g of EDTA (free acid) in ultrapure dH_2O to a final volume of 10 mL. Aliquot and store at −20 °C.
16. 1× NuPAGE Sample Buffer (Invitrogen).
17. 10× Reducing Agent (Invitrogen).
18. 1 M Iodoacetamide (IAA): (Sigma) 0.185 g of Iodoacetamide dissolved in 1 mL HPLC grade water stored at −20 °C.

2.4.2 Assessing Immunoaffinity Purification Efficiency

1. Saved samples from Subheading 3.4.1 (input supernatant, cell pellet, flow through, primary eluate, and secondary eluate).
2. 1× NuPAGE LDS Sample Buffer and 1× Reducing Agent (Invitrogen).
3. 2 % SDS.
4. Acetone.
5. Heat block at 70 °C.
6. NuPAGE 4–12 % Bis–Tris precast SDS-PAGE gel (10-well) (Invitrogen).
7. Xcell SureLock Mini-Cell electrophoresis system (Invitrogen).
8. 20× NuPAGE MOPS SDS Running Buffer (Invitrogen).

2.4.3 Optimization of Lysis Buffer Conditions

1. At least 3 Lysis Buffers at varying stringency (*see* Subheading 3.4.3).
2. Frozen cell powder (*see* Subheading 3.2.3).
3. Table top centrifuge rated to spin 20,000 ×g at 4 °C.
4. Sonicator.
5. Tris–HCl: (50 mM pH 7.4) containing 2 % SDS.
6. NuPAGE 4–12 % Bis–Tris precast SDS-PAGE gel (10-well) (Invitrogen).
7. 1× NuPAGE LDS Sample Buffer and 1× Reducing Agent (Invitrogen).
8. Xcell SureLock Mini-Cell electrophoresis system (Invitrogen).
9. 20× NuPAGE MOPS SDS Running Buffer (Invitrogen).

3 Methods

3.1 Epitope Selection

Numerous protein epitope and epitope-specific antibody combinations have been successfully used in isolating protein complexes for identifying their composition by mass spectrometry. Although many combinations are possible, we have limited our discussion to the tags that have been used specifically with the described method. These tags include TAP (tandem affinity purification) [6, 11], GFP (green fluorescent protein) [2, 5–7], and FLAG epitopes [8, 9]. Starting from the amino terminus of the bait protein, the TAP tag [17] (Table 1) consists of a calmodulin binding peptide (CBP), a tobacco etch virus protease (TEV) cleavage site and two IgG-binding units of protein A (ProtA) from *Staphylococcus aureus*. The TAP tag was originally used for sequential purification steps with IgG followed by TEV cleavage and re-isolation by Calmodulin-coated beads [17]. Our use of the TAP tag was limited to isolation with IgG to increase the identification of weak binders that might be lost following sequential purification steps [6, 11]. Although the tag is effective in isolating complexes, one limitation to consider is the large size of the epitope at 20 kDa. Another epitope successfully used in immunoisolating protein complexes is the fluorescent protein GFP (Table 1). Although GFP is also relatively large, at 27 kDa, it allows for both visualization and complex isolation [1] as demonstrated by Cristea et al. [2] using the Sindbis viral protein, nsP3, during infection. Immunoisolation is accomplished using an antibody against GFP [2, 5–7]. For several applications, GFP was inserted at the carboxyl terminus of the bait and the starting methionine of GFP was removed to avoid the possibility of internal initiation and expression of GFP alone (Table 1). Finally, immunoisolation has been completed using the FLAG tag (Table 1). The FLAG epitope consists of eight hydrophilic amino acids (AspTyrLysAspAspAspAspLys) [19] and the advantage of the tag is its small size at 1 kDa. Isolation of protein complexes has been completed using three repeats of the FLAG sequence expressed in-frame with the bait protein [8, 9]. Several antibodies are available against the FLAG epitope including the monoclonal Anti-FLAG M2 antibody (Sigma Aldrich). For all three epitopes discussed, it is possible to use either amino- or carboxyl-terminal tags on your bait, or on an internal flexible loop of the protein, provided that it does not interfere with protein function.

Introducing a unique epitope into the bait has the potential to alter the protein and associated functions. Therefore, it is critical to evaluate the known aspects of the bait to ensure wild-type protein function. For most of the examples referenced in this chapter, epitopes were introduced into the viral genomes and protein complexes were identified during infection. These recombinant viruses were evaluated for their ability to replicate with wild-type kinetics [2, 5, 6, 8, 11]. A failure to replicate efficiently suggests that the tag has disrupted an important function of your bait. For proteins which have been previously demonstrated to be nonessential during replication *in vitro*, alternative approaches will be needed to assess functionality. For example, the HCMV protein pUL27 is nonessential for replication but is necessary for the full antiviral activity of an antiviral agent [9]. Therefore, the recombinant virus was evaluated for its sensitivity to the compound. When a protein-specific antibody is available, it is also possible to evaluate the tagged protein using immunofluorescence analysis to ensure wild-type protein localization during infection. In the end, the existing phenotypical data must be intact in the tagged version in order to effectively link the identified interacting proteins to viral protein function during infection.

3.2 Selection, Preparation, and Lysis of Infected Cells

With the understanding that every viral protein is unique, successful isolation of target proteins requires optimization. Two steps requiring optimization include (1) selection of an appropriate time for isolation during infection and (2) identification of an optimal lysis condition (*see* Subheading 3.4.3). The goal of this protocol is to effectively isolate and identify protein complexes containing the viral protein of interest with the long-term goal of determining the functional relevance of the interactions.

Identifying the appropriate time to isolate the protein complexes during infection is critical. The time can often be determined based upon the timing of expression, as well as upon phenotypical data using viruses containing mutations in the protein of interest and loss of function. Confocal microscopy can be used to visualize the temporal expression and subcellular localization of the viral proteins of interest. If the viral protein is tagged with a fluorescent tag (e.g., GFP), time-lapse microscopy using direct fluorescence can provide an in-depth view of the dynamics of protein expression and localization. Viral protein abundance should also be considered in the decision. In general, highly abundant viral proteins are easier to isolate and analyze for identification of binding partners. Isolation of low abundant proteins can also be achieved with optimization. While the increasing sensitivity of mass spectrometric configurations allows the conclusive identification and quantification of low abundance proteins, the analysis will only be as good as the isolation. When analyzing a complex, one has to consider not only the abundance of the bait protein (i.e., the protein targeted for immunoaffinity purification) but also the stoichiometry within the isolated protein complex. Therefore, the ability to identify interactions present at low levels increases if the isolation is effective and performed with a sufficient amount of starting material. The efficiency of isolation for low abundance proteins can be improved by increasing the multiplicity of infection, as well as by increasing the number of cells used to isolate protein complexes. Additionally, one can increase the number of cells in a stepwise manner to account for the differential expression level of viral proteins at different time points of infection. A good starting point will be to infect approximately 1×10^8 cells, which for most viral proteins would provide sufficient amounts of affinity purified proteins for mass spectrometry analyses.

3.2.1 Preparing Infected Cells

1. Infect approximately 1×10^8 cells using conditions defined by the specific virus and biological system being used. For most adherent cells, using 10× 15 cm dishes of cells will provide sufficient material (*see* **Note 1**).
2. Harvest infected cells at the optimal time postinfection by aspirating off the media and washing 2× with 5 mL of cold PBS (4 °C) to remove culture serum.
3. For adherent cells, add 5 mL cold PBS and collect cells by scraping and transferring into a 250 mL cone tip centrifuge bottle on ice (*see* **Note 2**).
4. Wash the dish with 5 mL cold PBS to collect the residual cells and combine with the first harvest.
5. Pellet cells by centrifugation at $300 \times g$ for 10 min at 4 °C and aspirate the PBS.
6. Resuspend the cell pellet in 10 mL cold PBS and transfer it to a 15 mL screw-capped tube.
7. Pellet cells by centrifugation at $300 \times g$ for 10 min at 4 °C and aspirate off the PBS.
8. Determine the weight of the cell pellet by subtracting the weight of the tube from the sample. Using 1×10^8 cells will result in approximately 1.0 g of sample material (*see* **Note 3**).
9. Add 100 μL Cryo Buffer to 1.0 g of sample and mix with the pipet tip. Add protease inhibitor cocktail (PIC) at a concentration of 1:100 (v/v) to the Cryo Buffer and mix by gentle inversion of the tube just before adding the buffer to the sample. The cell pellet should have the appearance of a thick paste (*see* **Note 4**).

3.2.2 Freezing Cell Pellet

1. Place a 50 mL screw-capped tube into a metal tube rack and place it in an empty ice bucket.
2. Add ~40 mL of liquid nitrogen into the 50 mL screw-capped tube and then also fill the ice bucket with liquid nitrogen to just above the rack height. Follow appropriate safety precautions when using liquid nitrogen.
3. Using a 18-gauge needle, puncture the bottom of the 15 mL sample tube and, with the plunger from a 10 mL syringe, slowly push the sample in Cryo Buffer out of the 15 mL tube and let the sample drop as pellets into the 50 mL tube containing liquid nitrogen (*see* **Note 5**).
4. Collect the remaining material using a wide bore pipette tip and drop the material into the liquid nitrogen filled 50 mL tube (*see* **Note 6**). Carefully, puncture the lid of the 50 mL tube several times with an 18-gauge needle and then cap the tube. About 4–5 holes are enough to vent the liquid nitrogen and gas (*see* **Note 7**).

5. Empty out liquid nitrogen from the capped 50 mL tube by reversing the tube and letting the liquid nitrogen drop in the ice bucket.
6. The remaining frozen pellets can be stored inside the 50 mL conical tube at −80 °C until you are ready to proceed with the rest of the steps. Ensure that there is no liquid nitrogen remaining in the 50 mL conical tube before storing the sample in the −80 °C freezer.

3.2.3 Cryogenic Disruption of Cell Pellets

After the infected cells are frozen into small pellets, it is critical that the sample remains frozen at all times through the steps described below, all the way until the lysis step (*see* Subheading 3.4.1). If the material partially thaws, it could cause decreased efficiency of extraction and increased nonphysiological protein associations. If working with a high risk infectious virus, *see* Subheading 3.2.4, or with very small amounts of material, *see* Subheading 3.2.5.

1. Clean the grinding materials (spatula, Retsch Mixer mill jars, and grinding balls) stepwise with dH_2O, Windex, ultrapure dH_2O, and 100 % methanol, and then allow them to dry in the chemical hood. It is important that the jars are completely dry at the end of **step 1**, because any water remaining in the lids will freeze during **steps 2** and **3** and impede the closing and opening of the jar.
2. Prechill the jars and balls in an ice bucket or Styrofoam container filled with liquid nitrogen. Once the liquid nitrogen no longer has a bubbling appearance, the jars and balls have reached the proper temperature. Carefully remove them from the liquid nitrogen using forceps and deposit the frozen cell pellets inside the jars. For optimal grinding, the jars should be filled to no more than one-third of the volume of the jar (e.g., max. of 3 g cell pellets per 10 mL jar, and 7 g cell pellets per 25 mL jar).
3. Place a chilled ball on top of the cell pellets, close the jar, and return the jar to the liquid nitrogen container for additional cooling.
4. Place the filled jars in the holders of the Retsch Mixer Mill and set the mixer for grinding at 30 Hz. If only one jar is needed, use an empty jar without the ball as a balance. Perform 20 cycles of 2.5 min each with cooling of the jars following each grinding cycle. To cool the jars between the cycles, remove the jars and chill in the liquid nitrogen container. The jars may loosen during the grinding process so check that the lids are tightly closed before cooling the jars in liquid nitrogen.
5. Once all grinding cycles are finalized, cool the jars in liquid nitrogen. Once cooled, remove the jars from the liquid nitrogen, open the jar, and remove the cell powder with a chilled

spatula, placing the frozen cell powder into a 50 mL conical tube on dry ice. Work in this step must be done quickly to avoid the thawing of cell powder. It is helpful to keep the liquid nitrogen on hand for chilling the spatula several times as the power is removed from the jars. Store the sample at −80 °C.

3.2.4 Alternative Cryogenic Grinding-Premixing Lysis Solution

The goal of this procedure is to minimize contact with the infectious cell powder.

1. Prepare frozen cell pellets by adding ~200 μL Cryo Buffer per 1 g cells, and freezing the cell pellets by slow pipetting of the cell solution into a 50 mL conical tube containing liquid nitrogen. In contrast to Subheading 3.2.2, do not use needles and syringe plungers for freezing the cell pellets.
2. Prepare an appropriate Lysis Buffer for the immunoaffinity purification, optimizing its composition as described in Subheading 3.4.3, allowing for 5 mL of buffer per 1 g of cells.
3. Freeze the Lysis Buffer by pipetting drop wise into a 50 mL conical tube containing liquid nitrogen.
4. Store separately the frozen cell pellets and the frozen Lysis Buffer pellets until ready to perform the cryogenic grinding and immunoaffinity purification.
5. To perform the cryogenic disruption, combine the frozen cell pellets and the frozen Lysis Buffer pellets and carry out the grinding as described in Subheading 3.2.3.
6. When the grinding is completed, place the jars in a bucket filled with ice, and allow the ground cell-lysis mixture to slowly thaw. Transfer the lysate to a 50 mL conical tube, add additional protease inhibitor cocktail (1/100 v/v), and continue directly with the immunoaffinity purification procedure (Subheading 3.4.1).

3.2.5 Alternative Cryogenic Grinding-Smaller Sample Size

If using <0.5 g of sample as well as requiring the samples to be maintained in a sealed tube, it is possible to grind samples within 2.0 mL tubes using adaptors for the Retsch Mixer Mill and 3 mm grinding balls.

1. Clean the grinding materials (spatula, tweezers, and grinding balls) stepwise with dH_2O, Windex, ultrapure dH_2O, and 100 % methanol, and then allow them dry in the chemical hood.
2. Prepare frozen cell pellets by adding ~200 μL Cryo Buffer per 1 g cells, and freezing the cell pellets by slow pipetting of the cell solution into a 50 mL conical tube containing liquid nitrogen. In contrast to Subheading 3.2.2, do not use needles and syringe plungers for freezing the cell pellets.
3. Prechill the adaptor, 2.0 mL tubes and balls in an ice bucket or Styrofoam container filled with liquid nitrogen. Once the

liquid nitrogen no longer has a bubbling appearance, the tubes and balls have reached the proper temperature.

4. Carefully remove tubes from the liquid nitrogen using forceps and empty residual liquid nitrogen from the tubes.
5. Using prechilled tweezers, deposit the frozen cell pellets inside the tubes and do not fill more than one-third of the volume. Add one prechilled 3 mm grinding ball and cap tubes, returning them to liquid nitrogen.
6. Place tubes with sample into chilled adaptor and return to liquid nitrogen.
7. Place adaptor in the holders of the Retsch Mixer Mill and set the mixer for grinding at 30 Hz. If only one adaptor is needed, use an empty adaptor without tubes as a balance. Perform 6 cycles (or until completely disrupted) of 2.5 min each with cooling of the adaptor in the liquid nitrogen following each grinding cycle. It is helpful to keep the liquid nitrogen on hand for chilling the adaptor.
8. Store the sample within the 2.0 mL tubes at −80 °C. The optimized Lysis Buffer can be added directly to each 2.0 ml tube, *see* Subheading 3.4.1.

3.3 Antibody Conjugation

This protocol has been optimized for use with Epoxy-coated magnetic beads of 2.8 μm diameter (Invitrogen). However, the same procedure can be utilized with other sizes of magnetic beads and other types of beads by adjusting the amount of antibodies to account for the difference in the bead surface of binding. This conjugation protocol can be utilized for conjugating antibodies against tags (e.g., GFP, FLAG), antibodies raised directly against the protein of interest, as well as IgG (for isolation of proteins tagged with Protein A or with a TAP tag containing a Protein A component) (Table 1). The antibodies utilized should be affinity purified, and the solutions in which the antibodies are stored are critical. As the binding to epoxy occurs through primary amines, ensure that the antibodies used are not stored in Tris Buffer. Also, check that the antibody solution does not contain other proteins, such as BSA, that would compete with the antibody for binding to the beads and would significantly affect the identity of co-isolated proteins. It is best to do **steps 1–8** of the conjugation in the afternoon and continue **steps 9–17** the following morning. Do not allow the beads to dry in between the wash steps. All steps can be performed at room temperature.

When considering the amount of beads needed, one must consider the amount of cellular material being used and the abundance of the protein of interest. A small scale pilot experiment using less than 1×10^7 cells may require only 1–2 mg of beads, while a single isolation followed by comprehensive mass spectrometry analyses would require 5–10 mg of beads. If the protein of interest is not highly abundant, then using 10–20 mg of beads could be appropriate. A good starting point is 10 mg of beads to 1 g of ground sample.

1. Weigh out the appropriate amount of Dynabeads into a 2 mL round-bottom tube.
2. Wash the beads by adding 1 mL of Sodium Phosphate Buffer (pH 7.4) and mix using a vortex for 30 s.
3. Place tube containing the beads in the Sodium Phosphate Buffer on a shaker for additional mixing for 15 min. A TOMY shaker or an alternative shaker with vigorous mixing (e.g., medium shaking setting on a vortex) is recommended to agitate the tube and keep the beads from settling on the bottom (*see* **Note 8**).
4. Place the tube on a magnetic rack to separate the beads from the solution. Once the solution becomes clear and the beads are collected on the magnet side, remove and discard the buffer using a pipette.
5. Remove the tube containing only the beads from the magnet and add 1 mL of fresh Sodium Phosphate Buffer. Mix by vortexing for 30 s. Return the tube to the magnetic rack and remove the buffer as described above.
6. Remove the tube from the magnet and begin conjugating the antibody (*see* **Note 9**).
7. To conjugate 10 mg of beads, the total volume of the conjugation reaction will be 200 μL. The volume of 0.1 M Sodium Phosphate Buffer equals 200 μL—volume of antibody—66.66 μL 3 M Ammonium Sulfate. Ammonium Sulfate solution is added last for a final concentration of 1 M (*see* **Note 10**).
8. Seal the tube with parafilm and rotate the slurry overnight at 30 °C. Use an end-over-end rotation of the tube to help keep the beads suspended in the antibody solution throughout the conjugation period, and avoid the settling of the beads at the bottom of the tube.
9. The following morning, place the tube on the magnetic rack. Remove the supernatant using a pipette, and save it to test the efficiency of conjugation.
10. Wash the beads in 1 mL of Sodium Phosphate Buffer by gentle pipetting. Place the tube on the magnetic rack. To avoid losing beads inside the pipette tip, once the beads attach to the magnet wall, wash the tip using the buffer inside the tube, and allow again for the beads to settle on the magnet. Once the solution is clear, remove and discard the buffer.
11. Wash the beads with 1 mL 100 mM Glycine–HCl, pH 2.5 using the same procedure and cleaning of the tip introduced in the step above. However, it is critical that this is a FAST WASH (*see* **Note 11**).

12. Wash the beads with 1 mL of 10 mM Tris–HCl, pH 8.8 using the same procedure as the one described in **step 10**.
13. Wash the beads in 100 mM Triethylamine. This is a FAST WASH, so, remove the solution as soon as the beads are attached to the magnet.
14. Wash the beads four times with DPBS using 1 mL for each wash. Remove the buffer following each wash.
15. Add DPBS containing 0.5 % Triton X-100 to the beads and place the tube on a shaker (e.g., a TOMY shaker) for 15 min.
16. Wash the beads twice with DPBS using 1 mL for each wash. Remove the buffer following each wash.
17. Suspend the beads in DPBS with 0.02 % NaN_3 and store at 4 °C until their use in Subheading 3.4.1 (*see* **Note 12**).

3.4 Immunoaffinity Purification: Strategy, Optimization, and Assessment of Efficiency

3.4.1 Immunoaffinity Purification of Protein Complexes

1. Prepare two Lysis Buffers: (a) IP Buffer for suspending the cell powder and (b) Wash Buffer for washing the magnetic beads prior to and after the immunoaffinity purification. The composition of the IP Buffer is optimized as described in Subheading 3.4.3. Prepare 5–8 mL of IP Buffer per gram of frozen cell powder (*see* **Note 13**).
2. Bring the frozen cell powder from the −80 °C storage, and keep the cells on dry ice until ready to proceed with the experiment.
3. Place the tube containing the cell powder in a 4 °C ice bucket, and wait 20–30 s (*see* **Note 14**). Slowly add the IP Buffer to the frozen cell powder and periodically swirl the tube until the powder becomes solubilized (*see* **Note 15**).
4. Prepare a Polytron for the homogenization of the cell lysate. First, rinse the Polytron with ultrapure dH_2O, then fill a beaker with ultrapure dH_2O and allow the Polytron to run inside the beaker for 10 s.
5. To homogenize the cell lysate, run the Polytron for 2× 15 s at a speed of 22.5k with cooling on ice between the two steps. For this homogenization step, the cell lysate can be present in either a 15 or 50 mL conical tube. As foam is formed during this step, ensure that the volume of the cell lysate does not occupy more than 1/3 of the tube to avoid over flow.
6. If processing more than one sample, perform a rinse and a wash step in ultrapure dH_2O (*see* **step 4**) to clean the Polytron and avoid cross-contamination of samples (*see* **Note 16**).
7. Place the homogenized cell lysate on a slow rotor at 4 °C for 5 min.
8. To separate the insoluble fraction, subject the homogenized cell lysate to centrifugation at 8,000 ×g at 4 °C for 10 min.

9. While waiting for the completion of the centrifugation step, place the tube containing the conjugated magnetic beads on a magnetic rack to remove and discard the storage solution. Wash the beads three times with 1 mL cold Wash Buffer by gentle pipetting. To avoid drying of the beads, suspend the washed beads in 100–200 μL of Wash buffer and store them in the 4 °C ice bucket.
10. Once the centrifugation (**step 8**) is completed, retain the pellet and 40 μL of the supernatant for later analysis (*see* Subheading 3.4.2).
11. Move the clarified cell supernatant to a new conical tube. If using a centrifuge with swinging buckets, pour the cell supernatant into the new conical tube. If using a centrifuge with a fixed-angle container, then move the clarified cell supernatant to the new conical tube by pipetting. It is recommended to use a 50 mL conical tube if the cell supernatant is >10 mL, a 15 mL conical tube for cell lysates of 4–10 mL, and a 6 mL tube if the supernatant is <4 mL.
12. Carefully analyze the appearance of the cell supernatant, and ensure that there are no floating or mixed insoluble particles in the supernatant. If such particles are observed, remove them by carefully pipetting before proceeding. Particles may clog the beads and interfere with the affinity purification.
13. Add the necessary amounts of washed beads to the cell supernatant by pipetting (*see* Subheading 3.3).
14. Place the tube containing the cell lysate and the beads on a rotor at 4 °C and allow for 1 h of slow mixing (*see* **Note 17**).
15. Once the 1 h incubation is completed, separate the beads from the cell lysate by placing the tube into contact with a magnet (large bar magnet); secure the tube on the magnet using a rubber band. Wait 2–5 min (as necessary) for the beads to be fully captured on the magnet. Remove the flow through supernatant by pipetting, and save it into a new conical tube for later analysis (*see* Subheading 3.4.2).
16. Add 1 mL Wash Buffer to the beads in the large conical tube and transfer them to a 1.5 mL tube. Check that no beads are left in the large conical tube. If beads are still remaining in the conical tube, use some additional Wash buffer to transfer them to a 2.0 mL tube.
17. Place the tube in the magnetic rack and remove the Wash Buffer by pipetting.
18. Wash the beads three times with 1 mL Wash Buffer (*see* **Note 18**). After the third wash, move the bead slurry to a new eppendorf tube.
19. Once the beads are transferred to the new tube, wash the beads two more times with 1 mL Wash Buffer, and remove the Wash solution by pipetting.

20. Add 1 mL of PBS and transfer the beads to a new 1.5 mL round bottom tube. Wash once more with 1 mL PBS to remove any residual detergent and to improve the efficiency of elution.
21. Add 750 μL base Elution Buffer to the beads and place the tube on a shaker with vigorous shaking that precludes the beads from settling for 20 min at room temperature (*see* **Note 19**).
22. Place the tube on the magnetic rack and, once the beads are separated, transfer the eluate to a microcentrifuge tube (e.g., Axygen).
23. Flash freeze the eluate by dipping the tube in liquid nitrogen.
24. Place the tube in a vacuum centrifuge and evaporate the liquid overnight.
25. To test for the efficiency of elution, perform a second elution by adding 40 μL of a solution containing 1× LDS and 50 mM DTT to the beads remaining from **step 22**. Incubate at 70 °C for 10 min, then place the tube on a shaker at room temperature for another 10 min. Remove this second elution and reserve 10 % (i.e., 4 μL) for future analysis. Freeze the remainder of the second elution and store at −20 °C. Store the remaining beads at 4 °C to test the efficiency of the elution (*see* Subheading 3.4.2).
26. The next day, as **step 24** is completed, the primary eluate should be completely dry and can be prepared for analysis by mass spectrometry. One of the following steps should be performed depending on which of these bottom-up strategies is selected for preparing the samples for mass spectrometry analysis.
27. For in-gel digestion, resolve the co-isolated proteins by SDS-PAGE. Suspend the dried primary eluate in 40 μL of 1× NuPAGE Sample Buffer containing 1× Reducing Agent and heat for 10 min at 70 °C. Set aside 10 % (4 μL) of the eluate in a clean microcentrifuge tube for future analysis (*see* Subheading 3.4.2). Add 4 μL of 1 M Iodoacetamide to the remaining 90 % of primary eluate, and then incubate 30 min at room temperature in the dark. If not continuing to sample preparation for mass spectrometry, freeze in liquid nitrogen and store at −20 °C.
28. For in-solution digestion, suspend the dried primary eluate in 40 μL of 1× LDS Elution Buffer containing 50 mM DTT and heat at 70 °C for 10 min. Set aside 10 % (4 μL) of the eluate in a clean microcentrifuge tube for future analysis (*see* Subheading 3.4.2). If not continuing to sample preparation for mass spectrometry, freeze remaining 90 % of primary eluate in liquid nitrogen and store at −20 °C.

3.4.2 Assessing the Efficiency of Immunoaffinity Purification

1. To assess the efficiency of isolation of the targeted protein, it is necessary to compare, using Western blot, equivalent amounts (5–10 %) of the cell pellet (remaining after the centrifugation), the flow through (remaining after the immunoaffinity purification), the primary elution (obtained after the first elution), and the secondary elution (obtained after the second elution). A fraction of the input supernatant can also be compared to provide a point of reference for the molecular mass of the targeted protein.
2. For the input supernatant, dilute 40 μL of the saved supernatant (*see* Subheading 3.4.1, **step 10**) to a final volume of 60 μL containing 1× NuPAGE LDS Sample Buffer and 1× Reducing Agent.
3. To prepare the cell pellet (*see* Subheading **3.4.1**, **step 10**): (a) homogenize the pellet in 1 mL of 2 % SDS, (b) heat the mixture at 70 °C for 10 min, and then (c) subject it to centrifugation for 5 min max speed at room temperature (if the sample is viscous, the solubilization can be aided by sonication), (d) from the solubilized pellet, remove an aliquot that corresponds to the equivalent amount of the utilized input, and (e) dilute the aliquot in 1× NuPAGE LDS Sample Buffer and 1× Reducing Agent.
4. To prepare the flow through (*see* Subheading 3.4.1, **step 15**): (a) add 4 volumes of ice-cold acetone to a volume equivalent of 10 % of the total flow through, (b) mix briefly by vortexing, and (c) incubate on ice for at least 1 h, (d) separate the precipitated fraction by centrifugation at 3,000 ×*g* for 10 min at 4 °C, (e) discard the supernatant and wash the remaining precipitated pellet with 4 volumes of 80 % acetone/20 % dH_2O, and (f) allow the pellet to air dry and then resuspend it in 40 μL of 1× NuPAGE LDS Sample Buffer and 1× Reducing Agent.
5. For the primary eluate (*see* Subheading **3.4.1**, **step 21**), dilute the equivalent of 10 % of the elution (i.e., 4 μL) into a final volume of 40 μL 1× NuPAGE Sample Buffer and 1× Reducing Agent. If performing the "alternative elution" (*see* **Note 19**), the primary eluate can be prepared the same day as opposed to the base elution that must be prepared on the next day.
6. Prepare the secondary eluate (*see* Subheading 3.4.1, **step 25**) similar to the first eluate, by diluting the equivalent of 10 % (i.e., 4 μL) into a final volume of 40 μL 1× NuPAGE Sample Buffer and 1× Reducing Agent.
7. Heat all samples (prepared in **steps 1–5** above) for 10 min at 70 °C, and either directly proceed with the comparison by Western blotting or store them at −20 °C.
8. The comparison of these fractions using Western blot analysis should show the presence of the targeted protein significantly enriched in the primary eluate.

3.4.3 Optimization of Lysis Buffer Conditions

Optimization Prior to the Immunoaffinity Purification

Multiple factors can contribute to the properties of the targeted protein, such as subcellular localization and the presence of post-translational modifications. It is recommended to perform optimization experiments utilizing small scale sample amounts, such as ~1×10^7 cells or one 15 cm cell culture plate, prior to proceeding with large scale proteomics experiments. Different combinations and concentrations of salts and detergents (such as NaCl and Triton X-100) determine the stringency of the buffer, which must be stringent enough to allow for an efficient solubilization and isolation of the protein of interest, and mild enough to prevent the loss of interactions. It is recommended that at least three lysis buffer conditions with varying degrees of stringency are tested. Tables 2 and 3 list examples of Lysis Buffer conditions that have been successfully used to isolate proteins from various viral infections and cell types. The following steps are recommended for optimizing the Lysis Buffer.

1. Prepare cryogenically ground cell powder as described in Subheading 3.2.3. Split the frozen cell powder equally into prechilled tubes.
2. Add Lysis Buffer to each sample using 5 mL of buffer per 1 g of cells. The volume of Lysis Buffer per sample is usually ~0.5–1 mL if using adherent cells grown in a 15 cm culture dish.
3. Instead of the Polytron used to homogenize the larger scale purification (*see* Subheading 3.4.1, **steps 4–6**), perform the homogenization of the cell powder in the Lysis Buffers by vortexing 2× 30 s with cooling on ice in between the two steps.
4. Separate the insoluble fraction by centrifugation at 8,000 × g for 10 min at 4 °C. Remove and save the supernatant as the soluble fraction.
5. To extract the pellet, add 50 mM Tris–HCl, pH 7.4, containing 2 % SDS. Sonicate and then boil at 95 °C for 5 min. Centrifuge at 20,000 × g for 10 min and save the supernatant as representing the "pellet" fraction.
6. Compare the relative amounts of bait protein in the soluble and pellet fractions by loading equivalent amounts (5–10 %) for analysis by Western blot, probing with antibodies against the tag or the endogenous protein.
7. Select the lysis condition that yields the highest amount of bait protein in the soluble fraction (*see* **Note 20**).

Optimization Following the Immunoaffinity Purification

The comparison performed in the assessment of the efficiency of immunoaffinity purification (*see* Subheading 3.4.2) can inform on how to proceed with further optimizing the Lysis Buffer conditions.

1. The presence of the targeted protein in the cell pellet may suggest the need for a more stringent lysis buffer condition.

2. The presence of the protein in the flow through may indicate that (a) the amount of beads utilized is not sufficient for capturing all of the target protein, or (b) the tag or the epitope of the endogenous protein is not folded in an accessible manner or hindered by other interactions in this lysis condition.
3. The presence of the protein in the secondary elution suggests that the efficiency of primary elution is not sufficient and that either a larger elution volume or the alternative elution approach should be utilized.

The isolation conditions can be further optimized once the co-isolated proteins are analyzed by mass spectrometry. If a balance between an efficient isolation and reduced presence of nonspecific associations is difficult to achieve, the composition of the Wash Buffer (i.e., the buffer used to wash the beads following the immunoaffinity purification) can also be optimized. An experiment worth trying in this situation is to use a less stringent IP Buffer and a more stringent Wash Buffer (*see* Subheading 3.4.1, **step 1**).

Once the bait (targeted protein) and its interacting proteins are eluted from the beads, the resulting mixture of co-isolated proteins can be prepared for mass spectrometry analysis using several approaches. For example, the co-isolated protein mixtures can be digested either using an in-gel or an in-solution approach, as described in refs. 14, 20, 21. The in-gel approach involves the separation of proteins by SDS-PAGE prior to the enzymatic digestion, while the in-solution digestion involves the direct digestion of proteins while maintaining them in solution, and is usually followed by a chromatographic separation prior to mass spectrometry analysis. Since the in-gel digestion strategy already resolves the sample by 1-D or 2-D SDS-PAGE, this step may be directly followed by analysis by mass spectrometry. However, this will depend on the complexity of the sample (i.e., number of co-isolated proteins and posttranslational modifications). Therefore, it is common that either the in-gel or in-solution prepared samples are analyzed using a chromatographic separation (nLC, nano-liquid chromatography) coupled online with the mass spectrometer.

3.5 Selecting Controls

For every immunoaffinity purification experiment, it is important to select and perform the appropriate controls that will allow interpreting the results and distinguishing specific from nonspecific associations. Among the possible sources for nonspecific associations that have to be considered, the most prominent ones are nonspecific associations to (1) the tag and antibody utilized, (2) the resin selected for immunoaffinity purification (e.g., magnetic beads, agarose), or (3) the isolated proteins (i.e., the protein complex of interest).

3.5.1 Controlling for Nonspecific Association to the Tag, Antibody, and Resin

1. If a tag is being used to isolate the protein of interest, control samples are needed to ensure that the tag or the utilized antibodies are not the source of the identified protein interactors. It is recommended that the same system used to generate the tagged viral protein be used to generate a virus that expresses the free tag. For example, a control virus triggering the pancellular expression of GFP should be used in parallel experiments to immunoaffinity purifications of GFP-tagged viral proteins [1, 2]. These control viral infections should be analyzed at the same time points of infection as those used for the isolation of the viral protein of interest.
2. To rule out the possibility that proteins are interacting with the antibody or the beads, empty beads or antibody (or purified IgG)-conjugated beads can be used for parallel experiments.
3. When compared side by side, proteins identified by mass spectrometry in these controls but not in the experimental samples may correspond to nonspecific associations.
4. In some cases, an interacting protein may be identified in both the experimental sample and the control. It is possible that these interactions are still specific if they show a significant enrichment relative to the control. To help determine the specificity of interaction, the nLS (nano-liquid chromatography)-MS/MS data obtained from the mass spectrometry analysis of the co-isolated proteins can be used in conjunction with a computation program called SAINT (Significance Analysis of INTeractome) [22] to assign confidence scores to the protein interaction data.

3.5.2 Controlling for Nonspecific Associations to the Isolated Protein Complex of Interest

1. The nonspecific associations that are most difficult to control are those that occur with the isolated protein complex of interest. Once a viral protein is isolated, its true virus and host interacting partners can act as binding sites for nonspecific associations with other proteins in the cell lysate. These types of interactions frequently occur with highly abundant, "sticky" proteins. For example, the solubilization of a protein complex may trigger some unfolding that can result in the association with heat shock proteins. Strategies involving metabolic labeling with stable isotopes can be utilized to test for the presence of such nonspecific interactions.
2. When using a tagged protein, the I-DIRT (isotopic differentiation of interactions as random or targeted) approach can be utilized [14, 23]. Briefly, the cells infected with the tagged virus are grown in regular (light) media, and the cells infected with the wild-type virus are grown with media containing heavy amino acid isotopes. The cells are harvested and frozen as described above, and the frozen cell pellets are mixed prior

to the cryogenic cell lysis. The immunoaffinity purification from the mixed cell population is performed as described above, and the co-isolated proteins are analyzed by mass spectrometry. Upon analysis by mass spectrometry, the presence of single peaks representing peptides containing only the light amino acid isotopes would represent specific interactions. In contrast, the presence of doublet peaks representing peptides containing both the heavy and light isotopes will signify a nonspecific association or a fast-exchanging association.

3. A similar approach can be used when isolating endogenous proteins. shRNA-mediated knockdown can be performed on the protein of interest prior to the differential metabolic labeling.

3.5.3 Controlling for Spatial Preference of Nonspecific Interactions

1. In many circumstances, the protein of interest is localized to a distinct subcellular compartment, such as the nucleus or mitochondria. Upon lysis, the protein of interest and its interacting partners will be brought into contact with other subcellular compartments (e.g., the cytoplasmic fraction), exposing the isolated protein to interactions previously prevented by spatial restriction. These associations can be assessed using the two approaches shown above (*see* Subheadings 3.5.1 and 3.5.2). However, if these associations are numerous, these proteins can substantially increase the complexity of the sample and hinder the identification of real interactions by mass spectrometry. To prevent these non-physiological interactions, subcellular fractions can be performed prior to the cryogenic lysis and immunoaffinity purification.

4 Notes

1. The recent increase in the sensitivity of mass spectrometry instrumentation has significantly decreased the requirement for sample amount, allowing efficient isolations of viral proteins from as little as 1× 15 cm dish. However, it is still preferred to start with ~10× 15 cm dishes to allow for optimization of lysis conditions.
2. It is also possible to collect the cells in multiple 50 mL conical tubes. If using 50 mL tubes, after the centrifugation in **step 5**, aspirate the PBS and combine the cell pellets into a common tube and then repeat **step 5**.
3. Cell types such as HeLa and HEK293 cells tend to result in more than 1 g of sample material (~1.2 g), while cell types such as HFFs tend to result in less than 1 g (i.e., ~0.9 g).

4. If working with smaller amounts of material, i.e., 1–3 plates/sample, it may be beneficial to add an increasing volume of the Cryo Buffer at ~200 μL Cryo Buffer per 1.0 g. In this case, Subheading 3.2.2 will involve the direct slow pipetting of the resulting cell solution into the 50 mL conical tube containing liquid nitrogen. Use a 200 μL pipette to obtain an appropriate size of the frozen pellets.
5. The goal during freezing the sample is to produce small ball-bearing size frozen pellets of your sample allowing for efficient cryogenic disruption.
6. If there is still residual material remaining in the tube, an additional 100 μL of Cryo Buffer can be used to wash out remaining material to be added drop wise into the liquid nitrogen. As they freeze, the cell pellets will sink at the bottom of the liquid nitrogen filled 50 mL tube.
7. If the mass of the cell pellet is low, i.e. <0.5 g, an alternative approach is to use a wide bore pipette tip or a cut 200 μL tip to directly withdraw the pellet and slowly add the material drop wise into the 50 mL tube containing the liquid nitrogen. If the sample freezes in the tube, thaw it out and continue the process.
8. The round-bottom tube further prevents the beads from settling and allows maximum exposure of the bead surface to the solution.
9. To calculate the necessary volume of antibody solution to use, for purified high-affinity antibodies (e.g., in-house generated anti-GFP antibodies) use 3–5 μg Ab/mg beads. For commercially available antibodies that may not have as high affinity for binding, use 5–8 μg Ab/mg beads, and for conjugation of IgG use 8–10 μg IgG/mg beads.
10. In certain situations, the total volume of ~20 μL/mg beads may not be feasible because of diluted concentrations of available antibodies. For these circumstances, a larger total volume of conjugation can be used, but it is recommended that this total volume is not increased beyond ~60 μL/mg beads, as the efficiency of conjugation can be affected.
11. Do not leave the beads in contact with this solution for an extended period of time. Remove the solution as soon as the beads are secured on the magnet.
12. It is recommended to use the beads within 2 weeks of conjugation, as their isolation efficiency will decrease significantly if the beads are stored for extensive periods of time (e.g., ~40 % decrease after 1 month storage).

13. In most cases, the Wash Buffer is the same as the Lysis Buffer, without the inclusion of protease and phosphatase inhibitors, or DNase. Prepare 10 mL of wash buffer per sample, and keep both IP and Wash Buffers on ice (4 °C).
14. The 20–30 s waiting step prior to the addition of the Lysis Buffer will ensure that the buffer does not freeze on contact with the sample.
15. If using a lysis buffer containing DNase, first incubate the sample on a rotor at room temperature for 5–10 min. If no DNase is used, place the tube containing the cell lysate on a rotor at 4 °C, and allow for gentle mixing for 5–10 min, until the cell powder is fully suspended in the buffer. This step is not necessary if the entire cell powder was already dissolved in the buffer in **step 3**, which is the case when working with small amounts of material.
16. When finished with the last sample, rinse and wash the Polytron in ultrapure dH_2O, followed by a rinse with methanol. Use a tissue to tap off the excess methanol and let the Polytron to dry.
17. Avoid using more than 1 h of mixing, as the increase in incubation time will lead to an increase in accumulation of nonspecific proteins and loss of weak binding partners [1].
18. For each wash, use gentle pipetting with the Wash Buffer on top of the beads until they are fully suspended, and place the beads back on the magnet support.
19. Alternative Elution: The base Elution Buffer (*see* **step 21**) is preferred because it leads to less background protein and IgG contamination. However, if low recovery of bait protein is observed, this alternative procedure can be used in place of the base elution. (a) Add 40 μL of 1× LDS Elution Buffer to beads and incubate for 10 min at 70 °C, then for 10 min at room temperature with shaking. (b) Remove the primary eluate and repeat the elution with 1× LDS Elution Buffer to obtain a second elution. (c) Add 2.0 μL 1 M DTT to each eluate and heat to 70 °C for 10 min. Set aside 10 % of both eluates for future analysis. (d) If performing in-gel digestion, add 4 μL of 1 M iodoacetamide to remaining 90 % of eluates and incubate at RT for 30 min in the dark. (e) Freeze remaining 90 % of eluates in liquid nitrogen and store at −20 °C or proceed immediately to proteomic analysis.
20. If the yields are equivalent across several conditions, select the condition with the lowest stringency. This will allow for an efficient extraction of the bait protein, while helping to preserve interacting partners.

References

1. Cristea IM, Williams R, Chait BT, Rout MP (2005) Fluorescent proteins as proteomic probes. Mol Cell Proteomics 4:1933–1941
2. Cristea IM, Carroll JW, Rout MP, Rice CM, Chait BT, MacDonald MR (2006) Tracking and elucidating alphavirus-host protein interactions. J Biol Chem 281: 30269–30278
3. Cristea IM, Chait BT (2011) Affinity purification of protein complexes, Cold Spring Harbor Protocols 2011, pdb prot5611.
4. Cristea IM, Chait BT (2011) Conjugation of magnetic beads for immunopurification of protein complexes. Cold Spring Harbor Protocols 2011, pdb prot5610.
5. Moorman NJ, Sharon-Friling R, Shenk T, Cristea IM (2010) A targeted spatial-temporal proteomics approach implicates multiple cellular trafficking pathways in human cytomegalovirus virion maturation. Mol Cell Proteomics 9:851–860
6. Cristea IM, Moorman NJ, Terhune SS, Cuevas CD, O'Keefe ES, Rout MP, Chait BT, Shenk T (2010) Human cytomegalovirus pUL83 stimulates activity of the viral immediate-early promoter through its interaction with the cellular IFI16 protein. J Virol 84:7803–7814
7. Fehr AR, Gualberto NC, Savaryn JP, Terhune SS, Yu D (2012) Proteasome-dependent disruption of the E3 ubiquitin ligase anaphase-promoting complex by HCMV protein pUL21a. PLoS Pathog 8:e1002789
8. Cristea IM, Rozjabek H, Molloy KR, Karki S, White LL, Rice CM, Rout MP, Chait BT, MacDonald MR (2010) Host factors associated with the Sindbis virus RNA-dependent RNA polymerase: role for G3BP1 and G3BP2 in virus replication. J Virol 84:6720–6732
9. Reitsma JM, Savaryn JP, Faust K, Sato H, Halligan BD, Terhune SS (2011) Antiviral inhibition targeting the HCMV kinase pUL97 requires pUL27-dependent degradation of Tip60 acetyltransferase and cell-cycle arrest. Cell Host Microbe 9:103–114
10. Youn S, Li T, McCune BT, Edeling MA, Fremont DH, Cristea IM, Diamond MS (2012) Evidence for a genetic and physical interaction between nonstructural proteins NS1 and NS4B that modulates replication of west nile virus. J Virol 86:7360–7371
11. Moorman NJ, Cristea IM, Terhune SS, Rout MP, Chait BT, Shenk T (2008) Human cytomegalovirus protein UL38 inhibits host cell stress responses by antagonizing the tuberous sclerosis protein complex. Cell Host Microbe 3:253–262
12. Terhune SS, Moorman NJ, Cristea IM, Savaryn JP, Cuevas-Bennett C, Rout MP, Chait BT, Shenk T (2010) Human cytomegalovirus UL29/28 protein interacts with components of the NuRD complex which promote accumulation of immediate-early RNA. PLoS Pathog 6:e1000965
13. Greco TM, Yu F, Guise AJ, Cristea IM (2011) Nuclear import of histone deacetylase 5 by requisite nuclear localization signal phosphorylation. Mol Cell Proteomics 10(M110):004317
14. Tsai YC, Greco TM, Boonmee A, Miteva Y, Cristea IM (2012) Functional proteomics establishes the interaction of SIRT7 with chromatin remodeling complexes and expands its role in regulation of RNA polymerase I transcription. Mol Cell Proteomics 11(M111):015156
15. Blanco MA, Aleckovic M, Hua Y, Li T, Wei Y, Xu Z, Cristea IM, Kang Y (2011) Identification of staphylococcal nuclease domain-containing 1 (SND1) as a Metadherin-interacting protein with metastasis-promoting functions. J Biol Chem 286:19982–19992
16. Li T, Diner BA, Chen J, and Cristea IM (2012) Acetylation modulates cellular distribution and DNA sensing ability of interferon-inducible protein IFI16. Proc Natl Acad Sci USA 109: 10558–10563.
17. Rigaut G, Shevchenko A, Rutz B, Wilm M, Mann M, Seraphin B (1999) A generic protein purification method for protein complex characterization and proteome exploration. Nat Biotechnol 17:1030–1032
18. Chalfie M, Tu Y, Euskirchen G, Ward WW, Prasher DC (1994) Green fluorescent protein as a marker for gene expression. Science 263:802–805
19. Hopp TP, Prickett KS, Price VL, Libby RT, March CJ, Cerretti DR, Urdal DL, Conlon RJ (1988) A short polypeptide marker sequence useful for recombinant protein identification and purification. Nat Biotechnol 6:1204–1210
20. Greco TM, Miteva Y, Conlon FL, Cristea IM (2012) Complementary proteomic analysis of protein complexes. Methods Mol Biol 917:391–407
21. Wisniewski JR, Zougman A, Nagaraj N, Mann M (2009) Universal sample preparation method for proteome analysis. Nat Methods 6:359–362
22. Choi H, Larsen B, Lin ZY, Breitkreutz A, Mellacheruvu D, Fermin D, Qin ZS, Tyers M, Gingras AC, Nesvizhskii AI (2011) SAINT: probabilistic scoring of affinity purification-mass spectrometry data. Nat Methods 8: 70–73

23. Tackett AJ, DeGrasse JA, Sekedat MD, Oeffinger M, Rout MP, Chait BT (2005) I-DIRT, a general method for distinguishing between specific and nonspecific protein interactions. J Proteome Res 4:1752–1756
24. Goldberg AD, Banaszynski LA, Noh KM, Lewis PW, Elsaesser SJ, Stadler S, Dewell S, Law M, Guo X, Li X, Wen D, Chapgier A, DeKelver RC, Miller JC, Lee YL, Boydston EA, Holmes MC, Gregory PD, Greally JM, Rafii S, Yang C, Scambler PJ, Garrick D, Gibbons RJ, Higgs DR, Cristea IM, Urnov FD, Zheng D, Allis CD (2010) Distinct factors control histone variant H3.3 localization at specific genomic regions. Cell 140: 678–691
25. Selimi F, Cristea IM, Heller E, Chait BT, Heintz N (2009) Proteomic studies of a single CNS synapse type: the parallel fiber/purkinje cell synapse. PLoS Biol 7:e83

Chapter 5

Screening for Host Proteins with Pro- and Antiviral Activity Using High-Throughput RNAi

Samantha J. Griffiths

Abstract

RNA interference (RNAi) describes the mechanism of posttranscriptional gene silencing by small (typically 18–24 nucleotides) RNA molecules and includes small-interfering RNAs (siRNAs) and microRNAs (miRNAs). As siRNAs and miRNAs are simple to use experimentally, they are easily adaptable to high-throughput methodologies and provide an ideal tool for genome-wide gene depletion studies. Over recent years RNAi has been used extensively to investigate the complex interactions between pathogen and host, and the identification of novel cellular factors and pathways influencing viral disease pathogenesis exemplifies the power of this technique. Here, the use of RNAi to investigate the functional role of cellular proteins in herpesvirus (Herpes Simplex Virus Type I; HSV-1) replication and how to identify novel antiviral and proviral host proteins is described.

Key words RNA interference, Herpes simplex virus type 1 (HSV-1), High-throughput, siRNA, miRNA, 384-Well format

1 Introduction

Viruses are obligate parasites and as such they are dependent upon the host cell for replication and dissemination. Herpesviruses are a class of large double-stranded DNA viruses which infect and affect not only humans but also many other mammalian and fish species. A common feature of all members of the *Herpesviridae* family is the ability to establish both a productive, lytic infection and an asymptomatic, latent infection which may undergo periodic reactivation. The equilibrium between these two infection states requires a fine balance between innate and adaptive immune responses, and viral immune evasion mechanisms. The sheer size of herpesviruses makes the relationship between pathogen and host highly complex. For example, Herpes simplex virus Type 1, HSV-1, has a genome size of around 150 kb and expresses 80–85 proteins. A great deal of information has been gained over the years on herpesvirus pathogenesis by investigating protein–protein interactions [1–3],

Susanne M. Bailer and Diana Lieber (eds.), *Virus-Host Interactions: Methods and Protocols*, Methods in Molecular Biology, vol. 1064, DOI 10.1007/978-1-62703-601-6_5, © Springer Science+Business Media, LLC 2013

but whilst this provides information on physical interactions, biological experimentation is required to confirm a functional role in virus replication.

RNA interference describes a mechanism of posttranscriptional gene silencing, and includes both small-interfering RNAs (siRNAs) and microRNAs (miRNAs). siRNAs are short (21–23 nucleotides) RNA molecules which complementarily bind to the target mRNA sequence and target it for degradation by the RNA-induced silencing complex (RISC). miRNAs are also short (~22 nucleotides) RNAs that have a 6-bp "seed" sequence, usually the 5′ bases 2–7, which predominantly binds the 3′ untranslated region (3′UTR) to induce transcript degradation (full complementarity of seed sequence) or a translational stop (partial complementarity). Both miRNAs and siRNAs are relatively simple to use experimentally and are thus easily adaptable to genome-wide methodologies, making them an ideal tool to investigate specific gene depletion *in vitro* at the mRNA level. siRNA libraries have been exploited extensively to investigate a range of pathogens, including HIV [4–6], Influenza [7–9], West Nile Virus [10], Dengue virus [11], Hepatitis C virus [12], and Herpes Simplex virus type 1 (Griffiths et al., 2013, in press). Numerous studies over recent years have identified cellular miRNAs as significant (positive or negative) regulators of viral replication either directly by targetting viral-specific sequences [13–15], or indirectly, by targetting cellular genes involved in replication [16–18]. The use of miRNA agonist (enabling over-expression) and antagonist (inhibiting expression of cellular miRNAs) libraries lends a high-throughput, rapid approach to identify cellular miRNAs involved in a range of cellular processes, including virus replication [19]. These studies have identified many previously unknown cellular miRNAs, genes and pathways involved in virus replication and disease pathogenesis, and highlight the power of RNAi as a method to unveil the remaining mysteries surrounding virus–host interactions [20, 21].

This chapter describes a protocol to carry out an siRNA depletion screen in order to identify antiviral and proviral factors involved in HSV-1 replication. Whilst siRNAs are referred to specifically, the same protocol can be used for miRNA library screens in order to investigate the effect of miRNA over-expression (using miRNA "mimics") or depletion (using miRNA antagonists, or "inhibitors"). The method describes siRNA library preparation and generation of Library Master Plates, as well as culture of and preparation of HeLa cells used in this method. Two downstream phenotypic assay outputs are described here: (1) the effect of siRNA depletion on cell viability, by the addition of a dye expressing a fluorescent signal relative to the rate of cell metabolism, and (2) the effect of siRNA depletion on HSV-1 replication by utilizing a recombinant HSV-1 strain (C12) expressing the enhanced green fluorescent protein (eGFP) reporter gene from the US5 locus during the late stage of virus replication [22]. Reporter virus strains enable the analysis of virus growth over multiple rounds of replication; however, the same RNAi screening protocol can be applied to almost any phenotypic assay for which a reporter system is available. It is thus an incredibly powerful tool enabling investigations into many other facets of virus–host interactions, such as promoter activation using luciferase reporter genes, and reactivation of latent infection from quiescent reporter cell lines. Finally, this chapter provides some basic data analyses to enable initial identification of siRNA "hits," in this case cellular proteins with proviral and antiviral activity, are described. There are extensive statistical analyses that can be, and should be, applied to siRNA screen data and for accuracy and reliability of data a more thorough statistical analysis should be applied [23].

2 Materials

The protocol below details the equipment, quantity of consumables and volumes of reagents required to resuspend the Dharmacon Human Druggable (DHD) Genome siRNA library (Dharmacon, 0.5 nmol, 76 × 96-well plates) and generate 19 × 384-well Master Plates. Whilst the Master Plates generated with this protocol contain sufficient volume (100 μL) to carry out three complete screens, the volumes and quantities of consumables detailed for cell culture, siRNA transfection, and subsequent infection with HSV-1 are sufficient to carry out one complete screen. One complete screen consists of 19 × 384-well plates (50 nM final siRNA concentration) transfected in triplicate (57 plates), where one replicate (19 × 384-well plates) is used for a cell viability assay and two replicates (38 × 384-well plates) are used for an HSV-1 infectivity assay (*see* Fig. 1 for overview of siRNA plate management). Such repetition of the screen, including three biological replicates of technical duplicates, provides six individual data points and should produce a statistically robust dataset. It should be noted that, with advancements in genome annotation, siRNA libraries are regularly updated and that current versions of the Dharmacon Druggable Genome Library are not the same as used here (Version G004600-05). Take care to calculate the reagents and consumables required for your own library, based on the number of plates and replicates required. All assay preparation steps are carried out at room temperature and in sterile conditions unless otherwise specified.

2.1 siRNA Library Resuspension and Assay Plate Preparation

1. siRNA druggable genome library: siGENOME SMARTpool (mix of 4 siRNAs per gene) Druggable Genome siRNA library (Dharmacon Thermo Scientific), supplied as 0.5 nmol lyophilized siRNA per well in 76 × 96-well plates.

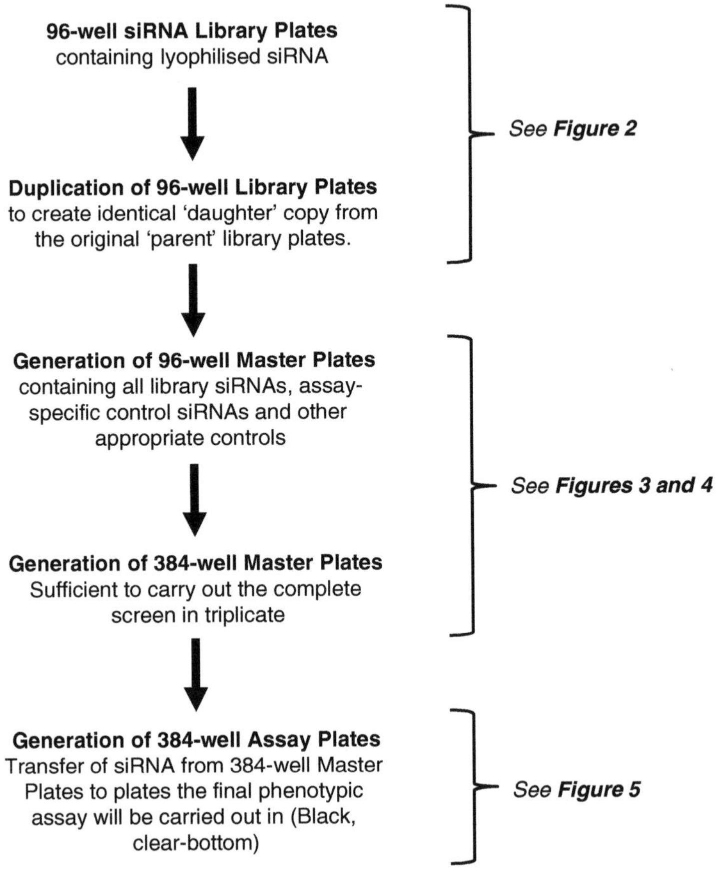

Fig. 1 Overview of siRNA Library plate management. Schematic flow chart describing plate management for large-scale siRNA libraries

2. 1× siRNA buffer: Dilute 400 mL 5× siRNA buffer (Thermo Scientific) in 1.6 L ultrapure, sterile, RNase- and DNase-free water. This will make enough 1× siRNA buffer to resuspend the library and prepare the 96-well Master Plates and 384-well Master Plates. Store the buffer at 4 °C until needed.
3. Robotic liquid handler: QIAGEN RapidPlate 384, or equivalent. Must be capable of liquid transfer between 96- and 384-well plates.
4. Pipette tips for robotic liquid handler (Zymark tips, Axygen).
5. 8-Channel multichannel pipette, 2–10 μL volume.
6. Automatic liquid dispenser (Thermo Scientific MultiDrop 384 or equivalent). Must be capable of dispensing liquid into 96- and 384-well plates.
7. 96-Well storage plates with lids (Thermo Scientific).
8. Plastic 96-well plate covers (Thermo Scientific).

9. Aluminum heat-sealing film for storage microplates.
10. Microplate heat-sealing device.
11. Adhesive plate seals.
12. siRNA controls: Assay-specific control siRNA, resuspended in 1× siRNA buffer to 20 μM concentration. For controls used in the HSV-1 screen (*see* Fig. 3 and **Note 8**).
13. 384-Well storage plates: Deep-well 384-well storage plates (120 μL capacity; Thermo Scientific).
14. 384-Well assay plates: Black plastic, clear- and flat-bottomed, sterile, tissue culture-treated 384-well plates (Corning or equivalent).
15. Barcode printer, such as the Brady® LABXPERT™ printer, and labels suitable for low temperature (−80 °C).
16. Low speed centrifuge with rotor for microtiter plates.
17. Microplate managing software, such as MDL Plate Manager.

2.2 Cell Culture, Transfection and Infection

1. Low-passage HeLa cells: 10× T75 sterile tissue culture flasks, 100 % confluent with low-passage (~p14–25) HeLa cells grown in Dulbecco's modified Eagle's medium (DMEM) supplemented with 4 mM L-glutamine, 5 % fetal calf serum (FCS), and penicillin–streptomycin (50 U/mL) (*see* **Note 1**).
2. Growth medium: DMEM supplemented with 4 mM L-GLUTAMINE, 5 % FCS, and penicillin–streptomycin (50 U/mL), 1.5 L.
3. Phosphate-buffered saline: Sterile phosphate-buffered saline (PBS) (Lonza), 400 mL.
4. Trypsin: Trypsin-Versene (EDTA) (Lonza), 200 mL.
5. Transfection medium: Phenol red-free DMEM/F-12 1:1, 5 % FCS supplemented with L-glutamine and 15 mM HEPES buffer (Gibco Invitrogen), 1 L.
6. Infection medium: Phenol red-free DMEM/F-12 1:1/5 % FCS supplemented with penicillin–streptomycin, L-glutamine, and 15 mM HEPES buffer (Gibco Invitrogen), 1 L.
7. T75 tissue culture flasks, 10.
8. T175 tissue culture flasks, 30.
9. Hank's Balanced Salt Solution: HBSS (Lonza), 750 mL.
10. siRNA transfection reagent: Lipid-based transfection reagent Dharmafect 1 (Dharmacon), 1.5 mL.
11. Sterile disposable serological pipettes, all volumes.
12. 384-Well siRNA Assay Plates: As prepared.
13. Cell incubator: Humidified incubator with 5 % CO_2 and maintained at 37 °C.

14. Automatic liquid dispenser (Thermo Scientific MultiDrop 384 or equivalent) capable of dispensing liquid into 96- and 384-well plates.
15. 8-Channel dispensing cassette for MultiDrop 384, two needed.
16. Sterile water: Sterile RNase and DNase-free water, 27× 50 mL tubes.
17. 70 % ethanol: Ethanol, diluted in sterile RNase- and DNase-free water, 25× 25 mL tubes.
18. Gas-permeable adhesive microplate seals: Plastic gas-permeable adhesive seals (Abgene, or equivalent) cut to 125 mm × 78 mm, 38 in total.
19. Cell viability reagent, such as CellTiter-Blue® (Promega), 45 mL.
20. HSV-1 reporter virus expressing eGFP (*see* **Note 2**).
21. Fluorescence plate reader, such as POLARstar OPTIMA (BMG Labtech).
22. Robotic stacker for a fluorescence plate reader equipped with a barcode scanner.

3 Methods

RNAs are temperature sensitive and the following steps should be carried out at as cool a temperature as possible, within the limitations of the laboratory. This process describes the resuspension of a lyophilized siRNA library to generate two duplicate copies, enabling a reduction in the overall cycles of freeze:thawing in between multiple uses of the library. Subheading 3.2 describes the generation of a 96-well Master Plate, which includes all the appropriate control siRNAs, and a 384-well Master Plate, at the required siRNA concentration, to generate the 50 nM Assay Plates, described in Subheading 3.3. Due to the temperature-sensitivity of siRNAs, it is recommended to limit the freeze:thaw cycles to fewer than 5. One way to achieve this is to prepare all assay plate replicates (9 in total; *see* Subheading 3.3 and **Note 9**) at the same time. Plates for subsequent screens (Screens 2 and 3) can be stored at −80 °C until needed. Each screen involves handling a large number of plates simultaneously (57) and, as there are several timing-essential steps such as a 20 min incubation of transfection complexes and an hour-long incubation for virus infection, it is important that plates are processed in the same sequential order. The use of a liquid handling equipment makes dispensation into plates very rapid (~12 min to dispense 10 μL transfection reagent to all 57 plates), and provided plates are kept in the correct order, the timing should remain the same for all plates.

3.1 siRNA Library Resuspension

1. Input siRNA plate details into a Microplate management software, such as MDL Plate Manager.
2. Using a label maker, print text or barcode labels for the siRNA library "Daughter" Plates, detailing the plate name, number, and siRNA concentration, for example "Druggable Genome Daughter Plate 1–3 μM."
3. Adhere the labels to the outer edge of 76 96-well storage plates.
4. Remove lyophilized siRNA library plates from −80 °C storage and equilibrate to room temperature.
5. Centrifuge plates for 10 min at 200×*g* to pellet any well contents.
6. Carefully remove plate seals and replace with clean plastic lids.
7. Clean an 8-channel dispensing cassette for the liquid handler (MultiDrop 384 or equivalent) by cleaning three times with 25 mL sterile 70 % ethanol and rinsing three times with 50 mL sterile DNase- and RNase-free water.
8. Place the ends of the 8-channel cassette tubes into a sterile glass bottle containing 1.1 L of 1× siRNA buffer.
9. Prime the cassette tubing with 1× siRNA buffer (*see* **Note 3**).
10. Dispense 165 μL 1× siRNA buffer to columns 3–12 of the 76 siRNA 96-well library plates to resuspend the 0.5 nmol lyophilized siRNA to 3 μM concentration (*see* **Note 4**) (Fig. 2).
11. Using the robotic liquid handler (RapidPlate 384) with a new box of tips (Axygen) per plate, transfer 85 μL from the siRNA library plates to the appropriately labelled 96-well storage plate to generate a duplicate "Daughter" siRNA library (*see* **Note 5**).
12. Use a manual or automatic heat-sealing device and aluminum heat-seals to seal the original siRNA library plates. Cover with the provided plastic lid and transfer to a −80 °C freezer for long-term storage.
13. Update plate manager software with details of the siRNA Daughter Plates.

3.2 Preparation of 96-Well and 384-Well Master Plates

1. Prepare a set of text or barcode labels for 19 384-well Master Plates, specifying the plate number and siRNA concentration, for example, "384-well Master Plate A1 300 nM" (*see* **Note 6**).
2. Adhere the labels to the outer edge of 19 384 deep-well storage plates (the 384-well Master Plates).
3. Clean an 8-channel dispensing cassette for the liquid handler (MultiDrop 384 or equivalent) by cleaning three times with 25 mL sterile 70 % ethanol and rinsing three times with 50 mL sterile DNase- and RNase-free water.
4. Dispense 85 μL 1× siRNA buffer to empty column 1 of each of the 76 DHD Library "Daughter" plates (*see* **Note 7**).

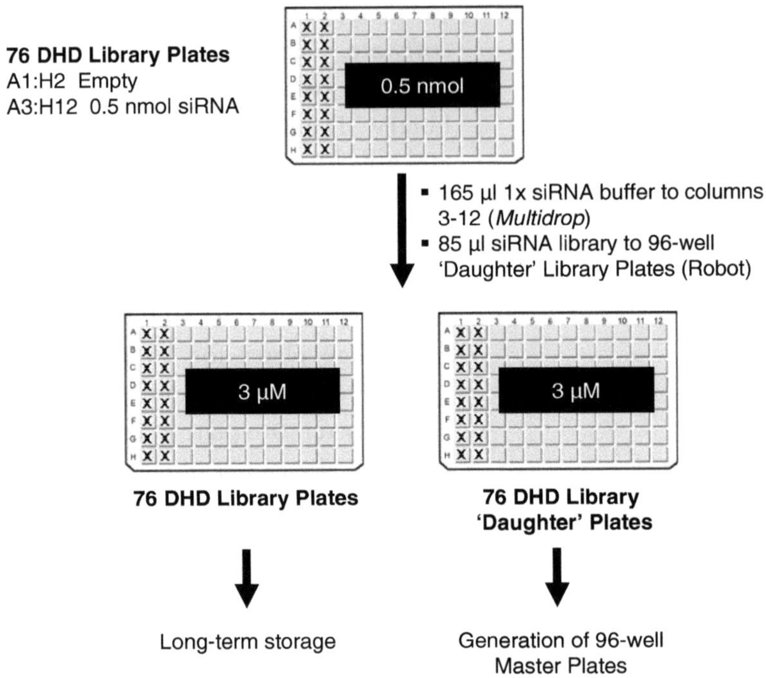

Fig. 2 siRNA library resuspension. Schematic diagram showing resuspension of the Dharmacon Human Druggable (DHD) siRNA library and generation of an identical duplicate "Daughter" Library (Subheading 3.1) in the 96-well format

5. Prepare the assay-specific control siRNA by diluting 600 µL of 20 µM siRNA stock in 3.4 mL 1× siRNA buffer to a final concentration of 3 µM (*see* **Note 8**).
6. Manually transfer 85 µL siRNA Control Pattern 1 (3 µM) to 96-well Master Plates 1, 2, 5, 6, 9, 10, etc. (384-well plate Quadrants A1 and A2), and 85 µL siRNA Control Pattern 2 (3 µM) to 96-well Master Plates 3, 4, 7, 8, 11, 12, etc. (384-well plate Quadrants B1 and B2) using an 8-channel pipette (Fig. 3) (*see* **Note 9**).
7. Using a liquid handler (MultiDrop 384 or equivalent), dispense 100 µL 1× siRNA buffer to all wells in the 19 pre-labelled 384-well Master Plates.
8. Using the robotic liquid handler, transfer 11 µL from each 96-well Master Plate (3 µM) to the appropriate Quadrant of the 384-well Master Plates (resulting concentration 300 nM) (Fig. 4).
9. Use a manual or automatic heat-sealing device and aluminum heat-seals to seal the 384-well Master Plates. Cover additionally with a plastic lid and transfer to a −80 °C freezer for long-term storage.
10. Update plate management software with details of the 96- and 384-well Master Plates.

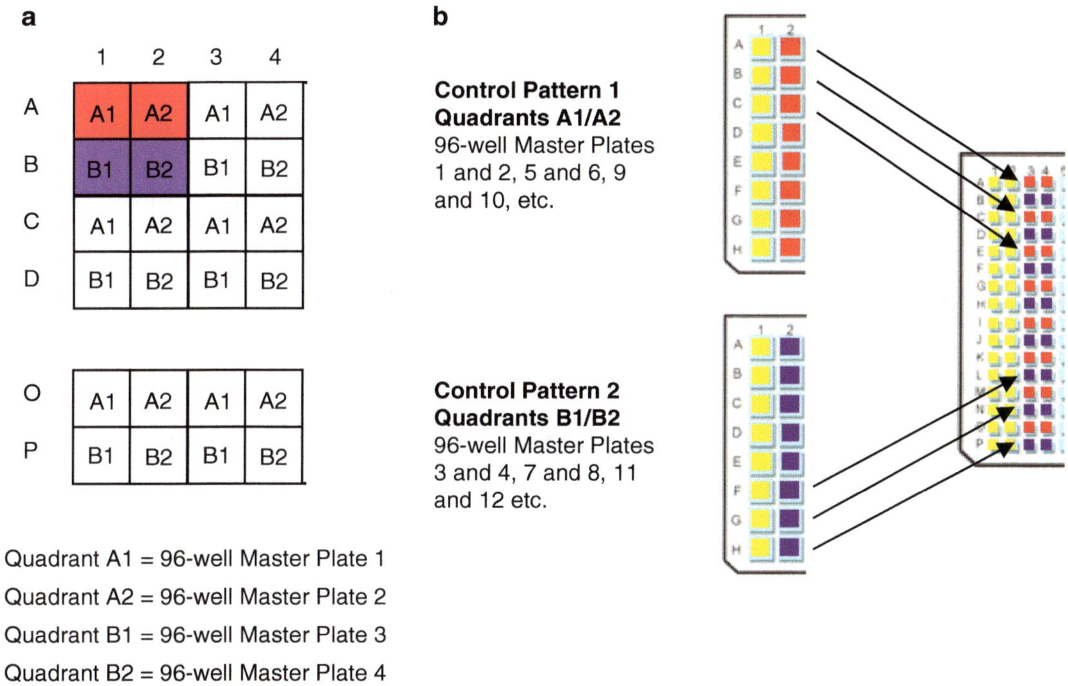

Fig. 3 96- to 384-Well pipetting schematic for siRNA control distribution. (**a**) Four 96-well plates are transferred onto one 384-well plate. Here, 96-well Master Plate 1 is translocated to Quadrant A1 in 384-well Master Plate 1; 96-well Master Plate 2 is translocated to Quadrant A2; 96-well Master Plate 3 is translocated to Quadrant B1; and 96-well Master Plate 4 is translocated to Quadrant B2. Subsequently, 384-well Master Plate 2 will contain 96-well Master Plates 5–8. (**b**) To generate siRNA controls in horizontal duplicates it is necessary to prepare two siRNA Control Patterns in alternate duplicate 96-well Master Plates. Control Pattern 1 (Quadrants A1 and A2) will be added to 96-well Master Plates 1 and 2, 5 and 6, 9 and 10, etc. whilst Control Pattern 2 (Quadrants B1 and B2) will be added to 96-well Master Plates 3 and 4, 7 and 8, 11 and 12, etc. This will result in 16 individual siRNA controls in horizontal duplicates on each 384-well Master Plate

3.3 Assay Plate Preparation

1. Using a label maker, prepare and print adhesive barcodes for 19× 384-well Assay Plates in triplicate (57 plates), specifying the plate number and assay output (replication, cell viability) (*see* **Note 10**).
2. Affix barcodes firmly to the side of the 57 black, clear-bottomed 384-well Assay Plates appropriate for the location of the barcode scanner.
3. Thaw the 384-well Master Plates at room temperature. Once thawed, centrifuge for 1 min at 200 *g* to collect any droplets from the lid of the plates.
4. Using a robotic liquid handler, transfer 10 μL siRNA from the 384-well Master Plates to three replicate Assay Plates (*see* **Note 11**).
5. Cover Assay Plates with adhesive seals, stack and seal in plastic bags/boxes and store at −80 °C until needed (*see* **Note 12**).
6. Cover Master Plates with heat-sealable adhesive seals and plastic lids, and store at −80 °C.

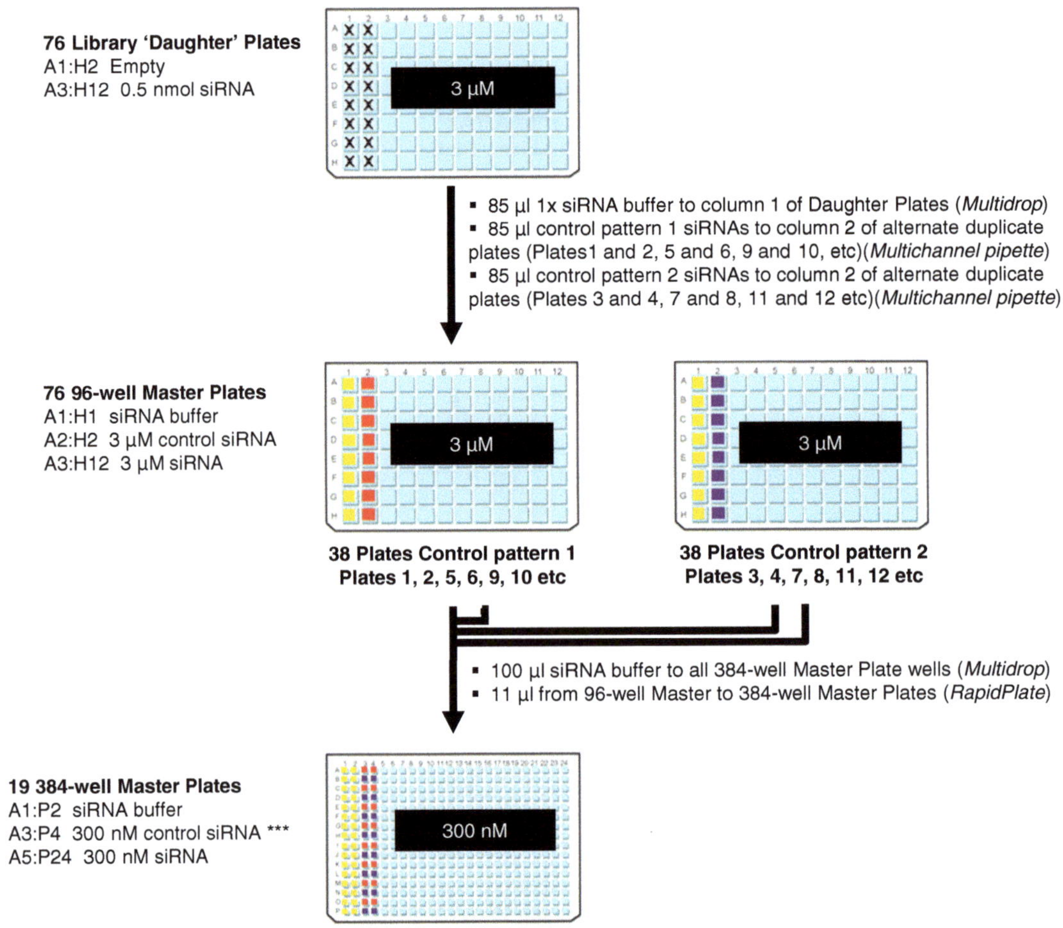

Fig. 4 Preparation of 96- and 384-well Master Plates. 96-Well Master Plates are prepared from siRNA "Daughter" Library plates by adding 1× siRNA buffer to Column 1 of all plates, to act as untransfected and uninfected control wells, and by adding assay-specific siRNA controls (3 µM) to Column 2 of each plates. 384-well Master Plates are prepared by the addition of 100 µL 1× siRNA buffer to each well in a 384-well storage plate (120 µL maximum volume) and the subsequent transfer of 11 µL from each 96-well Master Plate to the appropriate Quadrant in the 384-well Master Plate

3.4 Preparation of Cells for siRNA Transfection

1. Take 10× 100 % confluent T75 flasks of HeLa cells, remove medium and rinse each flask with 5 mL warmed sterile PBS.
2. Remove PBS and add 3 mL pre-warmed trypsin to each flask.
3. Leave for 5 min in a 37 °C incubator for cells to dislodge, before inactivating trypsin by adding 7 mL growth medium.
4. Pool cells from each flask into one, mix well and transfer 3.5 mL cell suspension to 30 T175 flasks.
5. Add a further 46.5 mL growth medium to each flask and incubate overnight at 37 °C in a humidified incubator (*see* **Note 13**).

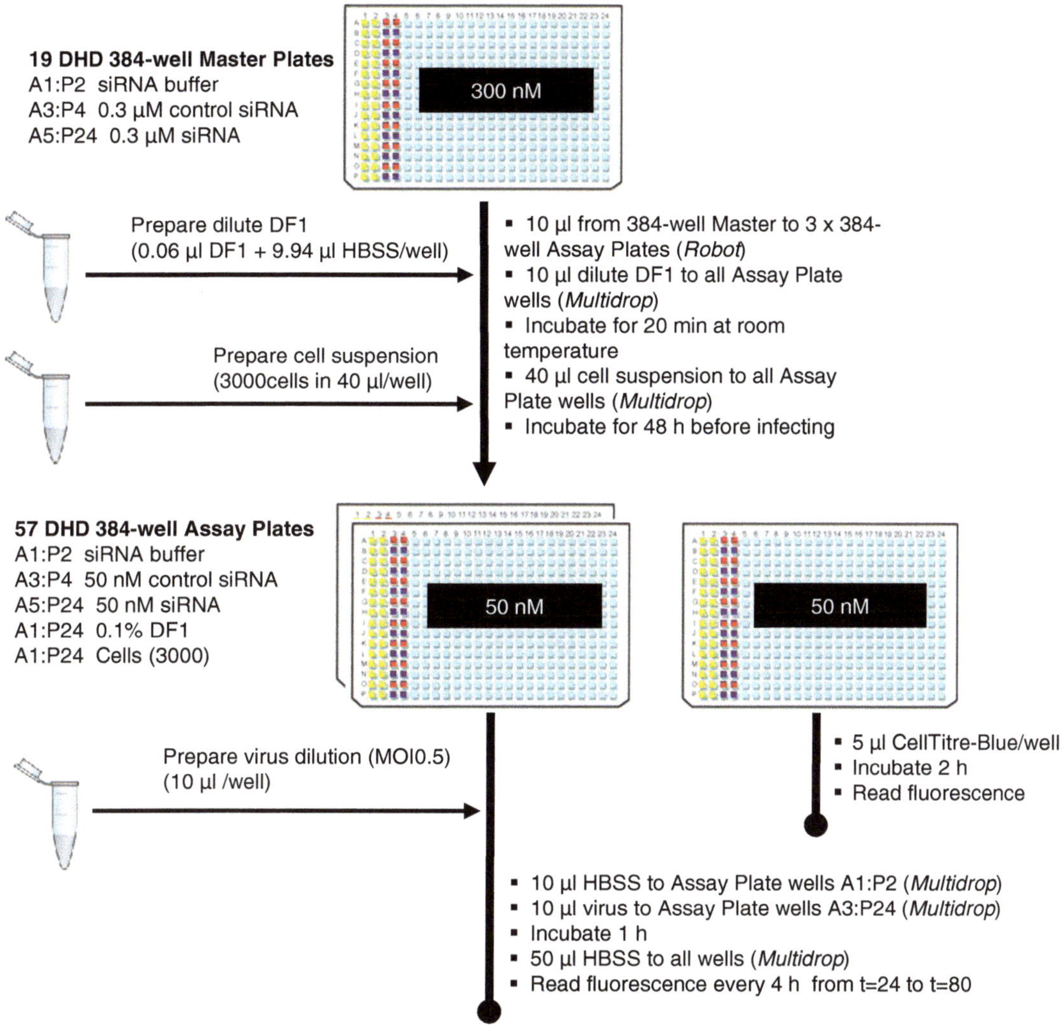

Fig. 5 siRNA transfection and Phenotypic Assays. A schematic representation of 384-well Assay Plate preparation, siRNA transfection, and Phenotypic Assays (Subheadings 3.3–3.6)

3.5 siRNA Transfection

1. Thaw the 0.3 µM 384-well Assay Plates and centrifuge for 5 min at 200 *g* to collect contents to the bottom of the well. Remove adhesive plate seals and recover with plate lids (*see* **Note 14**).
2. Dilute 1,446 µL Dharmafect 1 (DF1) transfection reagent in 239.5 ml HBSS to a concentration of 0.6 % (Fig. 5).
3. Clean an 8-channel dispensing cassette for the liquid handler (MultiDrop 384 or equivalent) by cleaning three times with 25 mL sterile 70 % ethanol and rinsing three times with 3× 50 mL sterile DNase- and RNase-free water.
4. Prime the dispensing cassette with the diluted transfection reagent and add 10 µL to all wells in each Assay Plate.

5. Centrifuge Assay Plates for 1 min at 200 *g* and incubate for 20 min at room temperature to allow transfection complexes to form.
6. Rinse the liquid handler dispensing cassette by cleaning three times with 25 mL sterile 70 % ethanol and rinsing three times with 50 mL sterile DNase- and RNase-free water.
7. Whilst transfection complexes are forming, prepare the HeLa cell suspension. Remove medium from the 30 T175 flasks, rinse in 10 mL sterile PBS, and dislodge cells in 5 mL trypsin by incubating at 37 °C for 5 min. Add 10 mL transfection medium to each flask and transfer all cells into a sterile 500 mL bottle.
8. Count the number of cells in your suspension and calculate the average cell density. Dilute the appropriate volume of cells in transfection medium to generate 1,000 ml of cell suspension at 7.5×10^4 cells/mL (*see* **Note 15**).
9. Using a liquid handler, dispense 40 μL cell suspension to all wells of each Assay Plate, swirling the bottle to ensure cells remain in suspension and equal cell numbers are dispensed in each well (*see* **Note 16**).
10. Transfer plates to a humidified incubator at 37 °C, taking care not to shake plates and cause uneven distribution of cells, and leave for 48 h to permit gene depletion at the mRNA level.
11. Clean the dispensing cassette for the liquid handler by cleaning three times with 25 mL sterile 70 % ethanol and rinsing three times with 50 ml sterile DNase- and RNase-free water.

3.6 Phenotypic Assay 1: Determination of siRNA Cytotoxicity

1. Thaw the CellTiter-Blue® (Promega) reagent.
2. Clean an 8-channel dispensing cassette for the liquid handler (MultiDrop 384 or equivalent) by cleaning three times with 25 mL sterile 70 % ethanol and rinsing three times with 50 ml sterile DNase- and RNase-free water.
3. Prime the cassette tubing with CellTiter-Blue reagent, and add 5 μL to each well of the 19 Cell Viability Assay Plates.
4. Return the plates to the incubator for 2 h at 37 °C before measuring the relative fluorescence signal (485 nm) in each well using a plate reader, such as the POLARstar OPTIMA (BMG Labtech).
5. Determine the effect of siRNA depletion on cell viability by comparing fluorescence to the average fluorescence of mock-transfected cells (Columns 1 and 2) (*see* **Note 17**).

3.7 Phenotypic Assay 2: HSV-1 Replication Assay

1. Clean an 8-channel dispensing cassette for the liquid handler (MultiDrop 384 or equivalent), set aside for HSV-1 use only, by cleaning three times with 25 mL sterile 70 % ethanol and rinsing three times with 50 mL sterile DNase- and RNase-free water.

2. Thaw an aliquot of HSV-1-eGFP in a 37 °C water bath until only just thawed, dilute in infection medium (*see* Subheading 2.2) to make 160 mL of virus inoculum at a multiplicity of infection (MOI) of 0.5. Prime the liquid handler cassette with diluted virus (*see* **Note 18**).
3. Remove the medium from one Assay Plate by inversion over a suitable liquid waste container and manually add 10 μL infection medium to columns 1 and 2 using a low-volume (2–10 μL) 8-channel pipette for uninfected control wells.
4. Transfer the plate to the liquid handler and dispense 10 μL diluted HSV-1-eGFP to each well in columns 3–24. Repeat for the remaining 37 plates.
5. Centrifuge Assay Plates for 5 min at 200 *g* and incubate at 37 °C for 1 h (*see* **Note 19**).
6. Clean the HSV-1 dispensing cassette by cleaning three times with 25 mL sterile 70 % ethanol and rinsing three times with 50 mL sterile DNase- and RNase-free water.
7. Clean a dispensing cassette for the liquid handler used for cells and/or medium, as above, and prime the cassette with infection medium.
8. After 1 h incubation with virus, add 50 μL infection medium to all wells in each of the 38× 384-well plates (*see* **Note 20**).
9. Cover plates with gas-permeable seals, stack and cover the top plate with a plastic lid (*see* **Note 21**).
10. Return plates to the 37 °C incubator and monitor virus replication as a measure of GFP fluorescence from 20 to 80 h postinfection, ensuring enough measurements are taken to generate an accurate replication growth curve (every 3–4 h) (*see* **Note 22**).

3.8 Data Analysis and Identification of Antiviral and Proviral Cellular Factors

1. For each time point, calculate the background fluorescence of uninfected cells (Columns 1 and 2) on each plate and subtract from the fluorescence observed in all siRNA-transfected cells on the same plate.
2. Use these background-subtracted values to plot a growth curve of mock-transfected cells (relative fluorescence versus time postinfection) from a selected number of plates to identify the linear phase of replication, and calculate the gradient of the slope of replication for all wells individually.
3. On each plate, calculate the average replication slope of negative controls [mock-transfected cells (no siRNA), RISC-free (RSCF)-transfected cells (an siRNA which is not processed by the RISC machinery), and nontargeting siRNA-transfected cells (a scrambled siRNA which does not target known human genes) or, for miRNA screens, *Caenorhabditis elegans* miRNA

mimic or Inhibitor, depending on the miRNA library] and normalize the slope of siRNA- or miRNA-transfected cells to the average slope of negative controls on a per plate basis, to adjust for inter-plate variations (*see* **Note 23**).

4. Determine a suitable cut-off point for a significant effect on the replication slope phenotype based on the effect of controls (for example, in comparison to positive controls known to have a strong, moderate, or slight effect on virus replication).
5. Remove or highlight those siRNAs which had a cytotoxic effect on cells (from the Cell Viability phenotypic assay).
6. Categorize the siRNAs based on their positive (increased replication slope) or negative (decreased replication slope) effect on virus replication. Those siRNAs leading to an increase in replication will target an antiviral gene (removal of an antiviral gene will enhance normal replication levels), whereas those siRNAs leading to an inhibition of virus replication will target proviral genes (removal of a gene required for virus replication will reduce normal replication levels) (*see* **Note 24**).
7. Document the details of your siRNA screen (including target gene, siRNA sequence, controls, raw data from each phenotypic screen and transformed data) using appropriate recording guidelines (www.miare.org) [24].

4 Notes

1. The passage number of HeLa cells refers to the number of times a cell line has been "split." The cell line used here was purchased from the European Collection of Cell Cultures (ECACC) and Passage 14, but how this relates to the number of passages since original isolation (in 1951) is not clear. What is important is to keep an accurate account of cell passage number since acquisition within your own laboratory. As cells kept in permanent culture will evolve over time, their phenotype can often change extensively with respect to growth kinetics and thus the phenotype of virus replication may not be reproducible. It is important to work with cells of as low a passage as possible, so that they best represent the phenotype of the original isolate, and so that subsequent phenotypic assays are reproducible and reliable.
2. This screen makes use of the HSV-1 strain C12, modified to contain the fluorescent reporter gene eGFP in the US5 locus, expressed during the late stage of virus replication kinetics [22]. Fluorescence changes are indicative of changes in virus replication.
3. When priming the 8-channel dispensing cassette for the liquid handler, watch the streams of liquid carefully to ensure it is

dispensing directly to the center of the well. If the liquid is dispensing out of alignment there may be a blockage in the channel. Using a sterile fine-gauge needle, carefully clear any obstruction and re-prime the cassette to ensure a direct flow of liquid.

4. Resuspending the library to 3 μM concentration results in a final siRNA concentration in the assay plates of 50 nM. The library plates are diluted 1:10 to generate the 384-well Master Plates at 300 nM. 10 μL siRNA is taken from the 384-well Master Plate to generate an assay plate, which has a further 50 μL added. This results in a final siRNA concentration of 50 nM. If a higher/lower final concentration of siRNA is required, the library resuspension or 384-well Master plate dilution step should be adjusted accordingly so that the final 1:6 dilution results in the desired concentration. We have found a total volume of 60 μL in Corning black 384-well plates is optimal.
5. siRNAs are temperature sensitive and it is not recommended to subject them to multiple freeze–thaw cycles. Generation of a duplicate "Daughter" library provides one copy (the original plates) for long-term storage at low temperature (−80 °C) and one (the "Daughter" plates) for immediate use, and use in subsequent screens.
6. The resuspended siRNA library plates have sufficient siRNA for multiple copies of identical 384-well Master Plates. When labelling Master Plates it is best to include a letter with the plate number to differentiate between additional copies, such as "384-well Master Plate A1," indicating it is Plate 1, replicate A. Subsequent Master Plates would then be labelled B1, C1, etc.
7. Columns 1 and 2 of the siRNA 96-well library plates are empty to provide the user space for assay-specific controls. It is important to add controls to every plate in order to determine the extent of and account for inter-plate variations in the assay, which can occur due to the preparation and handling of large numbers of multi-well plates. Column 1 of plates contains 1× siRNA buffer alone, which will translate to columns 1 and 2 in each 384-well plate to provide untransfected controls. In our virus replication assay, these columns will be left uninfected to provide "background" fluorescence readings (*see* further information in **Note 22**).
8. In this protocol Column 2 of the 96-well Library Plates is used for assay-specific control siRNAs. These will translate to Columns 3 and 4 of the 384-well plates. *See* Fig. 2 for a pipetting schematic and example siRNA control layout. In general, it is useful to include a range of controls including positive controls (siRNAs which will show a strong phenotype in your

selected assay) and negative controls (siRNAs or treatments known to have no effect in your assay). In addition, it is often helpful to target cellular and viral genes which result in moderate effects on your phenotypic assay. In our screen we exploit mock-transfection (transfection reagent treated with no siRNA), nontargeting siRNA (a scrambled sequence which targets no human gene), and RISC-free siRNA (which is not processed by the RISC machinery) as negative controls. Positive controls include eGFP, which whilst leaving HSV-1 replication unaffected does deplete the eGFP reporter gene, and viral genes essential for replication (HSV-1 VP16 and ICP4), as well as cellular genes known to be important for HSV-1 replication (chaperone protein BAG3, and HCF, required for viral transcription). The controls used will need to be modified according to each phenotypic assay. For miRNA libraries, it is important to use miRNA controls, such as unspecific *C. elegans* mimic and inhibitor sequences.

9. Here we describe the addition of control siRNAs to the 96-well Master Plate. As the 96-well Master Plates contain sufficient siRNA for multiple assays it may be more suitable to add assay-specific control siRNA directly into the 384-well Master Plate. Be aware that the siRNA concentration added to a 384-well plate would be 300 nM, whilst concentration added to the 96-well Master is 3 μM.

10. The output of our phenotypic assays is fluorescence. For ease of analyzing large numbers of plates we use a fluorescence plate reader (POLARstar OPTIMA, BMG Labtech) with an additional stacker machine equipped with a barcode scanner for automated reading of multiple plates. The barcode should contain all information required to identify the plate, for example, "DHD 1 HSV-A," "DHD HSV-B," and "DHD 1 CV," signifying it as the *D*harmacon *H*uman *D*ruggable siRNA library 384-well Plate 1, HSV-1 infection replicates A and B, and *C*ell *V*iability replicate.

11. Most robotic liquid handlers permit adjustments to enable liquid aspiration from and dispensation to a range of heights. Our 384-well Master Plates contain 111 μL. If the tips are lowered too far the contents of the wells will be displaced into neighboring wells. It is therefore crucial to adjust the height of the tips according to the volume of siRNA in the wells. Using the reagents and equipment detailed in Subheading 2, we find that a height of 0.95 cm is suitable to aspirate 30 μL from a full 384-well Master Plate. The height should be lowered to 0.65 cm then 0.35 cm for the second and third use of the Master Plate. This will need to be adjusted and optimized based upon your own equipment and consumables.

12. If the assay plates are going to be used within a relatively short period of time (for example, the day after preparation) then storage at −20 °C should suffice if storage space within a −80 °C freezer is limited. As prepared Assay Plates contain a very low volume of 10 μL, long-term storage is not recommended due to the risk of evaporation.
13. Virus replication is affected by the growth cycle of the cells being infected. We find transfection and virus replication are optimal when cells are undergoing active growth, and use cells when they are 50 % confluent. We find splitting cells the day before transfection at a level to ensure 50 % confluence after 24 h to be best. For example, in our hands we split one confluent T75 flask of HeLa cells into 3 T175 flasks. These are usually 50 % confluent the next day.
14. Occasionally, liquid handling robots do not dispense liquid to a well. When assay plates are removed from the −80 °C freezer for use it is worth looking at the underside of each plate and noting wells which have no siRNA or those which appear to have a lower volume. Empty wells appear clear, whilst wells with liquid in appear opaque or white. Empty wells can be filled manually using the remaining volume (21 μL) from the 384-well Master Plate, or left providing a note is made to disregard that well in subsequent data analyses.
15. If you have one, use an automated cell counter for a more accurate cell count. If not, remove at least three individual samples of cells from the suspension and count cells from multiple squares in order to generate a more accurate average cell count. This cell number has been optimized for transfection of HeLa cells in a 384-well plate format and may need optimizing depending on the cell line being used.
16. Addition of 40 μL cells to 10 μL DF1 (0.6 %) and 10 μL siRNA (300 nM) in the well results in a final DF1 transfection reagent concentration of 0.1 % and a final siRNA concentration of 50 nM.
17. In our cell viability assay plate columns 1 and 2 are mock-transfected (treated with transfection reagent but no siRNA). When analyzing cell viability values we normalize the fluorescence of siRNA-transfected wells to an average of these 16 control wells on a per plate basis to avoid inter-plate variations. As per manufacturer's instructions, a fluorescence value of <70 % of the control cells is considered cytotoxic.
18. Before carrying out a large-scale siRNA screen to investigate the role of cellular genes in virus replication, it is important to test a range of virus dilutions (multiplicities of infection) to ensure you achieve the desired assay output. This assay aimed to monitor virus growth over multiple rounds of replication,

so a kinetic growth curve from 20 to 80 h postinfection was measured to provide information on the effect of gene depletion on multiple stages of the virus life cycle. An MOI of 0.5 (0.5 virus plaque-forming units per cell) generated the most appropriate growth curves with HSV-1 for this purpose. Use of a high MOI for such assays is not recommended, as if gene depletion results in a moderate inhibitory effect on virus replication it is possible that a higher level of initial infection would be sufficient to overcome this.

19. When removing culture medium from the plates by inversion, it is possible that air bubbles can form on the bottom of the well. It is important that plates are centrifuged for 5 min at 200 *g* after addition of medium/virus to ensure air bubbles are removed and cell death, due to dehydration, does not occur.

20. As far as possible, the number of steps involving liquid dispensation directly onto cells should be reduced, as cells can become dislodged, leading to an empty patch in the well center and inaccurate interpretation of fluorescence measurements. For this reason, medium is added to the plates without removing the virus inoculum. In our experience, the virus growth kinetics are unaffected by removal of the inoculum.

21. As mentioned, the fluorescent plate reader described here is equipped with a stacker to enable reading of up to 25 plates at a time. This stacker is not compatible with the plastic lids of the Corning 384-well plates used here. For this reason, plates are covered with a gas-permeable adhesive seal following infection. These seals are larger than the plate which can occasionally result in the seal sticking to the plate above. This will prevent the plates from passing through the stacker machinery and cause a jam. It is crucial to cut the seals to the size of the plate to prevent this. For our purposes, the seals from Abgene have the minimal impact on fluorescence readings, and these are pre-cut to 78 × 126 mm for our assay plate. As the seals are gas permeable, the top plate of each stack is covered with a plastic lid to minimize evaporation.

22. The stacker equipment with the BMG Labtech PolarSTAR plate reader is incompatible with plastic lids on multi-well plates, resulting in the need for adhesive seals (*see* **Note 21**). The seals used here are opaque, so the fluorescence during ongoing virus replication is measured from the underside of the plate. As well as deciding on top or bottom-reading settings, other parameters will require optimization, such as the level of gain setting (overall signal amplification, needs adjustment to maximize signal:noise ratio, where noise is the background fluorescence of the plate, media, cells, etc.). The plate reading protocol will need to be optimized based upon the plate reading equipment available.

23. Replication slopes were calculated using the *linest* formula in Microsoft Excel. This essentially calculates the gradient of the slope between two points, by dividing the change in y by the change in x, where y is the relative fluorescence and x is the time postinfection. For the HSV-1 assay described here, the slope during the linear phase of replication was calculated, from 24 to 80 h postinfection.
24. Classification of cellular miRNAs as antiviral or proviral can be more complex. Whilst over-expression of an miRNA may have a significant effect on virus replication, it is not necessarily physiologically relevant if the miRNA is not expressed in the cell type of information. Such data will only be available from the use of an miRNA inhibitor. To increase confidence in the role of cellular miRNAs in virus replication, it is advisable to select miRNAs whose over-expression (mimic library) result in the opposite phenotype as that observed upon inhibition (miRNA inhibitor).

Acknowledgments

The author is grateful to Professor Jürgen Haas and Professor Peter Ghazal for funding and support [MRC (G0501453 J.H.) and Scottish Enterprise (P.G.)], to Kim Martin for assistance with the figures, and to David Griffiths and Kai Kropp for critically reading the manuscript, and providing helpful comments and suggestions.

References

1. Calderwood MA, Venkatesan K, Xing L et al (2007) Epstein-Barr virus and virus human protein interaction maps. Proc Natl Acad Sci USA 104:7606–7611
2. Fossum E, Friedel CC, Rajagopala SV et al (2009) Evolutionarily conserved herpesviral protein interaction networks. PLoS Pathog 5:e1000570
3. Uetz P, Dong YA, Zeretzke C et al (2006) Herpesviral protein networks and their interaction with the human proteome. Science 311:239–242
4. Brass AL, Dykxhoorn DM, Benita Y et al (2008) Identification of host proteins required for HIV infection through a functional genomic screen. Science 319:921–926
5. Konig R, Zhou Y, Elleder D et al (2008) Global analysis of host-pathogen interactions that regulate early-stage HIV-1 replication. Cell 135:49–60
6. Zhou H, Xu M, Huang Q et al (2008) Genome-scale RNAi screen for host factors required for HIV replication. Cell Host Microbe 4:495–504
7. Brass AL, Huang IC, Benita Y et al (2009) The IFITM proteins mediate cellular resistance to influenza A H1N1 virus, West Nile virus, and dengue virus. Cell 139:1243–1254
8. Karlas A, Machuy N, Shin Y et al (2010) Genome-wide RNAi screen identifies human host factors crucial for influenza virus replication. Nature 463:818–822
9. Konig R, Stertz S, Zhou Y et al (2010) Human host factors required for influenza virus replication. Nature 463:813–817
10. Krishnan MN, Ng A, Sukumaran B et al (2008) RNA interference screen for human genes associated with West Nile virus infection. Nature 455:242–245
11. Sessions OM, Barrows NJ, Souza-Neto JA et al (2009) Discovery of insect and human dengue virus host factors. Nature 458:1047–1050
12. Li Q, Brass AL, Ng A et al (2009) A genome-wide genetic screen for host factors required

for hepatitis C virus propagation. Proc Natl Acad Sci USA 106:16410–16415
13. Jopling CL, Yi M, Lancaster AM et al (2005) Modulation of hepatitis C virus RNA abundance by a liver-specific MicroRNA. Science 309:1577–1581
14. Ahluwalia JK, Khan SZ, Soni K et al (2008) Human cellular microRNA hsa-miR-29a interferes with viral nef protein expression and HIV-1 replication. Retrovirology 5:117. doi:10.1186/1742-4690-5-117, 1742-4690-5-117 [pii]
15. Murakami Y, Aly HH, Tajima A et al (2009) Regulation of the hepatitis C virus genome replication by miR-199a. J Hepatol 50:453–460
16. Lagos D, Pollara G, Henderson S et al (2010) miR-132 regulates antiviral innate immunity through suppression of the p300 transcriptional co-activator. Nat Cell Biol 12:513–519
17. Wang P, Hou J, Lin L et al (2010) Inducible microRNA-155 feedback promotes type I IFN signaling in antiviral innate immunity by targeting suppressor of cytokine signaling 1. J Immunol 185:6226–6233
18. Song L, Liu H, Gao S et al (2010) Cellular microRNAs inhibit replication of the H1N1 influenza A virus in infected cells. J Virol 84:8849–8860
19. Santhakumar D, Forster T, Laqtom NN et al (2010) Combined agonist-antagonist genome-wide functional screening identifies broadly active antiviral microRNAs. Proc Natl Acad Sci USA 107:13830–13835
20. Cherry S (2009) What have RNAi screens taught us about viral-host interactions? Curr Opin Microbiol 12:446–452
21. Hirsch AJ (2010) The use of RNAi-based screens to identify host proteins involved in viral replication. Future Microbiol 5:303–311
22. Arthur JL, Scarpini CG, Connor V et al (2001) Herpes simplex virus type 1 promoter activity during latency establishment, maintenance, and reactivation in primary dorsal root neurons in vitro. J Virol 75:3885–3895
23. Birmingham A, Selfors LM, Forster T et al (2009) Statistical methods for analysis of high-throughput RNA interference screens. Nat Methods 6:569–575
24. Taylor CF, Field D, Sansone SA et al (2008) Promoting coherent minimum reporting guidelines for biological and biomedical investigations: the MIBBI project. Nat Biotechnol 26:889–896

Chapter 6

High Resolution Gene Expression Profiling of RNA Synthesis, Processing, and Decay by Metabolic Labeling of Newly Transcribed RNA Using 4-Thiouridine

Lars Dölken

Abstract

Gene expression profiling of changes in total RNA levels has provided invaluable knowledge on the regulation of gene expression. Studies on the kinetics of this regulation, however, have been limited by the fact that total cellular RNA is a poor template for revealing short-term changes in gene expression, alterations in RNA decay rates and the kinetics of RNA processing as well as the differentiation thereof. Here, we describe the metabolic labeling and purification of newly transcribed RNA with 4-thiouridine to study the molecular mechanisms governing RNA synthesis, processing, and decay.

Key words 4-Thiouridine, Biotin, Streptavidin, Newly transcribed RNA, RNA processing, RNA decay, Gene expression profiling, Microarray, RNA-seq

1 Introduction

Total RNA levels are subject to extensive regulation involving alterations in the rates of transcription, RNA processing, and decay [1–3]. Most studies focus on regulation at transcriptional level, however, changes in RNA degradation rates and RNA processing may also significantly alter gene expression of coding and noncoding RNAs [3–7].

Lytic virus infections result in profound changes in cell gene expression. High-throughput transcriptional analysis like microarrays or next-generation sequencing (RNA-seq) can provide important insights into the underlying molecular mechanism [8–11]. A major constraint of these powerful technologies is the poor temporal resolution for such kinetic changes. It is important to note that this is not a specific problem of these technologies but rather due to intrinsic properties of the involved biological samples. As such, total RNA levels of a transcript with an RNA half-life of 10 h will simply take 10 h to go down by twofold following a complete

Susanne M. Bailer and Diana Lieber (eds.), *Virus-Host Interactions: Methods and Protocols*, Methods in Molecular Biology, vol. 1064, DOI 10.1007/978-1-62703-601-6_6, © Springer Science+Business Media, LLC 2013

shutdown (e.g., >1,000-fold down-regulation) of transcription. The inability to readily differentiate changes in RNA synthesis rates from changes in RNA decay rates represents another major limitation. RNA decay rates can be studied by monitoring ongoing RNA decay over time following transcriptional arrest, e.g., by administrating actinomycin D [12]. However, this approach is not readily applied to virus infections due to its cell-invasive nature and detrimental impact on the natural course of infection. In addition, these kinds of measurements are based on very small differences in total RNA levels and thus inherently imprecise for more than half of all cellular transcripts. Finally, the contribution of changes in RNA processing remains elusive. As such, while the consequences of alternative splicing, i.e., the presence of alternatively spliced transcript, are easily revealed by RNA-seq, the kinetics of this regulation and thus the underlying molecular mechanisms are not depicted.

All these problems can be solved by metabolic labeling of newly transcribed RNA using 4-thiouridine (4sU-tagging). This approach provides direct access to newly synthesized transcripts with minimal toxic effects [13–16]. Metabolic labeling is started by adding 4sU to cell culture medium and stopped by cell lysis followed by isolation of total cellular RNA. As RNA from mammalian cells does not contain thiol-groups, the thiol-labeled newly transcribed RNA can be specifically biotinylated generating a disulfide bond between biotin residues and newly transcribed RNA molecules. This tag then allows rigorous purification of newly transcribed RNA using streptavidin-coated magnetic beads. Newly transcribed RNA is finally recovered from the beads by simply adding a reducing agent (dithiothreitol) which cleaves the disulfide bond and releases the newly transcribed RNA molecules from the beads. The purified newly transcribed RNA can then be directly subjected to qRT-PCR, microarray analysis, and RNA-seq [17–19]. Employing this approach, snap-shot pictures of the real-time kinetics of eukaryotic gene expression are obtained [4, 17, 19]. When applied to lytic herpesvirus (murine cytomegalovirus) infection discrete clusters of genes regulated with distinct kinetics are identified revealing the underlying molecular signaling events as well as viral counter-regulation thereof [20].

Under steady-state conditions, RNA synthesis compensates for ongoing RNA decay. Therefore, the ratios of newly transcribed/total or newly transcribed/unlabeled RNA allow measurements of RNA half-lives [18, 21, 22]. In this setting the impact of measurement errors on the error bars of the half-live measurements is substantially less than for measurements based on transcriptional arrest resulting in significantly more precise measurements. When all three RNA fractions, i.e., newly transcribed, total and unlabeled, preexisting RNA are analyzed, both data normalization and data quality control are intrinsically provided by a linear-regression analysis model [22]. In case of non-steady-state conditions, comprehensive analysis of the changes in transcription rates and total RNA levels allows computational modeling of kinetic changes in RNA synthesis and decay [17].

Finally, it is important to note that the mean age of newly transcribed RNA (4sU-RNA) depends on the duration of 4sU exposure. This increases with the duration of 4sU-labeling. Therefore, when 4sU-RNA following different durations of labeling is prepared and analyzed by RNA-seq (*see* **Note 1**), the kinetics of RNA processing are revealed at nucleotide resolution [19].

In conclusion, this method provides access to the dynamics of RNA synthesis, processing, and decay in eukaryotic cells including all major model organisms [4, 16, 23]. In this chapter, the methodology to metabolically label and purify newly transcribed RNA from total cellular RNA is described.

2 Materials

Prepare all solutions using nuclease-free water. Using in-house purified, deionized water has resulted in problems, e.g., if the water contains reducing agents. In one case, this resulted in the complete loss of labeled RNA. Therefore, we strongly recommend buying premade nuclease-free NaCl, Tris–Cl, EDTA, and water. Take all necessary precautions to ensure nuclease-free conditions at all times. All solutions/reagents should be stored at room temperature unless otherwise specified.

2.1 Metabolic Labeling and Isolation of Total Cellular RNA

1. 4-Thiouridine (Carbosynth, 13957-31-8 250 mg): dissolve in sterile water to 50 mM stock concentration, store in small aliquots of 50–500 μL at −20 °C, discard unused reagent after use (do not refreeze).
2. Trizol (Invitrogen), store at 4 °C.
3. 15 mL Polypropylene Tubes SuperClear™ Gatefree™, in contrast to standard falcon tubes, these tolerate up to 15,000 × *g* (VWR International, Cat. No. 525-0153).
4. Polypropylen adaptors for four 15 mL "Falcon" tubes 62 × 120 mm (Laborgeräte Beranek, Germany, Cat. No. 356964).
5. Chloroform.
6. Isopropanol.
7. DEPC (to inactivate RNases), store at 4 °C.
8. Sodium citrate (Sigma): make 1.6 M stock solution, treat with 0.1 % DEPC over night with shaking to eliminate RNases and autoclave thereafter, alternatively buy nuclease-free sodium citrate (Sigma) and dissolve using nuclease-free water.

9. 5 M nuclease-free NaCl (Sigma).
10. RNA precipitation buffer (50 mL): 1.2 M NaCl, 0.8 M sodium citrate.
11. Ethanol.
12. Nuclease-free H_2O.

2.2 Biotinylation of Thiol-Labeled, Newly Transcribed RNA

1. 1 M nuclease-free Tris–Cl, pH 7.5 (Sigma).
2. 500 mM nuclease-free EDTA, pH 8.0 (Sigma).
3. Nuclease-free 10× Biotinylation Buffer (BB): 100 mM Tris pH 7.4, 10 mM EDTA, store in aliquots of 1–1.5 mL at 4 °C.
4. Dimethylformamide (Sigma).
5. EZ-Link Biotin-HPDP (Pierce, 50 mg, Cat. No. 21341): 1 mg/mL stock concentration dissolved in dimethylformamide, gentle warming will ensure complete solubilization, store in aliquots of 1 mL at 4 °C (*see* **Note 2**).
6. Nuclease-free H_2O.
7. Chloroform.
8. Phase Lock Gel (2.0 mL) Heavy Tubes (Eppendorf).
9. Isopropanol.
10. Ethanol.

2.3 Streptavidin-Capture Assay

1. μMacs Streptavidin Kit (Miltenyi, *see* **Note 3**).
2. Magnetic Stand (Miltenyi, one stand holds 4 or 8 columns of the μMacs Streptavidin Kit).
3. Tween 20 (Sigma).
4. Washing Buffer (WB): 100 mM Tris pH 7.5, 10 mM EDTA, 1 M NaCl, 0.1 % Tween 20.
5. Dithiothreitol (DTT): 100 mM DTT in nuclease-free H_2O, always prepare fresh before use.
6. RNeasy MinElute Kit (Qiagen), store columns at 4 °C.

3 Methods

Carry out all procedures at room temperature unless otherwise specified.

3.1 Metabolic Labeling of Newly Transcribed RNA with 4-Thiouridine

As 4-thiouridine is very efficiently taken up by cells, using too little volume of cell culture media may reduce labeling efficiency. We therefore recommend using 5/10 mL of medium for a 10/15 cm dish, respectively. Before beginning the labeling, make a detailed time table plan of the whole experiment allowing for 5 min between each condition (usually comprising up to five technical replicates).

3.1.1 Beginning of Labeling

1. Thaw 4-thiouridine (4sU) just before use and pipet required amount of 4sU for each condition into a sterile falcon tube.
2. Only treat cells of one condition at a time. Try to handle cells as quickly as possible.
3. Take the required amount of medium off the plates, add this to the 4sU-containing falcon tube and mix well. Discard the remaining medium from the plates.
4. Reapply 4sU-containing medium back to cell culture plates.

3.1.2 End of Labeling

1. Carefully remove all cell culture medium from cells (one condition at a time, max. 3–5 plates) and immediately add Trizol to each plate (5 mL per 15 cm dish). For complex experiments including multiple time points, this step is best done by two people, one removing the medium and the other adding Trizol and harvesting the lysate.
2. Pipette up and down several times and incubate at room temperature for 5 min to facilitate complete cell lysis.
3. Transfer to polypropylen tubes. Samples can be stored at –20 °C for up to 1 month until RNA is prepared.

3.2 RNA Preparation Using Modified Trizol Protocol

To ensure complete removal of non-4sU-labeled RNA, it is important to obtain very clean total cellular RNA (*see* **Note 4**). Therefore, do not use too little lysis reagent. In case of *in vivo* samples preparation of poly-A RNA may be required.

1. Add 1 mL chloroform (0.2 mL per mL Trizol) and shake vigorously for 15 s. Incubate at room temperature for 2–3 min.
2. Centrifuge at 13,000 × *g* for 15 min at 4 °C.
3. Transfer aqueous upper phase (containing the RNA) to a new 15 mL polypropylen tube.
4. Add ½ of the reaction volume of both RNA precipitation buffer and isopropanol (e.g., to 3 mL supernatant add 1.5 mL RNA precipitation buffer and 1.5 mL isopropanol).
5. Mix well. Incubate at room temperature for 10 min.
6. Centrifuge at 13,000 × *g* for 10 min at 4 °C. Discard supernatant.
7. Spin down briefly and remove residual isopropanol with 200 μL pipet.
8. Spin down briefly and remove residual isopropanol with 20 μL pipet.
9. Add an equal volume of 75 % ethanol and shake carefully until the pellet detaches. Avoid extensive vortexing as disrupting the pellet too much may make the removal of residual ethanol challenging.

10. Centrifuge at 13,000 × *g* for 10 min at 4 °C. Immediately discard supernatant. Spin down briefly and remove remaining ethanol with a 200 μL pipet. Repeat step and remove remaining ethanol with a 20 μL pipet. After these two steps no further drying of the pellet should be performed.
11. Add 100 μL of H_2O per 100 μg expected RNA yield.
12. Dissolve RNA by heating to 65 °C for 10 min with shaking and immediately place on ice.
13. Measure RNA spectrophotometrically. This RNA can be stored at −80 °C for at least 1 month.

3.3 Biotinylation Assay

Carry out all procedures at room temperature. Avoid direct exposure to bright light as 4sU is light sensitive. Use 30–150 μg total RNA for biotinylation.

3.3.1 Labeling Reaction

Pipet in this order:

1 μL 10× Biotinylation Buffer per 1 μg RNA.

7 μL nuclease-free H_2O per 1 μg RNA (H_2O + RNA).

2 μL Biotin-HPDP (1 mg/mL DMF) per 1 μg RNA.

Always add the biotin-HPDP last and mix immediately by pipetting. In case the biotin precipitates, DMF content can be increased to a final concentration of 40 %.

1. Incubate at room temperature for 1.5 h with rotation.
2. Add an equal volume of chloroform. Mix vigorously. Incubate for 2–3 min until phases begin to separate and bubbles start to disappear (*see* **Note 5**).
3. Centrifuge at full speed (20,000 × *g*) for 5 min. Carefully transfer upper phase into new tubes.
4. Repeat **steps 2** and **3** once. This step can also be performed using Phase Lock Gel (2.0 mL) Heavy Tubes (Eppendorf).
5. RNA precipitation: add 1/10 the volume of 5 M NaCl and an equal volume of isopropanol.
6. Centrifuge at 20,000 × *g* for 20 min at 4 °C. Discard supernatant.
7. Add an equal volume of 75 % ethanol, centrifuge at 20,000 × *g* for 10 min, discard supernatant.
8. Spin briefly and remove residual ethanol by pipetting as described above.
9. Do not allow RNA to dry but immediately resuspend it in 100 μL H_2O.

3.4 Separation of Labeled and Unlabeled RNA Using Streptavidin-Coated Magnetic Beads

1. Heat washing buffer (3 mL per sample) in a 50 mL Falcon tube to 65 °C.
2. Prepare fresh elution buffer: 100 mM dithiothreitol (DTT) in nuclease-free H_2O.
3. Heat biotinylated RNA samples to 65 °C for 10 min and immediately place on ice for 5 min to denature.
4. Pre-equilibrate Miltenyi columns with 1 ml room temperature washing buffer. This will take about 15 min.
5. Add 100 μL of biotinylated RNA to 100 μL of streptavidin beads. Incubate with rotation for 15 min.
6. Place μMacs columns into magnetic stand. Do not process more than 12 samples at a time (6–8 samples are optimal).
7. Apply beads (RNA) to the columns. Discard the flow-through (unless unlabeled RNA is of interest; *see* Subheading 3.6).
8. Wash 3× with 0.9 mL 65 °C washing buffer (pipet tips shrink when pipetting buffers at 65 °C).
9. Wash 3× with 0.9 mL room temperature washing buffer.
10. Pipet 700 μL Buffer RLT (RNeasy MinElute Cleanup Kit) into new 2 mL tubes.
11. Elute RNA directly into Buffer RLT by placing the tubes underneath the columns and adding 100 μL elution buffer (100 mM DTT) to the columns.
12. Perform a second elution round into the same tubes 3 min later.

3.5 Recovery of Newly Transcribed RNA

Continue with the RNeasy MinElute (Qiagen) Cleanup Protocol following the manufacturer's instructions shown below (*see* **Note 6**).

1. Add 500 μL 96–100 % ethanol to the diluted RNA and mix thoroughly by pipetting. Do not centrifuge.
2. Apply 700 μL of the sample to an RNeasy MinElute Spin Column in a 2 mL collection tube. Close the tube gently and centrifuge for 15 s at >8,000 × *g*. Discard the flow-through.
3. Apply the remaining 700 μL and repeat the centrifugation. Discard the flow-through. Transfer the spin column into a new 2 mL collection tube.
4. Pipet 500 μL Buffer RPE onto the spin column. Close the tube gently and centrifuge for 15 s at >8,000 × *g* to wash the column. Discard the flow-through. Transfer the spin column into a new 2 mL collection tube (not supplied).
5. Add 500 μL of 80 % ethanol to the spin column. Close the tube gently and centrifuge for 2 min at >8,000 × *g* to dry the silica-gel membrane. Discard the flow-through and transfer

the spin column into a new 2 mL collection tube. Open the cap of the spin column and centrifuge at full speed for 5 min. Discard the flow-through.

6. To elute, transfer the spin column to a new 1.5 mL collection tube. Pipet 20 μL nuclease-free water directly onto the center of the silica-gel membrane. Close the tube gently and incubate for 1 min before centrifuging for 1 min at maximum speed to elute.
7. Measure RNA using a Nanodrop 1000 Spectrophotometer.
8. Store RNA at −80 °C.

3.6 Recovery of Unlabeled, Unbound RNA

In case the unbound RNA needs to be recovered, collect the flow-through and the first wash for subsequent precipitation (together these contain >90 % of the unbound RNA). Combine the two fractions and recover the unbound RNA by isopropanol/ ethanol precipitation as performed after the biotinylation reaction (no salt needs to be added as the washing buffer already contains 1 M NaCl).

4 Notes

1. Using this modified Trizol protocol by Chomczynski et al. [24] improves the removal of DNA and glycoproteins. 5 mL Trizol per 15 cm dish produces nice clean RNA. Reducing the amount of Trizol may result in incomplete removal of RNAses and subsequent RNA degradation. As higher centrifugal forces are used the RNA pellets are more solid and easier to handle. This requires the use of special polypropylene tubes as the regular 15 mL Falcon tubes do not survive more than 6,000 × *g*.
2. The chloroform extraction is required to remove unincorporated biotin-HPDP. To reduce template RNA loss during the chloroform extraction step Phase Lock Gel Heavy tubes (2.0 mL, Eppendorf) may be used following the manufacturer's instructions. Usually we only use the phase-lock tubes for the second round as >1 mL biotinylation volume is too much volume to fit into these tubes.
3. Recovery of newly transcribed RNA is highly quantitative. If you started with the same RNA concentration you can expect the same amounts of newly transcribed RNA. In case the yields of labeled RNA are lower than expected carefully look for signs of RNA degradation by electrophoretic analysis of the newly transcribed RNA. Newly transcribed RNA is of higher molecular weight than total cellular RNA due to the presence of large unspliced transcripts and a slightly reduced contribution of rRNA (*see* **Note 7**). $OD_{260/280}$ ratios below 1.7 (instead of ~2.0)

usually indicate the carryover of washing buffer from the Qiagen kit. While this does not pose a major problem in downstream analysis (e.g., cDNA synthesis) it confounds RNA measurements (overestimating the amount of purified RNA). This problem is reduced by changing the collection tube after each centrifugation step in the RNeasy Minelute collection step or by precipitating the newly transcribed RNA with glycogen and isopropanol/ethanol.

4. We would strongly recommend not changing the provider of the streptavidin beads as we tested beads from four different companies and found only these to work without background RNA carryover.
5. It is crucial to prevent the dimethylformamide from getting in contact with incompatible plastic materials. Otherwise, substances are eluted from the plastic ware and carried along through the chloroform extraction and isopropanol/ethanol precipitation steps which cause a substantial loss (>75 %) of newly transcribed RNA during the streptavidin-capture assay. Most likely, this is due to damage to the coating of the Miltenyi beads. This becomes more prominent when the duration of labeling is shortened to 30 min or less. The same problem may occur when cell scrapers are used to collect the Trizol samples from cell culture plates.
6. For qRT-PCR analysis use 2.5 μL of labeled RNA in 20 μL cDNA synthesis mix. Subject 1:10 dilutions to qRT-PCR.
7. For RNA-seq, rRNA depletion is of less importance as newly transcribed RNA contains substantially less rRNA (due to substantially more intronic sequences). Therefore, the gain in sequencing depth is not as great as for total RNA. On the other hand, when performing rRNA depletion input RNA amounts may need to be increased.

Acknowledgments

This work was supported by NGFN Plus grant #01GS0801, MRC fellowship grant G1002523, and NHSBT grant WP11-05 to L.D.

References

1. Kim HD, Shay T, O'Shea EK et al (2009) Transcriptional regulatory circuits: predicting numbers from alphabets. Science 325(5939): 429–432
2. Jing Q, Huang S, Guth S et al (2005) Involvement of microRNA in AU-rich element-mediated mRNA instability. Cell 120(5):623–634
3. Nilsen TW, Graveley BR (2010) Expansion of the eukaryotic proteome by alternative splicing. Nature 463(7280):457–463
4. Miller C, Schwalb B, Maier K et al (2011) Dynamic transcriptome analysis measures rates of mRNA synthesis and decay in yeast. Mol Syst Biol 7:458

5. Shalem O, Dahan O, Levo M et al (2008) Transient transcriptional responses to stress are generated by opposing effects of mRNA production and degradation. Mol Syst Biol 4:223 doi:10.1038/msb.2008.59msb200859 [pii]
6. Cazalla D, Yario T, Steitz JA (2010) Down-regulation of a host microRNA by a Herpesvirus saimiri noncoding RNA. Science 328(5985):1563–1566
7. Heo I, Ha M, Lim J et al (2012) Mono-uridylation of pre-microRNA as a key step in the biogenesis of group II let-7 microRNAs. Cell 151(3):521–532. doi:10.1016/j.cell.2012.09.022 S0092-8674(12) 01129-4 [pii]
8. Simmen KA, Singh J, Luukkonen BGM et al (2001) Global modulation of cellular transcription by human cytomegalovirus is initiated by viral glycoprotein B. Proc Natl Acad Sci USA 98(13):7140–7145
9. Zhu H, Cong JP, Mamtora G et al (1998) Cellular gene expression altered by human cytomegalovirus: Global monitoring with oligonucleotide arrays. Proc Natl Acad Sci USA 95(24):14470–14475
10. Browne EP, Wing B, Coleman D et al (2001) Altered cellular mRNA levels in human cytomegalovirus-infected fibroblasts: viral block to the accumulation of antiviral mRNAs. J Virol 75(24):12319–12330
11. Hertel L, Mocarski ES (2004) Global analysis of host cell gene expression late during cytomegalovirus infection reveals extensive dysregulation of cell cycle gene expression and induction of Pseudomitosis independent of US28 function. J Virol 78(21):11988–12011. doi:78/21/11988 [pii] 10.1128/JVI.78.21.11988-12011.2004
12. Yang E, van Nimwegen E, Zavolan M et al (2003) Decay rates of human mRNAs: correlation with functional characteristics and sequence attributes. Genome Res 13(8):1863–1872
13. Melvin WT, Milne HB, Slater AA et al (1978) Incorporation of 6-thioguanosine and 4-thiouridine into RNA. Application to isolation of newly synthesised RNA by affinity chromatography. Eur J Biochem 92(2):373–379
14. Cleary MD, Meiering CD, Jan E et al (2005) Biosynthetic labeling of RNA with uracil phosphoribosyltransferase allows cell-specific microarray analysis of mRNA synthesis and decay. Nat Biotechnol 23(2):232–237
15. Kenzelmann M, Maertens S, Hergenhahn M et al (2007) Microarray analysis of newly synthesized RNA in cells and animals. Proc Natl Acad Sci USA 104(15):6164–6169
16. Dölken L, Ruzsics Z, Radle B et al (2008) High-resolution gene expression profiling for simultaneous kinetic parameter analysis of RNA synthesis and decay. RNA 14(9): 1959–1972
17. Rabani M, Levin JZ, Fan L et al (2011) Metabolic labeling of RNA uncovers principles of RNA production and degradation dynamics in mammalian cells. Nat Biotechnol 29(5):436–442. doi: nbt.1861 [pii] 10.1038/nbt.1861
18. Schwanhäusser B, Busse D, Li N et al (2011) Global quantification of mammalian gene expression control. Nature 473(7347):337–342. doi: nature10098 [pii] 10.1038/nature10098
19. Windhager L, Bonfert T, Burger K et al (2012) Ultra short and progressive 4sU-tagging reveals key characteristics of RNA processing at nucleotide resolution. Genome Res doi: gr.131847.111 [pii] 10.1101/gr.131847.111
20. Marcinowski L, Lidschreiber M, Windhager L et al (2012) Real-time transcriptional profiling of cellular and viral gene expression during lytic cytomegalovirus infection. PLoS Pathog 8(9): e1002908. doi:10.1371/journal.ppat.1002908 PPATHOGENS-D-12-00468 [pii]
21. Friedel CC, Dölken L, Ruzsics Z et al (2009) Conserved principles of mammalian transcriptional regulation revealed by RNA half-life. Nucleic Acids Res 37(17):e115. doi: gkp542 [pii] 10.1093/nar/gkp542
22. Dölken L, Malterer G, Erhard F et al (2010) Systematic analysis of viral and cellular microRNA targets in cells latently infected with human gamma-herpesviruses by RISC immunoprecipitation assay. Cell Host Microbe 7(4):324–334
23. Miller MR, Robinson KJ, Cleary MD et al (2009) TU-tagging: cell type-specific RNA isolation from intact complex tissues. Nat Methods 6(6):439–441
24. Chomczynski P, Mackey K (1995) Short technical reports. Modification of the TRI reagent procedure for isolation of RNA from polysaccharide- and proteoglycan-rich sources. Biotechniques 19(6):942–945

Chapter 7

Isolation and Characterization of Pathogen-Bearing Endosomes Enable Analysis of Endosomal Escape and Identification of New Cellular Cofactors of Infection

Konstanze D. Scheffer, Ruth Popa-Wagner, and Luise Florin

Abstract

Many pathogens, including viruses, bacteria, as well as bacterial toxins, enter their target cells by endocytosis leading to accumulation of pathogenic and cellular proteins in endosomes. Here, we present detailed experimental instructions on isolation of endosomes after virus infection and their subsequent biomolecular characterization. The isolation of endosomes is based on discontinuous sucrose gradient centrifugation, where different endosomal compartments accumulate at a specific sucrose interface. This enables the enrichment and separation of the virus-interacting and co-internalized cell-surface receptors and membrane-associated proteins. The endosomal fractions can be further analyzed by Western blot or quantitative real-time PCR, which reveals changes in the viral protein or DNA content during the processes of endocytosis and endosomal escape. In addition, comparative quantitative mass spectrometry enables the identification of unknown host-cell factors required for infection.

Key words Pathogen, Virus, Endosomes, Host-cell factor, Endocytosis, Trafficking, Endosomal escape, Proteome, PCR

1 Introduction

A common feature of all viruses is their dependence on the host cell machinery and the action of multifunctional viral proteins. In many cases, early events of infection, such as binding to specific cell surface receptors, signaling, entry into the host cell, egress from cellular compartments, and trafficking to replication sites is mediated by the interaction of few viral proteins with a large number of cellular components [1, 2]. For many viruses these processes are poorly understood. Therefore, it is of utmost importance to identify and characterize viral entry pathways, including the involved cellular proteins, for gaining a deeper understanding of virus infection mechanisms. Knowing the identity and comprehending the function of cellular and/or viral key players is essential for the development of new antiviral drugs. Homogenization and

Susanne M. Bailer and Diana Lieber (eds.), *Virus-Host Interactions: Methods and Protocols*, Methods in Molecular Biology, vol. 1064, DOI 10.1007/978-1-62703-601-6_7, © Springer Science+Business Media, LLC 2013

subcellular fractionation are well-established methods to isolate intact cellular organelles under physiological pH conditions without any chemical treatment [3–6]. Organelles or cellular compartments that are enriched by subcellular fractionation can then be characterized by proteomic screening strategies and/or other analytical methods [7]. We used the recently published method of Aniento and Gruenberg [4] for subcellular fractionation of tissue culture cells to isolate early endosomes of noninfected and virus-infected cells. We analyzed the impact of VP1-specific protein sequence motifs on endosomal escape of Adeno-associated virus type 2 [8] and identified new host-cell factors involved in human papillomavirus type 16 entry (KS and LF unpublished data). For papillomavirus type 16, we identified a distinct set of plasma membrane proteins, which were described as components of the tetraspanin-enriched microdomains. In addition, we found components of the cytoskeleton, proteins for vesicle trafficking, for protein folding, and ion transport channels needed for vesicle acidification. These results were in line with our previously published data showing the tetraspanin CD151 as a cofactor for papillomavirus endocytosis [9, 10].

Isolation and characterization of virus-filled endosomes is suitable for all viruses, enveloped and non-enveloped, and possibly other pathogens such as bacteria or bacterial toxins that enter cells via endocytosis and can therefore be detected in endosomes. The isolation of endosomes is based on discontinuous sucrose gradient centrifugation, where different endosomal compartments accumulate at a specific sucrose interface and can be collected easily by gradient fractionation (Fig. 1a). This enables the enrichment, separation, and identification of the virus-interacting and co-internalized cell-surface receptors and membrane-associated proteins perfectly isolated out of the plasma membrane by the cell itself via the process of endocytosis (K.S. and L.F. unpublished data).

These endosomes can be further analyzed by quantitative mass spectrometry, which reveals changes in the protein content compared to endosomes of noninfected cells (Fig. 1b, c). The relevance of the identified proteins can be studied in detail with respect to their role in the infection cycle by functional analysis using siRNA technology, dominant negative mutants, and/or inhibitors.

Besides identifying cellular interaction partners of viral proteins, endosome preparation by density gradient centrifugation is a useful tool for studying virus budding at intracellular membranes [11] or endosomal escape [8]. Budding into endosomal compartments or escape from endosomes is either mediated by viral proteins themselves or dependent on the virus-induced activity of cellular factors. Inhibition or deletion of the involved components leads to reduction of virus import or export, respectively, and can be determined by quantification of viral genomes within the endosomal fractions. Another main advantage of endosome preparation

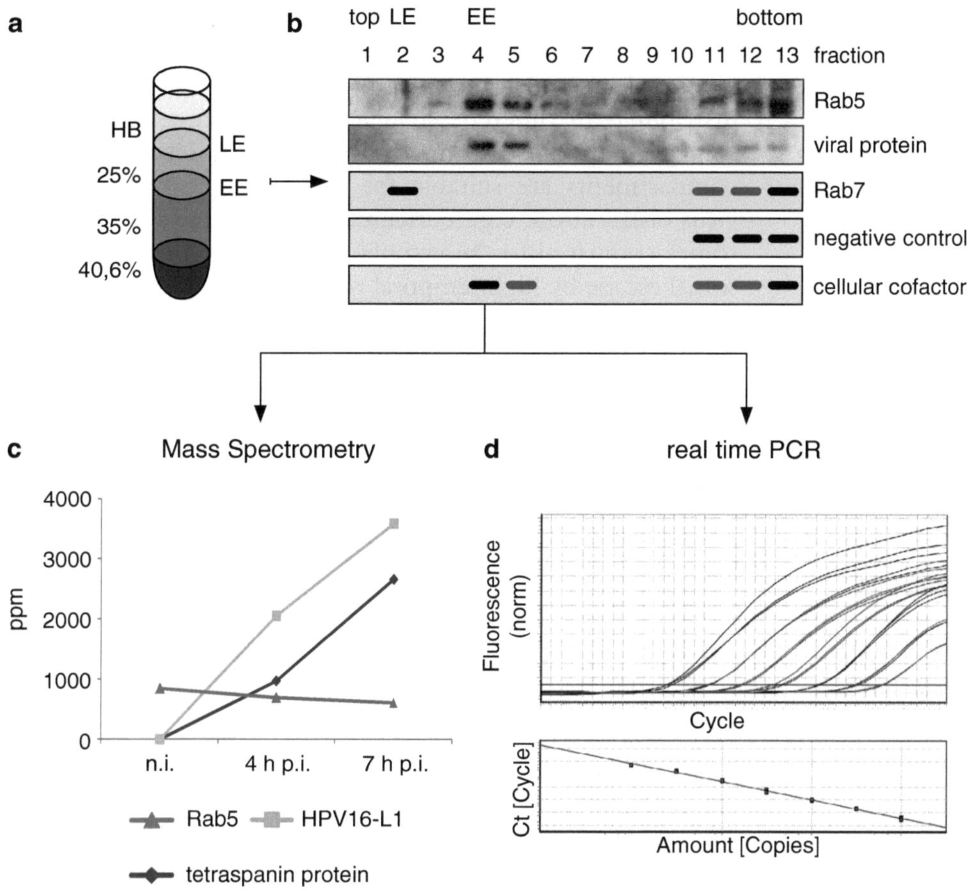

Fig. 1 Schematic illustration of in-depth analysis of virus-filled endosomes isolated by subcellular fractionation of tissue culture cells. (**a**) Infected cells are homogenized and the post-nuclear supernatant is prepared by discontinuous sucrose gradient centrifugation. Late endosomes focus at the 25 %/homogenization buffer interface and early endosomes are enriched at the 25 %/35 % interphase. (**b**) Diagram of Western blot analysis. The gradient is fractionated from the top into 13 fractions after centrifugation. Fractions are loaded on one SDS-Gel and are analyzed by Western blotting. Fractions can be probed for early endosomes (Rab5, *first panel*, lane 4 and 5), viral protein (*second panel*, lane 4 and 5), late endosomes (Rab7, schematic illustration *third panel*, lane 2), negative control (schematic illustration *fourth panel*, should not be detectable in the endosomal fraction), and putative cofactors for infection (schematic illustration *last panel*, lane 4 and 5). Lanes 11–13 show a significant proportion of protein input. (**c**) Analysis of isolated early endosomes by quantitative mass spectrometry-based proteomics. The enrichment of the papillomavirus capsid protein L1 and the co-enrichment of a tetraspanin molecule in early endosomes during the time course of infection (non-infected, 4, 7 h post infection) is shown. (**d**) Analysis of viral genomes within endosomal fractions by quantitative real-time PCR (qRT-PCR). Viral genomes are isolated from endosome fractions and quantified by qRT-PCR using specific primers and a specific fluorescence-labeled probe. For calculating the amount of viral genome copies per endosome fraction, a serially diluted DNA sample of known concentration is used as a standard. The *upper diagram* shows the increase in fluorescence of the DNA standard samples (each concentration in triplicate) with increasing cycle number. The *horizontal line* represents the cycle threshold (number of cycles at which the fluorescence exceeds a defined threshold; C_t). The *lower diagram* shows the standard graph, which is generated by plotting the C_t values against the logarithm of the viral genome copy numbers of the standard samples. Viral genome copies per endosome fraction are calculated according to the linear equation of the standard graph. *EE* early endosomes, *HB* Homogenization buffer, *LE* late endosomes, *n.i.* noninfected, *p.i.* post infection

by density gradient centrifugation as compared to subcellular fractionation using different detergents and buffer systems, is the purity of endosome fractions [11, 12]. As viruses are usually distributed in different cellular compartments, including endosomes, cytoplasm, and nucleus, only clearly separated subcellular compartments are suitable for determining small differences in virus distribution, e.g., enrichment of virus particles within endosomes compared to the cytoplasm (Fig. 1d). For analyzing endosomal escape by high temporal resolution imaging in real-time (*see* Martinez et al., Chapter 15 of this series).

In conclusion, the isolation of endosomes of infected cells allows various analytical possibilities to identify and characterize cellular cofactors of infection as well as virus trafficking strategies, which should lead to a better understanding of the early steps of viral infection.

2 Materials

2.1 Components for Homogenization and Fractionation

1. Cell line: Human cervix carcinoma cell line HeLa (German Resource Centre for Biological Material, DSMZ).
2. Cell culture dish, 145 × 20 mm (Greiner bio-one).
3. Cell culture medium: DMEM + GlutaMAX™-I (gibco® by *life*technologies™) supplemented with 10 % fetal bovine serum (Biochrom AG), 1 % modified Eagle medium nonessential amino acids (PAA), 1 % Penicillin/Streptomycin (PAA).
4. Phosphate buffered saline (PBS; 10×): 1.37 M NaCl, 27 mM KCl, 11.5 mM KH_2PO_4, 65 mM $Na_2HPO_4 \cdot 2H_2O$, pH 7.4.
5. Plastic tube, 15 mL (Greiner bio-one).
6. Rubber policeman: Rotilabo®-rubber wipers (Carl Roth).
7. Blue tips, 1 mL.
8. Homogenization buffer: 250 mM sucrose, 3 mM imidazole, pH 7.4, 10 mM Tris–HCl, pH 7.4 (*see* **Note 1**).
9. Protease inhibitor cocktail: Stock solution: 10 mM leupeptin, 10 mg/mL aprotinin, and 1 mM pepstatin. Working solution: dilute stock solution 1:1,000 in homogenization buffer. Stock solution can be stored at −20 °C for several months.
10. Sucrose solution in H_2O (25 %): 0.806 M sucrose, 3 mM imidazole, pH 7.4. The refractive index of this solution should be 1.3904 at 20 °C (*see* **Note 1**).
11. Sucrose solution in H_2O (35 %): 1.177 M sucrose, 3 mM imidazole, pH 7.4. The refractive index of this solution should be 1.3904 at 20 °C (*see* **Note 1**).

12. Sucrose solution in H_2O (62 %): 2.351 M sucrose, 3 mM imidazole, pH 7.4, 1 mM EDTA. The refractive index of this solution should be 1.4463 at 20 °C (*see* **Note 1**).
13. 22-G needle: 22G × 1¼″ Luer-Lock Sterican® 0.70 × 30 mm (Braun).
14. Syringe, 1 mL: Omnifix®-F 1 mL (Braun).
15. Refractometer.
16. Centrifuge tube, SW60: Open-Top Centrifuge Tube, Polyallomer, 11 × 60 mm.
17. Refrigerated centrifuge and rotor appropriate for cell sedimentation, Ultracentrifuge, SW60 rotor.
18. Low retention reaction tubes, 1.5 mL (Kisker Biotech).

2.2 Components for Mass Spectrometry

1. Iodoacetamide, 15 mM (Sigma).
2. Dithiothreitol, 5 mM (DTT, Sigma).
3. Ammonium bicarbonate, 25 mM containing 0.1 % RapiGest (Waters).
4. Porcine sequencing grade trypsin (Promega).

2.3 Components for Western Blotting

1. Protein sample buffer (5×): 0.3 M Tris–HCl pH 6.8, 5 % sodium dodecyl sulfate (SDS), a spatula tip of bromophenol blue, 50 % glycerol, 5 % β-mercaptoethanol. Leave one aliquot at 4 °C for current use and store remaining aliquots at −20 °C.
2. SDS gels: NuPAGE Novex 4–12 % Bis–Tris Mini gel, 1 mm, 15 wells (Invitrogen).
3. Pre-stained protein standard: Novex Sharp Protein Standard, 12 pre-stained protein bands in the range of 3.5–260 kDa (Invitrogen).
4. SDS-PAGE running buffer: NuPAGE MOPS SDS Running Buffer (Invitrogen).
5. Nitrocellulose (NC) membrane: PROTRan BA 85 (Schleicher & Schüll).
6. Whatman filter paper 3 MM (Schleicher & Schüll).
7. Western blot transfer buffer: 20 % methanol, 25 mM Tris–HCl pH 8.0, 19.2 mM glycine.
8. Phosphate-buffered saline (PBS).
9. Washing buffer (PBS-T): 0.5 % Tween-20/PBS.
10. Blocking solution: 6 % skimmed milk powder in PBS-T. Store at 4 °C.
11. Antibodies: anti-Rab5 (D-11; mouse monoclonal), anti-Lamin B (C-20; goat polyclonal), anti-IκB (C-21; rabbit polyclonal), anti-Vimentin (H-85; rabbit polyclonal) (Santa Cruz Biotechnology).

12. Enhanced chemiluminescence (ECL) substrate: Western Lightning-ECL (PerkinElmer).
13. Enhanced chemiluminescence films: Hyperfilm (Amersham).

2.4 Components for Quantitative Real-Time PCR

1. Viral DNA extraction: DNeasy Blood and Tissue Kit (Qiagen).
2. PCR master mix: TaqMan universal PCR Master Mix (Applied Biosystems), 0.2 μM TaqMan FAM-probe (specific for viral genome), 0.3 μM forward and reverse primers (specific for viral genome).
3. Viral DNA standard of known concentration.
4. qRT-PCR microtiter plate: twin.tec 96-well plate (Eppendorf).
5. Optical adhesive film: MicroAmp (Applied Biosystems).
6. Cycler: Mastercycler ep realplex (Eppendorf).

3 Methods

For isolation of virus-bearing endosomes, cells are mock-treated or infected with the virus of interest for different time points, depending on the specific infection kinetics of the virus. Cells are harvested and homogenized, and a post-nuclear supernatant is prepared. This supernatant is fractionated by a discontinuous sucrose density gradient and fractions containing early endosomes are collected at the 25 %/35 % interface (Fig. 1a). The fractions, which are enriched in early and late endosomes, can be identified by immunoblotting using specific endosomal markers such as Rab5 and Rab7, respectively (Fig. 1b) [8]. Viral proteins can also be detected in the respective fractions. Furthermore, known cellular interaction partners of viral proteins, which reside within endosomes (for instance the co-internalized receptor) can be probed in the endosomal fractions by immunoblotting. HeLa cells were used for the fractionation procedure of Anieto and Gruenberg [4] described below, however, this protocol is suitable for many kinds of tissue culture cells. The individual steps have to be altered with respect to the specific needs of the cells used (*see* Subheading 4 for details).

3.1 Cell Cultivation

Seed HeLa cells at a density of 3–5 × 10^6 cells/dish in a 145 × 20 mm culture dish in 25 mL of culture medium and incubate at 37 °C and 5 % CO_2 overnight (*see* **Notes 2** and **3**).

3.2 Infection

14–18 h after seeding remove the culture medium, add 15 mL of new culture medium and infect HeLa cells with the respective amount of virus up to the appropriate time points. Mock-infect one culture dish of HeLa cells as a control (*see* **Notes 4** and **5**). The following steps are all performed at 4 °C. The solutions should also have a temperature of 4 °C.

3.3 Homogenization of Tissue Culture Cells

1. Cool down cells on ice for 5 min, wash cells with PBS, and remove cells from the dish by scraping with a rubber policeman in 5 mL of PBS and transfer to a 15 mL plastic tube (*see* **Note 6**). Rinse the dish again with 5 mL of PBS and transfer to the same tube (*see* **Note 7**). Centrifuge for 5 min at 700 × *g* (*see* **Note 8**).
2. Remove the supernatant and resuspend the cell pellet gently in 1.8 mL of Homogenization buffer (HB) using a wide pore tip (blue tip cut to a diameter of 2 mm). Transfer homogenate to a 2 mL reaction tube and centrifuge for 10 min at 1,500 × *g* (*see* **Note 8**). Add 0.5 mL of HB + Protease-Inhibitor cocktail + 0.5 mM EDTA to the cell pellet and resuspend it using a blue (1 mL) tip. Quickly draw the homogenate ten times through a 22-G needle attached to a 1 mL syringe (*see* **Note 9**). Centrifuge homogenate for 10 min at 1,500 × *g* and 4 °C (*see* **Note 10**).
3. Collect the post-nuclear supernatant (PNS) and the nuclear pellet (NP). Resuspend NP in 1 mL of PBS and store it at −20 °C. Use 50 μL of the NP for SDS-PAGE and immunoblotting as a control. The PNS is ready for fractionation.

3.4 Gradient Preparation for Fractionation of Tissue Culture Homogenates

1. Dilute the PNS sample with a 62 % sucrose solution to obtain a final solution of 40.6 % sucrose/EDTA. Monitor the solution with a refractometer to control the exact sucrose concentration of 40.6 % (*see* **Note 11**).
2. Load the diluted PNS (approximately 1 mL) to the bottom of a SW60 centrifuge tube and overlay sequentially with 1.5 mL of 35 % sucrose solution (at the beginning in 150 μL steps). Then continue with overlaying 1 mL of 25 % sucrose solution. Add HB to fill the tube.
3. Place the tubes in a SW60 rotor and centrifuge for 90 min at 14,000 × *g* in an ultracentrifuge.
4. Fractionate the gradient carefully from the top in 300 μL steps and transfer every fraction to a low bind reaction tube. Early endosomes of infected as well as noninfected cells will be enriched at the 25 %/35 % interface (mainly fractions 4–6) and late endosomes will be located at the 25 %/HB interface (mainly fractions 2–4) (*see* **Note 12** and Fig. 1a). The isolated endosomes are ready for further investigation.

3.5 Analysis of Gradient Fractions for Endosomal, Cytoplasmic, and Nuclear Markers by Immunoblotting

1. Mix 14 μL of each gradient fraction (*see* **Note 13**) with 4 μL of 5× protein sample buffer and heat to 99 °C for 5 min. Centrifuge samples briefly (30 s at 3,000 × *g*) to bring down the condensate.
2. Load 8 μL of a pre-stained protein ladder and 17 μL of each gradient fraction, respectively, into wells of a 15-well 4–12 %

Bis–Tris mini gel. Electrophorese at 130 V until the dye front has reached the bottom of the gel (approximately 1.5 h).

3. For Western blot transfer, soak six Whatman paper, one NC membrane, and the SDS gel in Western blot transfer buffer and assemble them as a sandwich into a chamber for semi-dry blotting. Blot at 80 mA per mini gel for 1 h at 4 °C.
4. Rinse NC membrane with PBS-T. Then, incubate NC membrane with blocking solution for 1 h at room temperature (RT).
5. For detection of endosomal, cytoplasmic, and nuclear markers incubate NC membranes with anti-Rab5 (marker for early endosomes) (*see* **Note 14**), anti-IκB, anti-Vimentin (both markers for cytoplasmic components), or anti-Lamin B (marker for nuclear components) (each antibody diluted 1:1,000 in blocking solution) overnight at 4 °C (*see* **Note 15**).
6. Wash 5× with PBS-T at RT, 10 min each time. Incubate membranes with appropriate peroxidase-coupled secondary antibodies (anti-mouse, anti-goat, anti-rabbit peroxidase conjugates, diluted 1:5,000 in blocking solution) for 1 h at RT.
7. Wash 5× with PBS-T at RT, 10 min each time. Develop bands with ECL substrate, then absorb substrate with a Whatman paper and detect bands with ECL films (*see* **Note 16**).

3.6 Analysis of Endosomal Fractions by Quantitative Mass Spectrometry

1. Pellet the fractions containing early endosomes by ultracentrifugation at 100,000 × *g* for 1 h at 4 °C. Remove the supernatant completely.
2. Solubilize pelleted early endosomes in 25 mM ammonium bicarbonate containing 0.1 % RapiGest for 15 min at 80 °C.
3. Add 5 mM DTT to the pellet for reducing proteins for 45 min at 56 °C.
4. To alkylate free cysteins add 15 mM iodoacetamide for 1 h in the dark at 25 °C.
5. Digest the proteins with 0.2 μg porcine sequencing grade trypsin and incubate the samples overnight at 37 °C.
6. After digestion, hydrolyze RapiGest by adding 10 mM HCl for 10 min at 37 °C.
7. Remove the resulting precipitate by centrifugation at 13,000 × *g* for 15 min at 4 °C.
8. The supernatant is ready for defining the quantitative proteome of mock-infected and infected cells with a mass spectrometer.
9. Comparison of the proteomes shows enrichment of cellular proteins after infection (*see* **Note 17**), putative host cell factors required for virus endocytosis, trafficking, capsid disassembly, and other mechanisms involved in early steps of infection (Fig. 1c).

3.7 Analysis of Endosomal Escape by Western Blotting and Quantitative Real-Time PCR

For analyzing the amount of virus within endosomal fractions, the relative number of endosomes per endosome fraction has to be estimated. Then, viral genomes have to be extracted from endosome fractions and quantified by means of quantitative real-time PCR (qRT-PCR) [13]. For obtaining the exact proportion of virus particles per endosome fraction, the relative number of endosomes per fraction is related to the amount of viral genomes per fraction [8].

1. For estimating the relative number of endosomes per fraction, run all endosome fractions to be analyzed on one single SDS page. Following Western blotting, quantify Rab5-positive bands by means of the *ImageJ* gel analysis software (*see* **Note 18**).
2. Extract viral DNA from 200 μL of each endosome fraction using the Qiagen DNeasy Blood and Tissue Kit following the manufacturer's instructions. Elute viral DNA from the column in a total volume of 150 μL elution buffer (first elution step with 80 μL of buffer, second elution step with 70 μL of buffer). Store the viral DNA at -20 °C.
3. Prepare a PCR master mix (on ice) sufficient for the number of samples to be analyzed (*see* **Note 19**). A master mix for one triplicate is composed of 40 μL TaqMan universal Master Mix, 2 μL of each forward primer, reverse primer and probe, and 14 μL of PCR-grade water.
4. Prepare a DNA standard by tenfold serially diluting a virus DNA sample of known concentration in PCR-grade water, resulting in sample concentrations ranging from 3.5×10^8 viral genomes/mL to 3.5×10^1 viral genomes/mL. Mix 10 μL of each standard sample with 60 μL of master mix and pipet 20 μL of this mixture into three wells of a qRT-PCR microtiter plate.
5. Pipet 17.1 μL/well of the master mix into a qRT-PCR microtiter plate and add 2.9 μL of extracted viral DNA per well.
6. Seal a qRT-PCR plate with an optical adhesive film and centrifuge for 5 min at $500 \times g$ to remove any air bubble.
7. Run the following PCR program: (a) 2 min at 50 °C (destruction of contaminating PCR product by AmpErase), (b) 10 min at 95 °C (polymerase activation/pre-denaturation), (c) 15 s at 95 °C (denaturation), (d) 1 min at 60 °C (annealing/extension); repeat steps (c) and (d) 40 times. It may be necessary to adjust the PCR program to the virus-specific probe and primers used.
8. A standard graph is generated by plotting the C_t values (PCR cycle number at which a defined fluorescence threshold level is crossed) against the logarithm of the viral genome copy numbers of the standard samples (Fig. 1d).

4 Notes

1. Store sterile stock solutions for 4–8 weeks at −20 °C.
2. The amount of cells to be seeded the day before fractionation depends on the size and growth kinetics of the respective cells. They should have a confluence of 80–90 % on the day of homogenization and fractionation. If there are not enough cells for isolation of endosomes, seed 2–3 culture dishes per approach and pool the cells.
3. Infection and subcellular fractionation can also be performed with cells grown in suspension.
4. Infection time depends on the particular infection kinetics of the virus to be analyzed [8].
5. For papillomavirus infection, we use pseudoviruses harboring a luciferase-encoding plasmid and we applied 100 luciferase-vector positive pseudovirions per cell in 3 mL/1,000,000 cells. Quantification of pcDNA3.1/Luciferase positive pseudovirions was performed by qPCR as described previously [14]. Pre-binding in the cold is not necessary for papillomaviruses because of the slow internalization kinetic, but could be advisable for other viruses. Four and seven hours post-infection have proven to be suitable for preparation of papillomavirus filled endosomes and subsequent qMS analysis.
6. HeLa cells can easily be removed from the culture dish by scraping with a rubber policeman. Some cells such as primary keratinocytes are sensitive to mechanical treatment and have to be removed from the dish with chemical treatment (e.g., trypsin).
7. If more than one culture dish per approach has been seeded, cells need to be pooled at this step of the protocol.
8. The optimal G force for centrifugation depends on the particular cells used for the experiment. Primary keratinocytes for example have to be pelleted at a much lower G force ($300 \times g$) than HeLa cells. Therefore, the rate of centrifugation has to be adjusted for the individual requirements of every cell line.
9. This step is of great importance for the isolation of intact endosomes and has to be monitored by phase-contrast microscopy. First, cells should be drawn 10 times through a syringe with a 22 gauge needle, verifying by phase-contrast microscopy that cells are broken (Fig. 2). If still intact, increase the number of needle strokes and, where required (e.g., smaller cells such as T-lymphocytes), reduce the diameter of the needle. Cells that are grown in suspension or are removed from the culture dish with trypsin usually require harsher homogenization conditions than cells removed mechanically. Caveat: Too harsh conditions can destroy the endosomes.

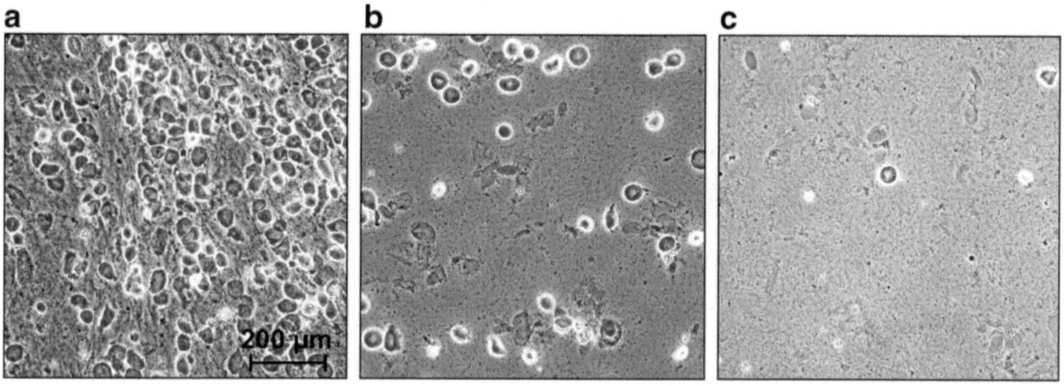

Fig. 2 Monitoring the homogenization process by phase-contrast microscopy. HeLa cells in Homogenization buffer without any mechanical treatment (**a**, cells are still intact) and after drawing 10 (**b**, incomplete homogenization) or 20 times (**c**, complete homogenization) through a 22-G needle attached to a 1 mL syringe

10. After centrifugation the supernatant should be cloudy. If the supernatant is clear, repeat **step 2** (*see* also **Note 9**) and draw the homogenate through the needle again.
11. PNS should first be mixed with 300 µL of a 62 % sucrose solution. Then monitor the resulting density with a refractometer to measure the exact sucrose concentration. Gradually increase the amount of sucrose solution in 50 µL steps until the sample is adjusted to approximately 40.6 %. It is important to gradually increase the amount of sucrose to avoid a too low or too high final sucrose concentration.
12. After successful isolation of endosomes and a proportionally high amount of early endosomes, a cloudy white ring should be visible at the 25–35 % boundary layer after ultracentrifugation. If this is not the case, detection of the early endosomal marker Rab5 by immunoblotting might be problematic, since the amount of isolated proteins might be too small for subsequent experiments. Reasons for a lack of isolated endosomes can be an insufficient number of cells at the beginning of the experiment, unbroken cells or that endosomes are destroyed at Subheading 3.3, **step 2** of the protocol (*see* also **Note 9**).
13. If the marker protein for endosomes (Rab5) cannot be readily detected by immunoblotting, the input amount can be increased accordingly.
14. The protein Rab5 is a well-established standard marker protein for early endosomes. However, besides Rab5 other marker proteins for endosomes (such as Annexin) are also suitable, but we recommend using Rab5. The protein disulfide isomerase (PDI) is a marker protein for the endoplasmic reticulum, the Rab2 protein for the intermediate compartment, and the cis/medial 58K protein for the Golgi complex and suitable as additional negative controls [11].

15. While it is sufficient to incubate NC membranes with antibodies anti-IκB and anti-Lamin B for 1 h at RT, anti-Rab5 needs to be left on membranes overnight, otherwise no bands could be detected for unclear reasons.
16. Wheras IκB and Lamin B bands are detectable on ECL films after short exposure of 10 s to 2 min, ECL films have to be exposed to Rab5-stained blots for at least 1.5 h in order to obtain clearly visible bands.
17. Compare different cell lines to study cell type-specific properties and cell type-specific protein content.
18. The relative amount of early endosomes of an endosome fraction corresponds to the thickness and intensity of Rab5-positive bands. Therefore, quantifying the intensity of Rab5-positive bands by means of the *ImageJ* gel analysis software reveals the relative amount of early endosomes within this fraction.
19. Standard samples are analyzed in triplicate and samples of interest are analyzed in duplicate.

Acknowledgments

This work was supported by grants to L.F. from the German Research Foundation (DFG, SFB490) and the Johannes Gutenberg University Mainz (intern-funding program). Thanks to Stefan Tenzer for mass spectrometry information.

References

1. Mercer J, Schelhaas M, Helenius A (2010) Virus entry by endocytosis. Annu Rev Biochem 79:803–833
2. Spoden GA, Sapp M, Florin L (2012) Host cell factors involved in papillomavirus entry. Med Microbiol Immunol 201(4):437–448
3. Khan MN, Savoie S, Bergeron JJ et al (1986) Characterization of rat liver endosomal fractions: in vitro activation of insulin-stimuable kinase in these structures. J Biol Chem 261(18):8462–8472
4. Aniento F, Gruenberg J (2004) Subcellular fractionation of tissue culture cells. Curr Protoc Protein Sci Chapter 4, Unit 4 3.
5. Gorvel JP, Chavrier P, Zerial M et al (1991) rab5 controls early endosome fusion in vitro. Cell 64:915–925
6. de Araujo ME, Huber LA, Stasyk T (2008) Isolation of endocitic organelles by density gradient centrifugation. Methods Mol Biol 424:317–331
7. Lee YH, Tan HT, Chung MC (2010) Subcellular fractionation methods and strategies for proteomics. Proteomics 10(22): 3935–3956
8. Popa-Wagner R, Porwal M, Kann M et al (2012) Impact of VP1-specific protein sequence motifs on AAV2 intracellular trafficking and nuclear entry. J Virol 86(17):9163–9174. doi: 10.1128/JVI.00282-12
9. Spoden GA, Freitag K, Husmann M et al (2008) Clathrin- and caveolin-independent entry of human papillomavirus type 16-involvement of tetraspanin-enriched microdomains (TEMs). PLoS One 3(10):e331
10. Scheffer KD, Gawlitza A, Spoden GA et al (2013) Tetraspanin CD151 mediates papillomavirus type 16 endocytosis. J Virol 87(6):3435–3446. doi: 10.1128/JVI.02906-12
11. Lambert C, Doring T, Prange R (2007) Hepatitis B virus maturation is sensitive to

functional inhibition of ESCRT-III, Vps4, and gamma 2-adaptin. J Virol 81:9050–9060
12. Cordwell SJ, Thingholm TE (2010) Technologies for plasma membrane proteomics. Proteomics 10(4):611–627
13. Veldwijk MR, Topaly J, Laufs S et al (2002) Development and optimization of a real-time PCR-based method for titration of AAV-2 vector stocks. Mol Ther 6(2):272–278
14. Spoden GA, Besold K, Krauter S et al (2012) Polyethylenimine is a strong inhibitor of human papillomavirus and cytomegalovirus infection. Antimicrob Agents Chemother 56(1):75–82

Chapter 8

Computational Analysis of Virus–Host Interactomes

Caroline C. Friedel

Abstract

High-throughput methods for screening of physical and functional interactions now provide the means to study virus–host interactions on a genome scale. The limited coverage of these methods and the large size and uncertain quality of the identified interaction sets, however, require sophisticated computational approaches to obtain novel insights and hypotheses on virus infection processes from these interactions. Here, we describe the central steps of bioinformatics methods applied most commonly for this task and highlight important aspects that need to be considered and potential pitfalls that should be avoided.

Key words Virus–host interactions, Yeast two-hybrid, RNA interference, Computational analysis, Databases, Functional enrichment analysis, Clustering, Interaction prediction

1 Introduction

Large-scale screens of virus–host interactions using either yeast two-hybrid (Y2H) or RNA interference (RNAi) now provide substantial resources for the computational analysis and modeling of processes involved in virus infection and proliferation [1, 2]. Following the first genome-wide Y2H screen of virus–host interactions in EBV [3], similar screens have been performed for hepatitis C virus (HCV) [4], vaccinia virus [5], H1N1 and H3N2 influenza virus [6], and HIV-1 [7]. An overview of these studies is provided in our recently published review [1]. Since then, numerous additional virus–host Y2H screens have been published including dengue virus [8], influenza virus polymerase [9], flavivirus NS3 and NS5 proteins [10], murine γ-herpesvirus 68 [11], SARS [12], chikungunya virus [13], papaya ringspot virus NIa-Pro protein [14] and human T-cell leukemia virus type 1 and type 2 [15].

In contrast to Y2H, which detects binary physical virus–host interactions, RNAi can also identify functional interactions of the virus with so-called host factors (HF), which are involved in protein complexes, signaling pathways, or cellular processes

Susanne M. Bailer and Diana Lieber (eds.), *Virus-Host Interactions: Methods and Protocols*, Methods in Molecular Biology, vol. 1064, DOI 10.1007/978-1-62703-601-6_8, © Springer Science+Business Media, LLC 2013

relevant for infection, as well as HFs binding to viral nonprotein components (e.g., nucleic acids) [16, 17]. Genome-wide RNAi screens of viral HFs were first performed in *Drosophila* systems for an insect picornavirus [18] and dengue [19] and influenza viruses [20]. Subsequently, genome-scale screens in human cells were published for HIV-1 [7, 21–23], West Nile virus (WNV) [24], HCV [25–27], and influenza virus [28–31] (see Table 2 in [1]). Most recently, a genome-scale study was published performing RNAi for 17 different viruses [32].

Not surprising for high-throughput methods, reproducibility of both Y2H and RNAi large-scale screens is extremely low resulting in very small overlaps between independent screens of virus–host interactions. This can be best assessed for RNAi screens as here several independent screens of the same viruses have been performed including HIV-1, HCV, and influenza virus [7, 21, 23, 25, 27, 28, 30, 31]. In all cases, overlaps were modest ranging from 3 to 6 % for HIV-1 [33], 3–16 % for HCV and 1–12 % for influenza virus. Overlaps are similarly low when comparing different Y2H screens of the same virus or between Y2H and RNAi screens as is exemplified by the case of dengue virus. In the most recent dengue-human Y2H screen by Khadka et al. [8], 188 interactions were identified involving 105 proteins. Only three of these had been identified as HFs in previous RNAi screens (<3 %) and only 1 of 20 (5 %) previously published interactions was detected. Reasons that have been suggested for these discrepancies are differences in the experimental setup, such as differences in cell culture systems, virus isolates, or siRNA pools, as well as different criteria to determine the final set of published interactions. These differences may lead to different subsets of targets and HFs identified such that a large number of interactions is missed in each screen (false negatives). In addition, many of the detected interactions may be false positives, i.e., wrongly detected, due to unspecific interactions of "sticky" proteins in case of Y2H and off-target effects in case of RNAi.

Both the large size and varying quality of the high-throughput screens make it difficult to directly obtain insights on virus infection processes from the screening results. Accordingly, computational and systems biology approaches are necessary to integrate results from different screens and additional data sources as well as identify general trends and connections among the targeted proteins such as common pathways and biological processes they are involved in. An overview of computational approaches used for these purposes was recently published [1]. In this chapter, the corresponding methods are described in more detail and potential pitfalls are highlighted.

2 Resources and Databases

The first step in the analysis of virus–host interactions generally is the compilation of both virus–host and cellular interactions from previously published studies. Although for large-scale screens these data are commonly provided as supplementary material, it is cumbersome to trawl the available literature and download all required supplementary tables. Furthermore, for small-scale studies, interaction data is mostly provided within the main text. To alleviate this problem, many interaction databases have been developed that collect and store interaction data. A number of such databases focus specifically on virus–host interactions, notably the HIV-1, human protein interaction database at NCBI [34], VirHostNet [35], VirusMINT [36], HPIDB [37], and ViPR[38]. In most of these cases, interactions were obtained based on extensive literature curation. While this increases the quality of the data, it requires a continuous curation effort to keep the data up-to-date. Unfortunately, most of these databases are no longer actively updated and currently none of them can be considered as a standard repository for virus–host interactions.

Alternatively, virus–host interactions can be obtained from protein interaction repositories with a more general focus, such as BioGRID [39] and BIND [40]. Both rely on a combination of manual curation and high-throughput submission. However, only BioGRID is still actively maintained as the most recent updates to BIND occurred in mid-2006. Despite its active status, BioGRID by far does not provide a complete picture of either viral or cellular interactomes as it depends on authors submitting their interaction data to BioGRID or the availability and the area of interest of manual curators. Unfortunately, virus–host interactions appear to be included only to a limited degree in BioGRID with many of the most recent studies not covered. In contrast, cellular interactions, in particular human interactions, are much better covered by BioGRID and other actively maintained protein interaction databases such as MINT [41], IntAct [42], or DIP [43]. Furthermore, the human protein reference database (HPRD) provides a large collection of manually curated human interactions but no new release has been published since 2010 [44].

In summary, none of the resources available on viral and cellular protein interactions likely covers all known interactions. Thus, the best strategy to capture as much information as possible is to combine data from all of these resources as they are often to a large degree complementary. For virus–host interactions an additional literature screen is generally necessary as little up-to-date information is contained in available databases. Furthermore, when compiling virus–host and cellular interaction networks, annotations with regard to the type of interaction—which are generally available in all discussed databases—should be taken into account. In particular, protein–gene interactions should be distinguished from protein–protein interactions and for the latter type of interactions, binary and indirect (via other proteins) physical interactions should be treated separately from functional interactions. In most cases, this is best done based on the annotated experimental methods as the interaction type annotation of most databases is not sufficiently fine-grained.

3 Virus–Host Interactions in the Context of the Cellular Interactome

One of the most commonly used approaches for the analysis of virus–host interactomes focuses on evaluating general characteristics of viral targets or HFs within the host networks. Despite their incompleteness and uncertain quality, networks of cellular interactions compiled from public databases as described in the previous section are generally used as an approximation of the true host interactome. Many previous studies have found interesting trends for viral (and also bacterial) targets and HFs mostly with regard to centrality and interconnectedness of these proteins [3, 4, 6, 28, 45–47]. Centrality measures aim to quantify the importance of a protein within the host interactome (Fig. 1a). The most well-known and most easily computable centrality measure is the degree of a protein, i.e., the number of its interactions. The motivation behind this centrality concept is that high-degree proteins (so-called hubs) likely interact with and influence many different pathways and processes and, thus, are important for the cellular system. Indeed, degree has been reported to be correlated to essentiality of a protein for cell survival [48, 49]. Several studies on virus–host interactions indicated a tendency for viruses to interact with or depend on highly connected host proteins [3, 4, 28, 45–47], suggesting that virus tend to target essential proteins or proteins involved in many different pathways.

Alternative centrality measures include distance and betweenness centrality, which focus on more global aspects. Both were found to be significantly increased for viral targets and HFs, mostly independent of the correlation between degree and distance or betweenness centrality [4, 45]. In case of distance centrality, proteins are considered central if the average distance, i.e., the length of the shortest path, to any other protein in the network is small. Distance centrality of a protein is then calculated as the sum of the reciprocals of the distances to the other proteins. For this purpose, shortest path lengths between any pair of proteins have to be calculated. In case of unweighted interaction networks, this can be done most easily using breadth-first searches starting from each protein in the network (Fig. 1b). Betweenness centrality of a protein *P*, on the other hand, sums up the fraction of shortest

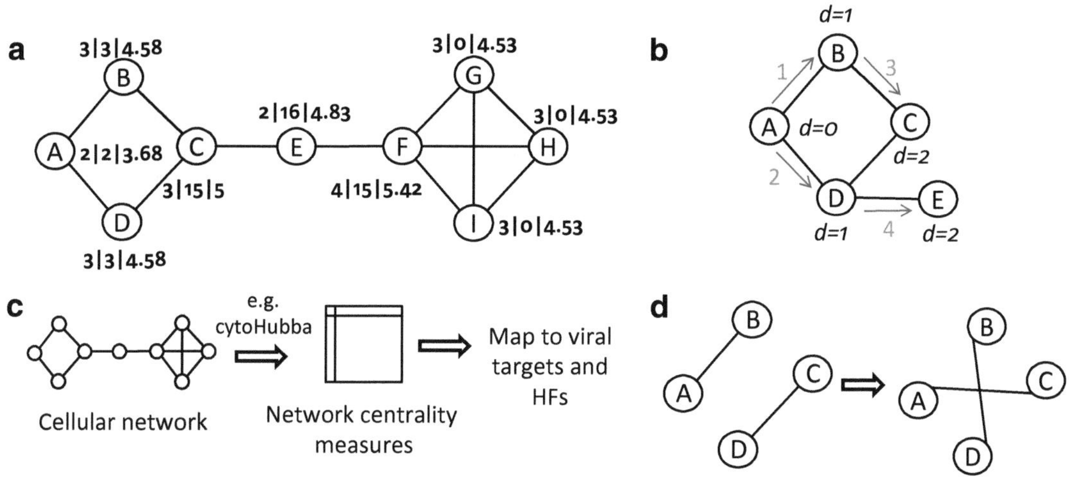

Fig. 1 Network centrality measures and network randomization. (**a**) Various network centrality measures were calculated for an example network (degree, betweenness, and distance centrality, annotated next to the proteins). The most central protein in terms of degree and distance is F, whereas betweenness centrality is highest for protein E, which has only two interactions. The reason for this is that any shortest path from one protein on the left (A–D) to any protein on the right (F–I), has to pass E. In contrast, proteins G–I have betweenness 0, as there is no shortest path going through them. Removal of any of the high betweenness-proteins (C–F) disconnects the networks, but not removal of the other proteins. (**b**) Calculation of distances (*d*) in an unweighted network using breadth-first search. In this example, all distances from A to any of the other proteins are calculated. The *gray arrows* indicate the order in which the interactions are traversed. In a breadth-first search, all interaction partners of a protein are visited before their interactions are traversed. (**c**) To determine centrality measures for viral targets and HFs, first centrality measures of all proteins in the cellular network are determined using existing tools (e.g., cytoHubba) and then values for the relevant proteins are selected. (**d**) Randomization of networks by rewiring of interactions. Afterwards, each protein has the same number of interactions as in the original network

paths between any pair of proteins that pass *P*. Thus, proteins are central if most shortest paths between many pairs of proteins go through them. Such proteins are called bottlenecks and the most extreme case of a bottleneck would be a protein whose removal disconnects the network (Fig. 1a). Betweenness centrality for unweighted interaction networks can be calculated most efficiently using Brandes' algorithm [50]. Unfortunately, no software is available so far for performing centrality analysis specifically for viral targets or HFs. However, existing tools for network analysis, e.g., the Cytoscape plugin cytoHubba (http://hub.iis.sinica.edu.tw/cytoHubba/), can be adapted to this task by first calculating centrality values for all proteins in the cellular network and then mapping them to the viral targets and HFs (Fig. 1c).

Although these trends are mostly confirmed with each new large-scale screen, the conclusions that can be drawn from these observations are limited and correlation is often mistaken for causation. Likely targeting of hubs and bottlenecks is not an end in itself but rather a consequence of the targeting of central pathways and biological processes that contain many highly interactive proteins due to their importance for the host. That viruses tend to target such important processes is certainly not surprising. Although it is tempting to speculate that the particular selection of hubs and bottlenecks allows targeting of these pathways and processes more efficiently with fewer interactions, scarcity of current data does not really allow confident conclusions in this respect. Nevertheless, knowledge of this trend—whatever its reason—is relevant for subsequent analyses performed on virus–host interactions as it serves to avoid some pitfalls. For instance, several groups have noted that viral interaction partners and HFs tend to be densely interconnected [3, 4, 6, 46]. This observation would not be remarkable if the density of the subnetwork were compared to any random subnetwork with the same number of proteins, as high-degree proteins tend to have a larger number of interactions between them by default. Instead, subnetwork density has to be compared against the random background of networks with the same number of interactions per protein. These random networks can be obtained by repeatedly switching end-points of two random edges (Fig. 1d). p-Values are obtained by repeating the random permutation several times (>100) and calculating the fraction of subnetworks among the viral targets or HFs in the random networks that have at least the same number of interactions as the true subnetwork. Similar strategies have to be applied whenever a pursued analysis approach might be biased by the increased degree and betweenness centrality of viral targets and HFs.

4 Evaluation of Targeted Pathways and Biological Processes

In order to better describe the mechanisms of virus infection and proliferation, it is crucial to understand which pathways and biological processes are specifically targeted by the virus. This is complicated by the following problems. First, the definition of pathways or processes is often ambiguous and may differ largely between experts or annotation resources. Second, many genes are involved in several processes or pathways and, thus, it may not always be possible to ascertain which of their functions is relevant for virus infection. Finally, for many proteins only some or even none of their functions may be known and, consequently, many pathways or processes have not been described at all or only incompletely. Essentially, there are two general approaches pursued for uncovering the involved pathways and processes in the context of virus infection. The first one focuses on identifying enriched pathways or processes based on existing functional annotations from public databases and statistical methods. The second one—which will be discussed in the next section—aims to identify novel functional modules based on protein interaction networks.

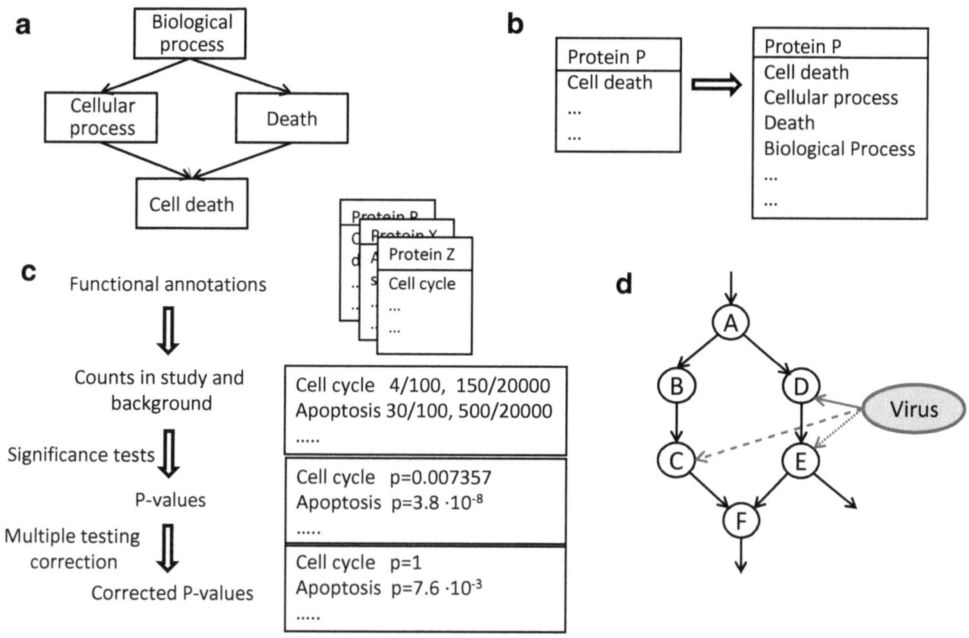

Fig. 2 Outline of functional enrichment analysis. (**a**) The most commonly used annotation resource for functional enrichment analysis is the Gene Ontology (GO), which is structured as a directed acyclic graph. Thus, a term may have more than one parent, e.g., *cell death* has the parents *cellular process* and *death*. Here, only the part of the *biological process* ontology above the term *cell death* is shown. (**b**) As only the most specific GO term is usually annotated directly to a protein, the annotation has to be extended for functional enrichment analysis by any superterm of the annotated terms. (**c**) Workflow for performing functional enrichment analysis. For this example, we assume 100 proteins identified as viral interactors or HFs and 20,000 proteins in the background population, e.g., all human proteins. p-Values were calculated with Fisher's exact test and multiple testing correction performed using the Bonferroni method. (**d**) Exemplary case illustrating the relevance of interactions when evaluating targeted pathways. The pathway consisting of proteins A–F is targeted by the virus at two proteins, either D and C (*solid* and *dashed line*) or C and E (*solid* and *dotted line*). In the first case, the complete pathway is influenced; in the second case, only the right-hand path from A to E but not the one on the left side. In functional enrichment analysis, however, the pathway is simply represented as a set of proteins and the two situations cannot be distinguished

Several resources on protein function annotation are currently available. The most commonly used resource is the Gene Ontology (GO) which provides three hierarchically structured vocabularies (called ontologies) to describe biological processes, molecular functions, and cellular components, respectively [51] (Fig. 2a). Pathway annotation can be obtained from the KEGG [52] and BioCarta (http://www.biocarta.com) databases which also provide interactions between proteins/genes. In principle, any type of annotation can be used for enrichment analysis, including also protein domains from Pfam [53], keywords from Uniprot [54] or disease annotations from OMIM (http://www.omim.org). The main distinction between annotation resources is whether they are hierarchically structured or not. In the first case, proteins will generally be annotated with the most specific term in the hierarchy but any superterm higher up in the hierarchy also applies to the protein. To address this problem, annotations are extended during enrichment analysis such that for any specific annotation term all superterms are also presumed to be annotated to the respective protein (Fig. 2b). Generally, this results in a large degree of redundancy in the results as superterms of significantly enriched terms often also tend to be enriched.

The most commonly used approach for identifying the relevant functional categories among the large list of annotated functions is based on assessing the statistical difference between the observed frequency of a function among the targets or HFs and the frequency in the background (Fig. 2c). The reason for using statistical testing is that neither the absolute counts nor the ratios of frequencies are informative on their own. For functional categories that are very frequent in the overall protein population, large counts among the targets are expected. In contrast, for a very infrequent category, a few hits among the targets may be sufficient for statistical significance. The most commonly used statistical tests for this purpose are Fisher's exact test and the hypergeometric test, which are both based on the hypergeometric null distribution and thus equivalent [55]. As these tests are applied individually to each functional category, one additional aspect becomes important, namely multiple testing correction. Essentially, a p-value quantifies the probability that a specific value of the test statistic is expected at random according to the null distribution. Thus, the standard cut-off of 0.05 for significance tests indicates that the probability of seeing this result at random is about 1 in 20 if only one significance test is performed. However, if thousands of tests are performed as in the case of functional enrichment analysis, this means that we can expect a lot of random results with this value. To address this problem multiple testing correction is applied. Here, the most rigorous and straightforward correction method is the Bonferroni method which simply multiplies all p-values by the number of significance tests. As this method is very stringent and discards many truly significant results, several other multiple testing correction methods have been developed. The most commonly used one is the method by Benjamini and Hochberg [56] for control of the false discovery rate (FDR), i.e., the number of results erroneously called significant. Most multiple testing correction methods are available in the statistical programming language *R*, for instance in the *multtest* package [57].

A large number of software tools and Web servers have been published so far for functional enrichment analysis (see, e.g., [55] for an overview), in most cases focused on the GO. Among these the DAVID Web server should be noted especially for its ease of use as it allows enrichment analysis for a wide range of annotation resources as well as protein identifier types (e.g., gene symbols,

Affymetrix IDs, Entrez Gene IDs). In addition to the classical view of enriched functional categories sorted by associated p-values, a clustering of categories based on the overlap of annotated proteins can be performed, which in light of the inherent redundancy within and between annotation resources provides a better overview of the relevant categories. One important feature generally provided by all tools is that a list of genes can be provided as background population by the user instead of the complete genome. This is important if only a nonrandom subset of the genome was selected for screening, such as the druggable genome. In this case, an enrichment analysis against the genome would generally pick up any functional category already enriched in the background population.

The advantage of functional enrichment analysis is that it is easy to perform using existing tools even without programming skills and provides a first "quick-and-dirty" overview which processes may be involved in virus infection. For instance, in our recently published study on SARS-host interactions [12], GO enrichment analysis provided the first clue that immunophilins might be suitable drug targets for coronavirus treatment. However, there are also several problems associated with enrichment analysis as it is standardly performed. First, it is based on gene lists, requiring a cut-off in case the readout from the experiment is continuous, as, e.g., for RNAi screens. This problem can be addressed by using statistical tests to compare distributions instead of frequencies, such as the Kolmogorov-Smirnov test, and some tools provide this option, e.g., GeneTrail [58]. Second, functional categories are assumed to be independent of each other, which is a very simplifying assumption as categories can overlap in many genes. This does not only result in a large redundancy in the output and affects its interpretability, but may also violate the assumptions behind the statistical tests and multiple correction methods. Despite this problem, more statistically sound approaches as discussed by Goeman and Bühlmann [59] have not gained wide-spread acceptance. Finally, when focusing on functional categories as gene sets only, interactions between genes and proteins are ignored and consistency of the results is not evaluated (Fig. 2d). As a consequence, results from the functional enrichment analysis should always be taken with a grain of salt and not be considered as an important finding by itself but rather be used to derive hypotheses that are followed up and validated by other means.

5 Identification of Novel Functional Modules Involved in Virus Infection

The approaches described in the previous section rely on existing knowledge of pathways and biological processes and predefined functional categories. As this knowledge is likely incomplete, several methods have been developed to identify previously undescribed

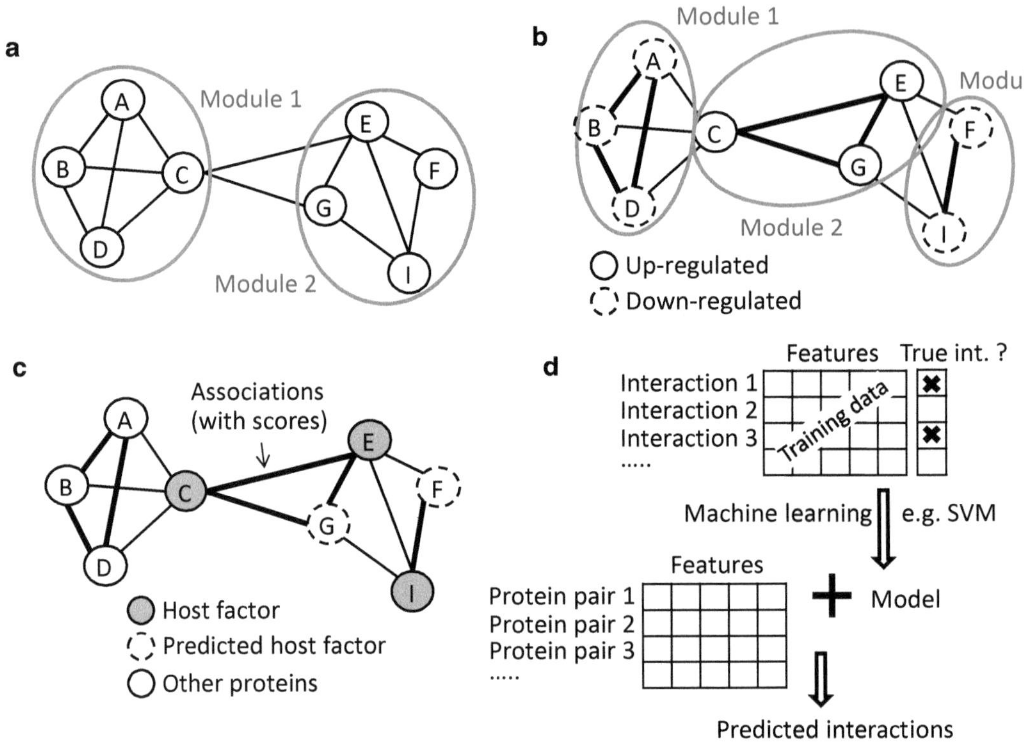

Fig. 3 Identification of targeted modules and prediction of virus–host interactions. (**a**) Clustering of viral targets and HF based on the density of cellular interactions between them. In this example, two clusters are identified in which the proteins are tightly connected. (**b**) Additional properties, such as co-expression, can be taken into account into the module identification by assigning weights to the interactions. In the example, proteins C, E, G are upregulated among infection and the others are downregulated resulting in stronger weights for some interactions (*thick lines*) and different identified modules compared to (**a**). (**c**) Prediction of novel host factors using the guilt-by-association concept. Associations in this case are the same as in (**b**) and calculated from protein interactions and co-expression. Proteins C, E, I have previously been identified as HFs. As G and F interact with all of these HFs and are co-expressed, they are predicted as HFs. The other proteins are only weakly associated with an HF and, thus, not predicted as a novel HF. (**d**) Outline of the approach for predicting direct virus–host interaction. To derive a prediction model, supervised machine learning algorithms (such as support vector machines (SVM)) require a set of known positive (i.e., interacting protein pairs) and negative interactions (i.e., pairs of proteins not interacting) with additional features describing these interactions, e.g., correlation of expression. This set is called the training data. The resulting model is then applied to a set of potential interactions for which the same features have been calculated and for each interaction the model predicts whether this is a true interaction or not

processes associated with virus infection. In general, these are based on identifying functional modules of virus targets or HFs from the network of cellular interactions using network clustering approaches. The clustering method most commonly used for this purpose is MCODE [60], which aims to detect densely connected subnetworks within large cellular networks (Fig. 3a). Density of a subnetwork is defined as the number of edges in the subnetwork divided by the maximum possible number of interactions among the involved proteins and ranges between 0 and 1. MCODE uses three steps, including protein weighting, determination of dense modules, and optional postprocessing of the detected modules. In the first step of MCODE, proteins are weighted based on the density of interactions among their neighbors in the network. In the second step, modules are extended starting from the highest weighted protein not yet contained in a cluster until the density in the module falls below a density threshold. This threshold is defined as a percentage of the weight of the seed protein. Using this percentage parameter, the number and density of the predicted modules can be adjusted. As MCODE is available as a Cytoscape plugin, it can be easily applied to the network of host interactions among the viral targets and HFs and the results can be immediately inspected visually. It is likely this ease of use that led to its predominance for the analysis of virus–host interactions (e.g., in [7, 28, 33]) and not necessarily a better performance in identifying functional modules. At least for the application of protein complex detection, other network clustering approaches have shown superior performance compared to MCODE [61], e.g., Markov Clustering (MCL) [62], but have yet to be applied for the analysis of virus–host interactomes.

The major challenge in the *de novo* detection of functional modules involved in virus infection is not the detection of these modules. Depending on the parameters, MCODE or other graph clustering algorithms will always identify some densely connected subnetworks among the viral targets and HFs. Accordingly, the difficulty consists in the assessment of the significance of the results and the biological interpretation of the modules. So far, the problem of module significance has been mostly ignored for this application and all focus has been put on the biological interpretation of the results. This is unfortunate as significance analysis not only serves to distinguish truly relevant results from mere random observations. It can also help to limit the list of identified modules to the most interesting ones for which more in-depth analysis is performed. As the number of identified modules can be large (e.g., 152 in case of the König et al. study on influenza virus HFs [28]), such detailed analysis is often omitted. Instead modules are commonly mapped to known processes and pathways, e.g., from the GO or KEGG, and only considered further if they are significantly enriched in at least one functional category. As a consequence, a large fraction of detected modules are often discarded (e.g., almost 50 % in the König et al. study mentioned above [28]), most notably the so far undescribed and likely novel functional modules. Accordingly, in most studies on virus–host interactions, network clustering so far provided only little incremental insights compared to a simple enrichment analysis. Thus, the advantage lies mostly in the extension of known processes by additional proteins as well as interactions between the proteins.

Apart from network clustering based only on the interactions between viral targets and HFs, additional approaches have been developed to identify modules which are not only connected by many interactions but also similar with regard to other properties, such as phenotype after RNAi knockdown (Fig. 3b). One straightforward way to do this is to assign edge weights to the interaction networks based on the other properties considered. This allows applying state-of-the-art weighted graph clustering approaches such as MCL or even standard distance-based clustering approaches such as average linkage clustering. The latter approach was used by Gonzalez and Zimmer [63] to identify clusters of interacting proteins that also show a similar phenotype in an RNAi screen. In this case, the challenging aspect is the definition of an appropriate weight function/distance metric to quantify different types of similarities between proteins. Given the edge weights, existing implementations of clustering algorithms for instance in *R* or *Matlab* can then be easily applied.

6 Prediction of Virus–Host Interactions

The small overlaps between screens of viral targets or HFs for the same species indicate that a large number of interactions are missed in each screen and, thus, a substantial number of interactions still remain to be detected. Accordingly, several methods have been developed to identify novel virus–host interactions or HFs. Just as for the large-scale screening methods, two objectives can be distinguished here: (1) the identification of proteins either interacting functionally with the virus (similar to RNAi) or (2) the identification of binary physical interactions between a viral and a host protein (similar to Y2H). Most approaches for the first application can be roughly subsumed by the term "guilt-by-association" (Fig. 3c). Accordingly, proteins are predicted as HFs if they are closely associated either functionally or physically with other HFs. What distinguishes the individual methods is the definition of the associations and the prediction of the HFs based on these associations. Usually, associations and confidence scores for these associations are calculated by integrating several types of evidence, such as co-expression, and domain co-occurrence, for instance using Bayesian methods [64]. Alternatively, functional associations including confidence scores for each type of evidence are also readily available from the STRING database, which integrates evidence from genomic context, high-throughput experiments, co-expression, and literature mining [65].

Using the association scores and information on known HFs, the likelihood of a protein to be an HF can be scored. The most straightforward way to do this involves a summing up of the association scores of this protein to known HFs, either with or without normalization to the total sum of association scores of this protein [64, 66]. A number of more sophisticated methods are described in a recent article by Murali et al. [66], including their novel SinkSource algorithm. As all of these methods only provide likelihood scores for a protein being an HF, a cut-off has to be applied to obtain the final predictions and the quality of the predictions depends strongly on the choice of the cut-off. Thus, to compare different methods, evaluation procedures should be used that are independent of a particular choice of cut-off, such as receiver operating characteristic (ROC) curves or precision-recall curves. In both cases, proteins are sorted by their confidence scores calculated based on all other proteins and all possible cut-offs are evaluated. For each cut-off, true positive rate (= fraction of HFs correctly predicted = recall) and false positive rate (= fraction of non-HFs wrongly predicted as HF), in case of ROC curves, or recall and precision (= fraction of predictions that are HFs), in case of recall-precision curves, are calculated and plotted against each other. If the curve for one method is always above the curve for another method, the first method is clearly superior. If no such clear trends are observed, the area under the curve (AUC) can be calculated which provides one single measure of performance. For ROC curves, the AUC quantifies the probability that a true HF is ranked before a random non-HF.

For the prediction of physical binding between a virus and a host protein, in principle the same methods can be used that have been developed for the prediction of intraspecies interactions. Generally, these approaches exploit similarities of a protein pair to known interacting protein pairs either from the same or a different species. These similarities may be quantified in terms of sequence or structural similarities between the proteins (e.g., [67, 68]) or other evidence as used for scoring associations for the prediction of HFs (e.g., [69, 70]). In the latter case, so-called supervised machine learning approaches are generally applied to learn a classification model that identifies true interactions based on certain features of the interaction. To learn the model, both known true interactions are required (positive examples) as well as protein pairs that do not interact (negative examples). Here, the challenging aspect is the selection and calculation of the interaction features and the collection of positive and negative examples (training data). Given this training data, any out-of-the-box supervised learning algorithm can be used, for instance support vector machines (SVM) or any other algorithm included in the WEKA software library [71].

The limitation of these approaches for the prediction of direct virus–host protein interactions consists in the scarcity of training data. For most viruses, the number of known interactions to the host is very small even when including closely related species. Accordingly, sequence and structure similarity to known interacting pairs is in most cases not large enough to confidently transfer interactions. Furthermore, other types of experimental evidence commonly used to infer interactions, such as gene expression studies, are also generally not available. Despite these difficulties efforts have been undertaken with some success to predict virus–host interactions mostly based on sequence homology and other sequence features but also protein centrality measures and GO annotations [66, 72, 73]. In all of these cases, however, predictions were focused on HIV-1–human interactions for which the largest amount of data is available. It remains to be seen how successful these approaches can be for less well-studied virus–host interactomes.

7 Conclusions

In summary, a large number of methods have been developed for the computational analysis of virus–host screens focusing either on the role of the viral targets and HFs within the host network or biological processes and pathways targeted by the virus. Mostly, however, these approaches are not readily available as software tools, thus limiting their applicability for biological users. Fortunately, at least in some cases the methods can be replicated using existing implementations for individual steps such that only little programming skills are required.

References

1. Friedel CC, Haas J (2011) Virus-host interactomes and global models of virus-infected cells. Trends Microbiol 19(10):501–508
2. Striebinger H, Kögl M, Bailer SM (2013) High-throughput analysis of virus-host interactions by yeast two hybrid assay. In: Bailer SM, Lieber D (eds) Virus-Host Interactions: Methods and Protocols, Methods in Molecular Biology, vol. 1064
3. Calderwood MA et al (2007) Epstein-Barr virus and virus human protein interaction maps. Proc Natl Acad Sci USA 104(18):7606–7611
4. de Chassey B et al (2008) Hepatitis C virus infection protein network. Mol Syst Biol 4:230
5. Zhang L et al (2009) Analysis of vaccinia virus-host protein-protein interactions: validations of yeast two-hybrid screenings. J Proteome Res 8(9):4311–4318
6. Shapira SD et al (2009) A physical and regulatory map of host-influenza interactions reveals pathways in H1N1 infection. Cell 139(7): 1255–1267
7. Konig R et al (2008) Global analysis of host-pathogen interactions that regulate early-stage HIV-1 replication. Cell 135(1):49–60
8. Khadka S et al (2011) A physical interaction network of dengue virus and human proteins. Mol Cell Proteomics 10(12):M111.012187
9. Tafforeau L et al (2011) Generation and comprehensive analysis of an influenza virus polymerase cellular interaction network. J Virol 85(24):13010–13018
10. Le Breton M et al (2011) Flavivirus NS3 and NS5 proteins interaction network: a high-throughput yeast two-hybrid screen. BMC Microbiol 11:234
11. Lee S et al (2011) An integrated approach to elucidate the intra-viral and viral-cellular protein interaction networks of a gamma-herpesvirus. PLoS Pathog 7(10):e1002297
12. Pfefferle S et al (2011) The SARS-coronavirus-host interactome: identification of cyclophilins as target for pan-coronavirus inhibitors. PLoS Pathog 7(10):e1002331
13. Bourai M et al (2012) Mapping of Chikungunya virus interactions with host proteins identified nsP2 as a highly connected viral component. J Virol 86(6):3121–3134
14. Gao L et al (2012) A set of host proteins interacting with papaya ringspot virus NIa-Pro

protein identified in a yeast two-hybrid system. Acta Virol 56(1):25–30
15. Simonis N et al (2012) Host-pathogen interactome mapping for HTLV-1 and 2 retroviruses. Retrovirology 9(1):26
16. Cherry S (2009) What have RNAi screens taught us about viral-host interactions? Curr Opin Microbiol 12(4):446–452
17. Griffiths SJ (2013) Screening for host proteins with pro- and antiviral activity using high-thoughput RNAi. In: Bailer SM, Lieber D (eds) Virus-Host Interactions: Methods and Protocols, Methods in Molecular Biology, vol. 1064
18. Cherry S et al (2005) Genome-wide RNAi screen reveals a specific sensitivity of IRES-containing RNA viruses to host translation inhibition. Genes Dev 19(4):445–452
19. Sessions OM et al (2009) Discovery of insect and human dengue virus host factors. Nature 458(7241):1047–1050
20. Hao L et al (2008) Drosophila RNAi screen identifies host genes important for influenza virus replication. Nature 454(7206):890–893
21. Brass AL et al (2008) Identification of host proteins required for HIV infection through a functional genomic screen. Science 319(5865):921–926
22. Yeung ML et al (2009) A genome-wide short hairpin RNA screening of jurkat T-cells for human proteins contributing to productive HIV-1 replication. J Biol Chem 284(29): 19463–19473
23. Zhou H et al (2008) Genome-scale RNAi screen for host factors required for HIV replication. Cell Host Microbe 4(5):495–504
24. Krishnan MN et al (2008) RNA interference screen for human genes associated with West Nile virus infection. Nature 455(7210): 242–245
25. Tai AW et al (2009) A functional genomic screen identifies cellular cofactors of hepatitis C virus replication. Cell Host Microbe 5(3):298–307
26. Ng TI et al (2007) Identification of host genes involved in hepatitis C virus replication by small interfering RNA technology. Hepatology 45(6):1413–1421
27. Li Q et al (2009) A genome-wide genetic screen for host factors required for hepatitis C virus propagation. Proc Natl Acad Sci USA 106(38):16410–16415
28. Konig R et al (2010) Human host factors required for influenza virus replication. Nature 463(7282):813–817
29. Bortz E et al. (2011) Host- and strain-specific regulation of influenza virus polymerase activity by interacting cellular proteins. mBio 2(4):e00151-11
30. Karlas A et al (2010) Genome-wide RNAi screen identifies human host factors crucial for influenza virus replication. Nature 463(7282): 818–822
31. Brass AL et al (2009) The IFITM proteins mediate cellular resistance to influenza A H1N1 virus, West Nile virus, and dengue virus. Cell 139(7):1243–1254
32. Snijder B et al (2012) Single-cell analysis of population context advances RNAi screening at multiple levels. Mol Syst Biol 8:579
33. Bushman FD et al (2009) Host cell factors in HIV replication: meta-analysis of genome-wide studies. PLoS Pathog 5(5):e1000437
34. Fu W et al (2009) Human immunodeficiency virus type 1, human protein interaction database at NCBI. Nucleic Acids Res 37(Database issue):D417–D422
35. Navratil V et al (2009) VirHostNet: a knowledge base for the management and the analysis of proteome-wide virus-host interaction networks. Nucleic Acids Res 37(Database issue):D661–D668
36. Chatr-aryamontri A et al (2009) VirusMINT: a viral protein interaction database. Nucleic Acids Res 37(Database issue):D669–D673
37. Kumar R, Nanduri B (2010) HPIDB–a unified resource for host-pathogen interactions. BMC Bioinformatics 11(Suppl 6):S16
38. Pickett BE et al (2012) ViPR: an open bioinformatics database and analysis resource for virology research. Nucleic Acids Res 40(Database issue):D593–D598
39. Stark C et al (2011) The BioGRID interaction database: 2011 update. Nucleic Acids Res 39(Database issue):D698–D704
40. Alfarano C et al (2005) The biomolecular interaction network database and related tools 2005 update. Nucleic Acids Res 33(Database issue):D418–D424
41. Licata L et al (2012) MINT, the molecular interaction database: 2012 update. Nucleic Acids Res 40(Database issue):D857–D861
42. Kerrien S et al (2012) The IntAct molecular interaction database in 2012. Nucleic Acids Res 40(Database issue):D841–D846
43. Salwinski L et al (2004) The database of interacting proteins: 2004 update. Nucleic Acids Res 32(Database issue):D449–D451
44. Keshava Prasad TS et al (2009) Human protein reference database–2009 update. Nucleic Acids Res 37(Database issue):D767–D772
45. Dyer MD, Murali TM, Sobral BW (2008) The landscape of human proteins interacting with viruses and other pathogens. PLoS Pathog 4(2):e32
46. Wuchty S, Siwo G, Ferdig MT (2010) Viral organization of human proteins. PLoS One 5(8):e11796

47. van Dijk D et al (2010) Identifying potential survival strategies of HIV-1 through virus-host protein interaction networks. BMC Syst Biol 4:96
48. Yu H et al (2004) Genomic analysis of essentiality within protein networks. Trends Genet 20(6):227–231
49. Jeong H et al (2001) Lethality and centrality in protein networks. Nature 411(6833):41–42
50. Brandes U (2001) A faster algorithm for betweenness centrality. J Math Sociol 25: 163–177
51. Ashburner M et al (2000) Gene ontology: tool for the unification of biology. The gene ontology consortium. Nat Genet 25(1):25–29
52. Kanehisa M et al (2010) KEGG for representation and analysis of molecular networks involving diseases and drugs. Nucleic Acids Res 38(Database issue):D355–D360
53. Punta M et al (2012) The Pfam protein families database. Nucleic Acids Res 40(Database issue):D290–D301
54. UniProt Consortium (2012) Reorganizing the protein space at the Universal Protein Resource (UniProt). Nucleic Acids Res 40(Database issue):D71–D75
55. Rivals I et al (2007) Enrichment or depletion of a GO category within a class of genes: which test? Bioinformatics 23(4):401–407
56. Benjamini Y, Hochberg Y (1995) Controlling the false discovery rate: a practical and powerful approach to multiple testing. J R Stat Soc Series B Stat Methodol 57(1):289–300
57. Pollard KS et al (2010) multtest: Resampling-based multiple hypothesis testing. R package version 2.6.0. http://CRAN.R-project.org/package=multtest
58. Backes C et al (2007) GeneTrail–advanced gene set enrichment analysis. Nucleic Acids Res 35(Web Server issue):W186–W192
59. Goeman JJ, Buhlmann P (2007) Analyzing gene expression data in terms of gene sets: methodological issues. Bioinformatics 23(8): 980–987
60. Bader GD, Hogue CW (2003) An automated method for finding molecular complexes in large protein interaction networks. BMC Bioinformatics 4:2
61. Brohée S, van Helden J (2006) Evaluation of clustering algorithms for protein-protein interaction networks. BMC Bioinformatics 7:488
62. van Dongen S (2000) Graph clustering by flow simulation. University of Utrecht, Utrecht, Netherlands
63. Gonzalez O, Zimmer R (2011) Contextual analysis of RNAi-based functional screens using interaction networks. Bioinformatics 27(19):2707–2713
64. Lee I et al (2011) Prioritizing candidate disease genes by network-based boosting of genome-wide association data. Genome Res 21(7):1109–1121
65. Szklarczyk D et al (2011) The STRING database in 2011: functional interaction networks of proteins, globally integrated and scored. Nucleic Acids Res 39(Database issue): D561–D568
66. Murali TM et al (2011) Network-based prediction and analysis of HIV dependency factors. PLoS Comput Biol 7(9):e1002164
67. Ng SK, Zhang Z, Tan SH (2003) Integrative approach for computationally inferring protein domain interactions. Bioinformatics 19(8): 923–929
68. Yu H et al (2004) Annotation transfer between genomes: protein-protein interologs and protein-DNA regulogs. Genome Res 14(6):1107–1118
69. Jansen R et al (2003) A Bayesian networks approach for predicting protein-protein interactions from genomic data. Science 302(5644): 449–453
70. Qi Y, Bar-Joseph Z, Klein-Seetharaman J (2006) Evaluation of different biological data and computational classification methods for use in protein interaction prediction. Proteins 63(3):490–500
71. Hall M et al (2009) The WEKA data mining software: an update. SIGKDD Explorations 11(1):10–18
72. Evans P et al (2009) Prediction of HIV-1 virus-host protein interactions using virus and host sequence motifs. BMC Med Genomics 2:27
73. Qi Y et al (2010) Semi-supervised multi-task learning for predicting interactions between HIV-1 and human proteins. Bioinformatics 26(18):i645–i652

Chapter 9

Interspecies Heterokaryon Assay to Characterize the Nucleocytoplasmic Shuttling of Herpesviral Proteins

Shuai Wang, Kezheng Wang, and Chunfu Zheng

Abstract

Nucleocytoplasmic trafficking of proteins plays important roles in processes of the viral life cycle. Interspecies heterokaryon assay is one of the most effective methods to investigate the nucleocytoplasmic trafficking properties of a protein. In our lab, the interspecies heterokaryon assay has been applied to identify a few herpesviral proteins with nucleocytoplasmic shuttling property. In this chapter, the detailed information and methods of the heterokaryon assay are presented.

Key words Heterokaryon assay, Nucleocytoplasmic trafficking, Herpes simplex virus type 1, Cell fusion, Leptomycin B

1 Introduction

In eukaryotic cells, genetic material and the transcriptional machinery in the nucleus are separated from the translational machinery and metabolic systems in the cytoplasm by the nuclear envelope. Nucleocytoplasmic trafficking of proteins is crucial for the life processes of eukaryotic cells [1–3]. The nuclear pore complexes (NPC) mediate the bidirectional transport of macromolecules between the cytoplasm and the nucleus [4]. Theoretically, small molecules, including ions, metabolites, and proteins with masses < 40 kDa, are allowed to passively diffuse through the nuclear pore complex (NPC) without energy consumption. However, an energy- and signal-dependent mechanism is required for large proteins (more than 40–60 kDa) and even most of the smaller proteins to pass through the NPC [5, 6].

Nuclear localization signals (NLSs) and nuclear export signals (NESs) within a protein mediate its shuttling between the nucleus and the cytoplasm [7, 8]. At the same time, a protein lacking an

Shuai Wang and Kezheng Wang contributed equally to this chapter.

Susanne M. Bailer and Diana Lieber (eds.), *Virus-Host Interactions: Methods and Protocols*, Methods in Molecular Biology, vol. 1064, DOI 10.1007/978-1-62703-601-6_9, © Springer Science+Business Media, LLC 2013

NLS or NES can shuttle by cotransport with another actively transported protein containing an NLS or NES. Nuclear import of proteins is mediated by NLSs, which are classified into four categories [9, 10]. The classical type of NLS consists of a short stretch of basic amino acids. The second category is the bipartite NLS, consisting of two stretches of basic amino acids separated by 10–12 amino acids. The third category of NLS contains charged/polar residues interspersed with nonpolar residues. The fourth category is the NLSs of arginine-rich domains from many viral proteins, including HIV-1 Rev [11] and Tat [12], and HSV-1 VP13/14 [13].

NES mediates export of a protein from the nucleus. A NES usually consists of an approximately 10-amino-acid leucine-rich domain, which has been found in both cellular and viral proteins executing quite heterologous biological functions. An increasing number of viral proteins have been identified to contain a leucine-rich NES sequence, including HIV-1 Rev [14], influenza virus NS1 [15], HSV-1 ICP27 [16], and γ34.5 [17]. However, atypical NESs that do not contain a leucine-rich domain have also been described for various proteins, such as hepatitis D antigen HDAg-L [18].

Shuttling transport factors, including importins and exportins, recognize proteins containing an NLS or NES and mediate their transport between the nucleus and the cytoplasm [19]. The classical NLS is recognized by the heterodimeric importin α/β complex with the importin α subunit binding directly to the NLS and the importin β subunit mediating the interaction with the cytoplasmic surface of the NPC [20]. In contrast, arginine-rich NLSs of many viral proteins directly interact with importin β in the absence of importin α [21, 22]. Chromosomal region maintenance 1 (CRM1; exportin 1) has been proven to interact with leucine-rich NESs directly and to be responsible for the export of NES-containing proteins [7, 23]. The CRM1-dependent nuclear export mechanism is well characterized; however, CRM1-independent export mechanisms have been found in some proteins [18, 24]. The energy for nuclear transport is provided by a small Ras family of GTPases, Ran [6, 25].

To investigate the nucleocytoplasmic shuttling property of a given protein which predominantly or at least partly locates in the cytoplasm, a simple approach is to use the chemical compounds to inhibit the nuclear export of the protein. Leptomycin B (LMB) is often used to identify the nucleocytoplasmic trafficking property of a protein. The pharmacological compound LMB directly interacts with CRM1 and inhibits CRM1-mediated protein export [26, 27]. For a nucleocytoplasmic trafficking protein containing a leucine-rich NES, LMB treatment will abolish its nuclear export, leading to nuclear accumulation of the protein. Alternatively, transcription inhibitors, such as actinomycin D (ActD), have also been used to investigate the nucleocytoplasmic shuttling of several proteins, whose nuclear export is required for RNA transport. However, LMB treatment does not alter the subcellular localization of

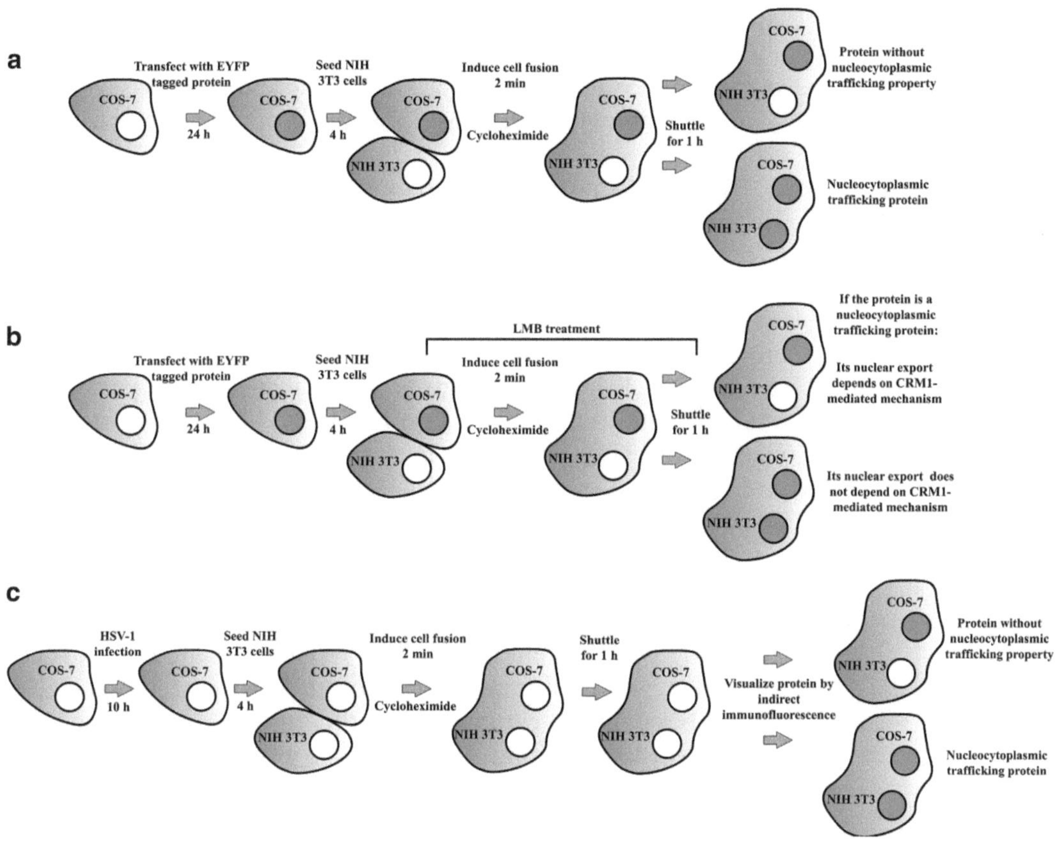

Fig. 1 Schematic diagram of interspecies heterokaryon assay. (**a, b**): The schematic diagram of interspecies heterokaryon assays by transfection of expressing plasmid. The nucleocytoplasmic trafficking properties allow the protein to be exported out of the nuclei of COS-7 cells and reimported into the nuclei of NIH 3T3 cells (**a**). If LMB treatment abolishes the transport of a nucleocytoplasmic trafficking protein, the nuclear export of the protein depends on CRM1- mediated mechanism (**b**). (**c**): Procedure for the heterokaryon assay in the context of HSV-1 infection, and the protein was visualize by indirect immunofluorescence

nuclear proteins. Instead, an interspecies heterokaryon assay is often applied to study the nucleocytoplasmic trafficking properties of a nuclear protein. Interspecies heterokaryon assays have been successfully applied to identify a few shuttling viral proteins in our lab [28–32]. In the heterokaryon assay, monkey COS-7 cells are transfected with a given protein or infected with viruses and subsequently fused to an equivalent number of murine NIH 3T3 cells in the presence of the protein synthesis inhibitor cycloheximide. The protein is visualized by tagging with a fluorescent protein or indirect immunofluorescence. If the protein is capable of nucleocytoplasmic trafficking, it will be exported out of the nuclei of COS-7 cells and reimported into the nuclei of NIH 3T3 cells (*see* Fig. 1a). Hoechst or DAPI dye staining allows differentiation between COS-7 and NIH 3T3 nuclei. COS-7 cells are stained diffusely throughout the nuclei, whereas NIH 3T3 cells nuclei are stained with a distinctive speckled pattern. The interspecies heterokaryon assay is sensitive and suitable to investigate the nucleocytoplasmic trafficking properties of transfected proteins and viral proteins in an infection context. However, the interspecies heterokaryon assay is not suitable for proteins located, or partly located in the cytoplasm.

Recently, a novel bipartite assay NEX-TRAP (Nuclear EXport Trapped by RAPamycin) was developed to investigate the protein nuclear export *in vitro* [33]. Rapamycin treatment induces dimerization of FRB (FK506-rapamycin (FR)-binding domain) and FKBP (FK506-binding protein-12) [34]. In the NEX-TRAP assay, a given protein fused to EYFP-NLS-FRB accumulates in the nucleus at steady state. FKBP is fused to an integral membrane protein that resides at the Trans-Golgi network (TGN). Rapamycin treatment mediates FKBP to trap FRB in the cytoplasm at the TGN. If the given protein is capable of nuclear export, rapamycin treatment will make the fused protein mainly locate in the cytoplasm [33]. NEX-TRAP is very sensitive, accurate, and easy to perform; however, it requires fusion of the given protein to EYFP-NLS-FRB, and is not suitable to study the viral protein in the context of viral infection.

Many viral genomes locate and replicate in the nuclei, therefore, an increasing number of viral proteins have been reported to shuttle between the cytoplasm and the nucleus to exert their specific biological functions. The nucleocytoplasmic shuttling mechanism benefits viruses by mediating viral mRNA transport [35, 36], viral particle assembly [37], etc. Furthermore, the nucleocytoplasmic transport property of viral proteins has been intensively characterized to develop new therapeutic strategies [38, 39].

Herpes simplex virus type 1 (HSV-1), a typical member of the alphaherpesvirus family, is a large, nuclear-replicating, dsDNA virus causing cold sores, herpes keratitis, and herpes encephalitis. HSV-1 VP19C is a structural protein of the viral particle, which is essential for the assembly of the capsid. In our recently published paper, a heterokaryon assay was performed to prove VP19C as a nucleocytoplasmic shuttling protein [40], making VP19C the first herpesviral capsid protein with nucleocytoplasmic shuttling property. In this chapter, the detailed methods of the heterokaryon assay of VP19C are described.

2 Materials

1. COS-7 cells.
2. NIH 3T3 cells.
3. pEYFP-VP19C plasmid: The open reading frame (ORF) of HSV-1 capsid protein VP19C was amplified with *EcoR* I and *BamH* I restriction sites inserted at the beginning and end of the ORF, respectively. The PCR products were then cloned

into the *EcoR* I and *BamH* I restriction sites of the pEYFP-C1 vector (Clontech) to generate plasmid pEYFP-VP19C.

4. Dulbecco's modified Eagle's medium with 4.5 g/L of glucose (DMEM, Gibco).
5. Fetal Bovine Serum (FBS, Gibco).
6. 0.2 % Trypsin (Gibco).
7. Opti-MEM® I Reduced Serum Medium without serum (Invitrogen).
8. Lipofectamine™ 2000 (Invitrogen).
9. Cycloheximide, 20 mg/mL (Sigma).
10. Leptomycin B, 540 μg/mL in ethanol (Invitrogen).
11. PBS: Weigh 0.24 g of KH_2PO_4, 1.44 g of Na_2HPO_4, 8.0 g of NaCl, 0.2 g of KCl, add H_2O to 1 L, pH 7.4, sterile.
12. 50 % polyethylene glycol 1,300–1,600 (w/w) (Sigma), sterile.
13. Hoechst or DAPI.
14. 37 °C 5 % CO_2 humidified cell incubator.
15. Six-well plate (Corning).
16. Fluorescence microscope (Zeiss).

Additional materials for the heterokaryon assay in the context of HSV-1 infection:

17. HSV-1.
18. 4 % paraformaldehyde in PBS.
19. 1 % Triton X-100.
20. BSA.
21. VP19C polyclonal antibody.
22. FITC-labeled Goat Anti-Rabbit IgG (CST).

3 Methods

To investigate the nucleocytoplasmic shuttling property of VP19C, interspecies heterokaryons of COS-7 and NIH 3T3 cells were generated as described previously [31, 32]. All procedures should be performed in a biosafety cabinet to avoid contamination unless otherwise specified.

3.1 Preparation of COS-7 Cells

1. Remove the medium from confluent monolayer COS-7 cells in 25 cm^2 cell culture flasks, and wash the cells with 5 mL PBS twice (*see* **Note 1**).
2. Add 200 μL 0.2 % trypsin (Gibco) to digest the cell at 37 °C for 2 min. Then add 5 mL DMEM (Gibco) with 10 % FBS and pipette evenly.

3. Seed the COS-7 cells onto six-well plates (about 2×10^5 cells/well), and add DMEM with 10 % FBS to 2 mL each well. Mix gently by rocking the plate back and forth. Incubate the cells at 37 °C in a 5 % CO_2 incubator overnight.

3.2 Transient Transfection of COS-7 Cells with pEYFP-VP19C

Transiently transfect COS-7 cells with 1.0–1.5 μg of pEYFP-VP19C. Lipofectamine™ 2000 is recommended (*see* **Note 2**).

1. Remove the DMEM from the wells and add 1 mL Opti-MEM® I Reduced Serum Medium.
2. For each well, dilute 1.0–1.5 μg of pEYFP-VP19C in 100 μL Opti-MEM® I Reduced Serum Medium, and mix gently.
3. Dilute 2 μL of Lipofectamine™ 2000 in 100 μL Opti-MEM® I Reduced Serum Medium. Mix gently and incubate the mixture for 5 min at room temperature.
4. Mix the diluted pEYFP-VP19C and diluted Lipofectamine™ 2000 gently and incubate for another 20 min at room temperature.
5. Add the plasmid-Lipofectamine™ 2000 complex into the wells containing COS-7 cells. Rock the plate back and forth to mix evenly. Incubate the cells at 37 °C in a 5 % CO_2 incubator overnight. Change the medium 6 h later.

3.3 Plating of NIH 3T3 Cells

After 24 h incubation (*see* **Note 3**), VP19C-EYFP can be observed in the nuclei of COS-7 cells by fluorescence microscopy (*see* **Note 4**). Then plate NIH 3T3 cells (about 6×10^5 cells/well) onto the COS-7 cells.

1. Remove the medium from confluent monolayer NIH 3T3 cells in 25 cm^2 cell culture flask, and wash the cells with 5 mL PBS twice. Digest the cells with 200 μL 0.2 % trypsin at 37 °C for 2 min. Then add 5 mL DMEM with 10 % FBS and pipette cells evenly.
2. Remove the medium from the wells containing COS-7 cells. Seed equal amounts of NIH 3T3 cells (about 6×10^5 cells/well) onto the transfected COS-7 cells. Add DMEM with 10 % FBS to 2 mL each well. Add cycloheximide to a final concentration of 50 μg/mL to block *de novo* protein synthesis. Mix gently and incubate the cells at 37 °C in a 5 % CO_2 incubator.

3.4 Formation of Heterokaryons

After 4 h incubation, NIH 3T3 cells will adhere to the surface of plate. Polyethylene glycol (PEG) 1,300–1,600 (Sigma) is used to induce the fusion of COS-7 and NIH 3T3 cells.

1. Remove the medium and wash the cells with PBS twice. Add 2 mL of 50 % PEG 1,300–1,600 (wt/wt) in PBS to the cells to induce cell fusion for 2 min (*see* **Note 5**).

2. After 2 min of incubation, the cells are washed extensively with PBS. Then add 2 mL DMEM with 10 % FBS and 50 μg/mL cycloheximide. Incubate the cells in a 5 % CO_2 incubator and allow the protein to shuttle for 1–2 h (*see* **Note 6**).
3. Remove the medium and add Hoechst or DAPI dye (1:1,000 in PBS) to stain the cells for 5 min.
4. Remove the dye and wash the cells intensively with PBS thrice. Then examine the nucleocytoplasmic shuttling of the protein by a fluorescence microscope (*see* **Note** 7).

3.5 Interspecies Heterokaryon Assay in the Context of Infection

Interspecies heterokaryon assays in the context of viral infection are similar to the transfection protocol (*see* Fig. 1c). The detailed methods are described as following. **Steps 1** through **6** describe infection and generation of heterokaryons. **Steps** 7 through **17** depict visualization of the protein of interest, VP19C, by immunofluorescence microscopy.

1. Preparation of COS-7 cells (about 2 × 10E5 cells/well) as in Subheading 3.1.
2. Dilute 4×10^5 HSV-1 to 0.5 mL DMEM without FBS for each well (*see* **Note 8**).
3. Remove the DMEM from the wells and add 1 mL diluted virus each well (Multiplicity of infection of 2) and incubate the cells at 37 °C in a 5 % CO_2 incubator for 2 h.
4. Remove the diluted virus and add 2 mL DMEM with 2 % FBS. Incubate the cells at 37 °C in a 5 % CO_2 incubator for 8 h (*see* **Note 9**).
5. Plate NIH 3T3 cells (about 6×10^5 cells/well) onto the COS-7 cells as in Subheading 3.3.
6. Induce the fusion of COS-7 and NIH 3T3 cells and allow VP19C to shuttle as in Subheading 3.4.
7. Wash the cells with PBS twice.
8. Add 2 mL 4 % paraformaldehyde in PBS to fix cells at 4 °C for 1 h.
9. Remove paraformaldehyde and wash the cells with 2 mL PBS for 10 min three times.
10. Permeabilize cells with 2 mL 1 % Triton X-100 for 10 min.
11. Remove 1 % Triton X-100 and wash the cells with 2 mL PBS for 10 min three times.
12. Add 2 mL 5 % BSA in PBS and incubate the cells at 37 °C for 1 h.
13. Dilute VP19C polyclonal antibody in 1 % BSA in PBS. Remove 5 % BSA and add the diluted VP19C antibody. Incubate the cells at 4 °C overnight.

14. Wash the cells with PBS containing 0.05 % Tween-20 for 10 min three times.
15. Dilute FITC-labeled Goat Anti-Rabbit IgG in 1 % BSA in PBS. Add the diluted antibody into the wells and incubate the cells at 37 °C for 1.5 h.
16. Wash the cells with PBS containing 0.05 % Tween-20 for 10 min three times.
17. Stain the cells with Hoechst or DAPI dye and examine the nucleocytoplasmic shuttling of VP19C as in Subheading 3.5.

4 Notes

1. Human cells, such as HeLa cells, are applied as alternative to COS-7 cells in interspecies heterokaryon assays. The nuclei of human cells, just like the nuclei of COS-7 cells, are also stained diffusely throughout the nuclei by DAPI or Hoechst, and are easily differentiated from the nuclei of mouse NIH 3T3 cells.
2. If the heterokaryon assay is performed using transfected protein without a fluorescent protein tag or using a viral protein during infection, an immunofluorescence assay is required to detect the protein after **step 2** of Subheading 3.4.
3. After transfection, cells can be incubated at 37 °C in a 5 % CO_2 incubator for a longer time until the fluorescence can be observed.
4. VP19C-EYFP locates in the nuclei. Interspecies heterokaryons are not suitable to identify the nucleocytoplasmic property of proteins which locate or partly locate in the cytoplasm.
5. The experimental conditions for PEG treatment should be controlled strictly, as prolonged or intensive treatment with PEG will disrupt the nuclear membrane. If necessary, use a nuclear protein without nucleocytoplasmic shuttling property as a negative control.
6. LMB, the inhibitor of CRM1, is usually used to determine whether the nucleocytoplasmic shuttling of a protein is dependent on CRM1. For LMB treatment, cells are incubated at 10 ng/mL LMB for 4 h before and after the cell fusion (*see* Fig. 1b).
7. The nuclei of COS-7 cells are stained diffusely throughout the nuclei and NIH 3T3 nuclei are stained with a distinctive speckled pattern. Proteins with nucleocytoplasmic trafficking property shuttle from the nuclei of COS-7 cells to the nuclei of NIH 3T3 cells. Abolishment of nucleocytoplasmic shuttling of the protein by LMB treatment suggests that the nuclear export

of the protein depends on a CRM1-mediated mechanism (*see* Fig. 1a, b).

8. The MOI of HSV-1 infection should be more than or equal to 2 to ensure that most COS-7 cells are infected.
9. Viral gene expression is regulated in distinct temporal phases. VP19C is a late gene of HSV-1 and is suitable to be detected at 15–16 h post-infection. However, the time for HSV-1 infection should be controlled to avoid cell death.

Acknowledgments

This work was supported by the grants from the Start-Up Fund of the Hundred Talents Program of the Chinese Academy of Sciences (20071010-141); the National Natural Science Foundation of China (81171584, 81101263); the Program for Changjiang Scholars and Innovative Research Team in Soochow University and the Program for Jiangsu Provincial Innovative Research Team.

References

1. Yoneda Y (2000) Nucleocytoplasmic protein traffic and its significance to cell function. Genes Cells 5:777–787
2. Guttler T, Gorlich D (2011) Ran-dependent nuclear export mediators: a structural perspective. EMBO J 30:3457–3474
3. Cheng F, McLaughlin PJ, Zagon IS (2010) Regulation of cell proliferation by the opioid growth factor receptor is dependent on karyopherin beta and Ran for nucleocytoplasmic trafficking. Exp Biol Med (Maywood) 235:1093–1101
4. Sorokin AV, Kim ER, Ovchinnikov LP (2007) Nucleocytoplasmic transport of proteins. Biochemistry (Mosc) 72:1439–1457
5. Goryaynov A, Ma J, Yang W (2012) Single-molecule studies of nucleocytoplasmic transport: from one dimension to three dimensions. Integr Biol (Camb) 4:10–21
6. Lui K, Huang Y (2009) RanGTPase: a key regulator of nucleocytoplasmic trafficking. Mol Cell Pharmacol 1:148–156
7. Guttler T, Madl T, Neumann P, Deichsel D, Corsini L, Monecke T et al (2010) NES consensus redefined by structures of PKI-type and Rev-type nuclear export signals bound to CRM1. Nat Struct Mol Biol 17:1367–1376
8. Oksayan S, Wiltzer L, Rowe CL, Blondel D, Jans DA, Moseley GW (2012) A novel nuclear trafficking module regulates the nucleocytoplasmic localization of the rabies virus interferon antagonist, P protein. J Biol Chem 287:28112–28121
9. Nigg EA (1997) Nucleocytoplasmic transport: signals, mechanisms and regulation. Nature 386:779–787
10. Gorlich D, Mattaj IW (1996) Nucleocytoplasmic transport. Science 271:1513–1518
11. Hammerschmid M, Palmeri D, Ruhl M, Jaksche H, Weichselbraun I, Bohnlein E et al (1994) Scanning mutagenesis of the arginine-rich region of the human immunodeficiency virus type 1 Rev trans activator. J Virol 68:7329–7335
12. Hauber J, Malim MH, Cullen BR (1989) Mutational analysis of the conserved basic domain of human immunodeficiency virus tat protein. J Virol 63:1181–1187
13. Donnelly M, Elliott G (2001) Nuclear localization and shuttling of herpes simplex virus tegument protein VP13/14. J Virol 75:2566–2574
14. Kalland KH, Szilvay AM, Brokstad KA, Saetrevik W, Haukenes G (1994) The human immunodeficiency virus type 1 Rev protein shuttles between the cytoplasm and nuclear compartments. Mol Cell Biol 14:7436–7444
15. Li Y, Yamakita Y, Krug RM (1998) Regulation of a nuclear export signal by an adjacent inhibitory sequence: the effector domain of the influenza virus NS1 protein. Proc Natl Acad Sci USA 95:4864–4869

16. Mears WE, Rice SA (1998) The herpes simplex virus immediate-early protein ICP27 shuttles between nucleus and cytoplasm. Virology 242:128–137
17. Cheng G, Brett ME, He B (2002) Signals that dictate nuclear, nucleolar, and cytoplasmic shuttling of the gamma(1)34.5 protein of herpes simplex virus type 1. J Virol 76: 9434–9445
18. Lee CH, Chang SC, Wu CH, Chang MF (2001) A novel chromosome region maintenance 1-independent nuclear export signal of the large form of hepatitis delta antigen that is required for the viral assembly. J Biol Chem 276:8142–8148
19. Sekimoto T, Yoneda Y (2012) Intrinsic and extrinsic negative regulators of nuclear protein transport processes. Genes Cells 17:525–535
20. Marfori M, Mynott A, Ellis JJ, Mehdi AM, Saunders NF, Curmi PM et al (2011) Molecular basis for specificity of nuclear import and prediction of nuclear localization. Biochim Biophys Acta 1813:1562–1577
21. Truant R, Cullen BR (1999) The arginine-rich domains present in human immunodeficiency virus type 1 Tat and Rev function as direct importin beta-dependent nuclear localization signals. Mol Cell Biol 19:1210–1217
22. Palmeri D, Malim MH (1999) Importin beta can mediate the nuclear import of an arginine-rich nuclear localization signal in the absence of importin alpha. Mol Cell Biol 19: 1218–1225
23. Koyama M, Matsuura Y (2011) Structural basis for CRM1-mediated nuclear export. Seikagaku 83:834–838
24. Chen T, Brownawell AM, Macara IG (2004) Nucleocytoplasmic shuttling of JAZ, a new cargo protein for exportin-5. Mol Cell Biol 24:6608–6619
25. Yudin D, Fainzilber M (2009) Ran on tracks–cytoplasmic roles for a nuclear regulator. J Cell Sci 122:587–593
26. Kudo N, Matsumori N, Taoka H, Fujiwara D, Schreiner EP, Wolff B et al (1999) Leptomycin B inactivates CRM1/exportin 1 by covalent modification at a cysteine residue in the central conserved region. Proc Natl Acad Sci USA 96:9112–9117
27. Fukuda M, Asano S, Nakamura T, Adachi M, Yoshida M, Yanagida M et al (1997) CRM1 is responsible for intracellular transport mediated by the nuclear export signal. Nature 390: 308–311
28. Zheng C, Brownlie R, Babiuk LA, van Drunen Littel-van den Hurk S (2005) Characterization of the nuclear localization and nuclear export signals of bovine herpesvirus 1 VP22. J Virol 79:11864–11872
29. Ding Q, Guo H, Lin F, Pan W, Ye B, Zheng AC (2010) Characterization of the nuclear import and export mechanisms of bovine herpesvirus-1 infected cell protein 27. Virus Res 149:95–103
30. Li M, Wang S, Cai M, Guo H, Zheng C (2011) Characterization of molecular determinants for nucleocytoplasmic shuttling of PRV UL54. Virology 417:385–393
31. Zheng C, Lin F, Wang S, Xing J (2011) A novel virus-encoded nucleocytoplasmic shuttling protein: the UL3 protein of herpes simplex virus type 1. J Virol Methods 177: 206–210
32. Zheng C, Brownlie R, Babiuk LA, van Drunen Littel-van den Hurk S (2004) Characterization of nuclear localization and export signals of the major tegument protein VP8 of bovine herpesvirus-1. Virology 324:327–339
33. Raschbichler V, Lieber D, Bailer SM (2012) NEX-TRAP, a novel method for in vivo analysis of nuclear export of proteins. Traffic 13:1326–1334
34. Klemm JD, Beals CR, Crabtree GR (1997) Rapid targeting of nuclear proteins to the cytoplasm. Curr Biol 7:638–644
35. Sandri-Goldin RM (2001) Nuclear export of herpes virus RNA. Curr Top Microbiol Immunol 259:2–23
36. Cullen BR (2003) Nuclear mRNA export: insights from virology. Trends Biochem Sci 28:419–424
37. Huang X, Liu T, Muller J, Levandowski RA, Ye Z (2001) Effect of influenza virus matrix protein and viral RNA on ribonucleoprotein formation and nuclear export. Virology 287: 405–416
38. Alvisi G, Rawlinson SM, Ghildyal R, Ripalti A, Jans DA (2008) Regulated nucleocytoplasmic trafficking of viral gene products: a therapeutic target? Biochim Biophys Acta 1784:213–227
39. Fulcher AJ, Jans DA (2011) Regulation of nucleocytoplasmic trafficking of viral proteins: an integral role in pathogenesis? Biochim Biophys Acta 1813:2176–2190
40. Zhao L, Zheng C (2012) The first identified nucleocytoplasmic shuttling herpesviral capsid protein: herpes simplex virus type 1 VP19C. PLoS One 7:e41825

Chapter 10

Detection of Integrated Herpesvirus Genomes by Fluorescence *In Situ* Hybridization (FISH)

Benedikt B. Kaufer

Abstract

Fluorescence *in situ* hybridization (FISH) is widely used to visualize nucleotide sequences in interphase cells or on metaphase chromosomes using specific probes that are complementary to the respective targets. Besides its broad application in cytogenetics and cancer research, FISH facilitates the localization of virus genomes in infected cells. Some herpesviruses, including human herpesvirus 6 (HHV-6) and Marek's disease virus (MDV), have been shown to integrate their genetic material into host chromosomes, which allows transmission of HHV-6 via the germ line and is required for efficient MDV-induced tumor formation. We describe here the detection by FISH of integrated herpesvirus genomes in metaphase chromosomes and interphase nuclei of herpesvirus-infected cells.

Key words Fluorescence *in situ* hybridization, Metaphase spread, Herpesvirus, Integration, Telomere, Telomeric repeats, Marek's disease virus, Human herpesvirus 6, Lymphoblastoid cell lines

1 Introduction

Fluorescence *in situ* hybridization (FISH) facilitates detection of specific DNA and RNA sequences in interphase cells or metaphase chromosomes. It is a widely used tool in various research areas including cancer research, pre- and postnatal diagnostics, cytogenetics, developmental biology, gene mapping and virology [1, 2]. In virology, FISH allows detection of viral genomic sequences in infected cells during various stages of the virus life cycle [3–7].

Several DNA viruses and retroviruses have been shown to integrate their genetic material into host chromosomes, which is required for efficient replication and maintenance of the viral genome in proliferating cells [8, 9]. Some herpesviruses, including human herpesvirus 6 (HHV-6) and the lymphoma-inducing Marek's disease virus (MDV), have been shown to integrate their genetic material into host chromosomes [10–12]. Integration of these two herpesviruses has recently been shown to occur in host telomeres, a

Susanne M. Bailer and Diana Lieber (eds.), *Virus-Host Interactions: Methods and Protocols*, Methods in Molecular Biology, vol. 1064, DOI 10.1007/978-1-62703-601-6_10,

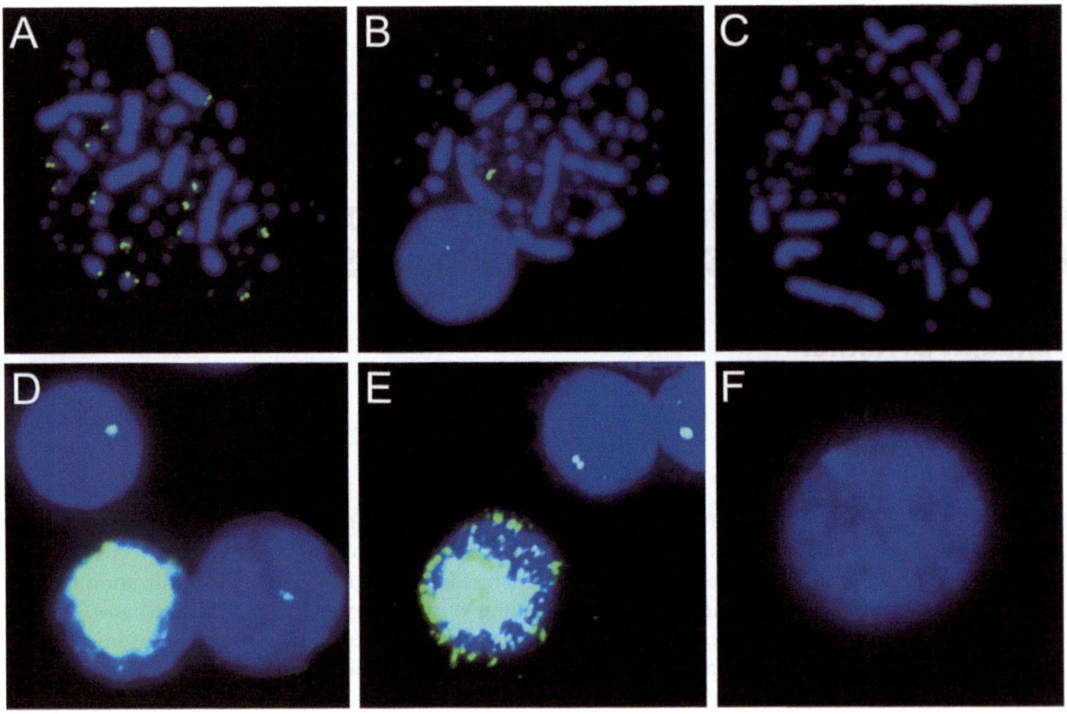

Fig. 1 Fluorescence *in situ* hybridization analysis detecting MDV integration sites (anti-DIG FITC, *green*) on metaphase chromosomes (DAPI stain, *blue*) performed on MDV-induced or control lymphoblastoid cell lines (LCLs). (**a** and **b**) MDV-induced LCLs harboring several (**a**) or only a single (**b**) MDV integration site. (**c**) Metaphase chromosomes of negative control cell line CU91. (**d** and **e**) MDV reactivation resulting in large amounts of viral genomic DNA detectable in some but not all interphase nuclei. (**f**) Interphase nucleus of CU91 negative control cell line

protective structure at the end of linear chromosomes [13, 14]. In case of HHV-6, integrated virus genomes are present in the germ line of about 1 % of the human population, a consequence and prerequisite for vertical transmission of the virus from parents to offspring [8, 14]. Intriguingly, HHV-6 and MDV genome integration is not a dead end as both viruses can reactivate from cells that contain only integrated virus genomes, resulting in lytic virus replication [14, 15]; however, the exact mechanism of and cues for genome mobilization from the integrated state are unknown.

Integrated herpesvirus genomes can be readily detected in metaphase chromosomes of MDV transformed cells by FISH (Fig. 1a, b), while no signal is detected in chromosomes of cells that do not harbor the MDV genome (Fig. 1c). In addition, FISH can be used to visualize viral DNA in lytically infected cells and upon reactivation of the virus from the latent stage of infection. During lytic replication, the virus genome is massively amplified and can be detected as a strong fluorescence signal all over interphase nuclei, while only defined spots can be detected in the

Table 1
Troubleshooting guide

Observation/problem	Potential cause	Suggestion
No or very weak staining	Not enough probe	Increase amount of probe during hybridization
	Insufficient probe labeling	Test the probe by dot blot hybridization
	Incomplete chromosome or probe denaturation	Control denaturation temperature and denaturation buffer composition
	Stringency of post-hybridization washes too high	Decrease temperature of stringency washes Increase salt concentration of SSC wash buffer (e.g., 2–4× SSC) Decrease duration of the wash steps
Background staining		
1. On chromosomes or nuclei	Low stringency of post-hybridization washes	Decrease salt concentration of SSC wash buffer (e.g., 0.5–1.0× SSC) Increase stringency of wash temperature Increase duration of the wash steps
	Insufficient blocking	Increase amount of salmon sperm or Cot-1 DNA
	Hybridization conditions	Increase hybridization temperature Decrease hybridization time
	Too much probe	Decrease probe concentration
2. Around the chromosomes and the nuclei, but not on the slide	Suboptimal cell preparation with a lot of cell debris	Increase hypotonic treatment Increase pepsin treatment
	DNA probe fragments are too long	Generate shorter probes
3. Generalized background on slides but not at chromosomes	Insufficient post-hybridization wash	Change wash buffer at least 3 times Shake coplin jar to ensure proper washing
	Antibody concentration too high	Reduce antibody concentration
4. Spotty background over the slide	Excessive slide *aging*	Less stringent *aging* conditions Try chemical *aging* with ethanol
	Probe gets trapped in protein aggregate	Increase pepsin treatment
	Antibody aggregates	Centrifuge antibody stock solutions to remove aggregates

nuclei of latently infected cells (Fig. 1d, e). No signal is detected in interphase nuclei that do not harbor the MDV genome (Fig. 1f).

This chapter describes the detection of herpesvirus genomes by FISH, including a detailed description of the preparation of metaphase spreads, generation of herpesvirus-specific probes, denaturation and hybridization of DNA probes to herpesvirus genomes, and the analysis of the samples via fluorescence microscopy. Potential problems are discussed in Subheading 4 and the troubleshooting guide (Table 1).

2 Materials

2.1 Metaphase Chromosome Preparation

1. LM-Hahn medium: Combine 500 mL McCoy's 5A medium, 500 mL Leibovitz L-15, 100 mL 10× tryptose phosphate broth, 2.6 mL 10 % Na-bicarbonate, 11 mL 1 mM β-mercaptoethanol, 11 mL 100× NaPyruvate (11 g/L H_2O), 11 mL Antibiotics-Antimycotics solution, 11 mL 200 mM L-Glutamine, 110 mL fetal bovine serum (FBS) and 88 mL chicken serum [16].
2. RPMI 1640 medium supplemented with 10 % FBS and 100 U/mL penicillin and 0.1 mg/mL streptomycin.
3. Ultrapure deionized water (ddH_2O; *see* **Note 1**).
4. Phosphate buffered saline (PBS). Dissolve 8.0 g NaCl, 0.2 g KCl, 1.44 g Na_2HPO_4 and 0.24 g KH_2PO_4 in 800 mL ddH_2O. Adjust the pH to 7.4 with 1 M HCl and bring to a final volume of 1 L with ddH_2O. Prepare aliquots, autoclave and store at room temperature.
5. Colcemid solution (10 μg/mL colcemid in PBS).
6. Hypotonic solution: 0.075 M KCl in ddH_2O.
7. Fixative: Methanol:acetic acid (3:1, v/v), store at −20 °C.
8. Precleaned microscope slides (e.g., Superfrost, Fisher Scientific).
9. Adjustable water bath (range: 20–80 °C).
10. Phase contrast microscope.

2.2 FISH Probes

1. Purified herpesvirus BAC, cosmid or viral DNA (*see* **Note 2**).
2. Frequently cutting restriction enzymes (4-bp recognition sites), e.g., *HaeIII* or *DpnI*.
3. DNA purification kit (e.g., QIAquick PCR Purification Kit, Qiagen).
4. Biotin or DIG-High Prime random priming labeling kit (Roche).

2.3 Fluorescence In Situ Hybridization

1. 22 × 22 mm glass coverslips.
2. Rubber cement.
3. Coplin jars (50 mL) optimally with a screw cap.
4. Glass bowl.
5. Humidified chamber (airtight container containing a damp paper towel or blotting paper).
6. Incubators at 37 and 80 °C.
7. Ethanol series: 70, 90 and 100 % ethanol in ddH_2O.
8. Pepsin solution: 0.005 % pepsin in 10 mM HCl.
9. 20× SSC buffer: Dissolve 175.3 g NaCl and 88.2 g Na-citrate in 800 mL ddH_2O. Adjust the pH to 7.0 with 1 M HCl and

bring to a final volume of 1 L with ddH_2O. Prepare aliquots, autoclave and store at room temperature.

10. 2× SSC wash buffer: Mix 100 mL of 20× SSC and 900 mL ddH_2O and store at room temperature.
11. 0.5× SSC wash buffer: Mix 25 mL of 20× SSC and 975 mL ddH_2O and store at room temperature.
12. Denaturation buffer: Mix 70 mL deionized formamide and 30 mL of 2× SSC and store at 4 °C (*see* **Note 3**).
13. Phosphate buffer: Mix 80 mL 500 mM sodium phosphate dibasic solution with 20 mL sodium phosphate monobasic solution to obtain pH 7.0. Prepare aliquots, autoclave and store at room temperature.
14. Dextran sulfate stock solution (50 %): Dissolve 25 g dextran sulfate in 40 mL ddH_2O by stirring over night at 4 °C. Adjust volume to 50 mL with ddH_2O, aliquot and store at −20 °C.
15. Hybridization buffer: Mix 50 mL deionized formamide, 10 mL 20× SSC, 10 mL phosphate buffer pH 7.0, 16 mL dextran sulfate stock solution (50 %) and bring to a final volume of 100 mL with ddH_2O. Aliquot the solution and store at −20 °C.
16. Sheared salmon sperm or Cot-1 DNA (20 mg/mL).
17. Stringency wash buffer: Mix 50 mL deionized formamide and 50 mL 2× SSC and store at 4 °C.
18. Detergent wash buffer: Mix 1 L of 4× SSC and 500 μL Tween-20 and store at room temperature.
19. Antibody solution: Dilute fluorescently labeled avidin or anti-digoxigenin antibody in detergent wash buffer 1:500 or according to the manufacturer's instruction (Sigma-Aldrich; *see* **Note 4**).
20. Mounting solution containing DAPI 4,6-diamidino-2-phenylindole (e.g., DAPI Vectashield, Vector).
21. Fluorescence microscope containing appropriate filter sets for all fluorophores used, a 60× or 100× oil immersion objective and a CCD (charge coupled device) camera.

3 Methods

3.1 Preparation of Cells

1. Propagate MDV or HHV-6 infected cells or lymphoblastoid cell lines (LCLs) in the appropriate medium and passage cells 1 day prior to metaphase preparation. Alternatively, cells can be directly isolated *ex vivo* (*see* **Note 5**).
2. Add 0.1 μg/mL colcemid to the cell suspension and incubate for 1–3 h for enrichment of metaphase cells (*see* **Note 6**).
3. Transfer 1×10^6 to 1×10^7 suspension cells into a 15 mL conical tube and centrifuge at $400 \times g$ for 8 min at room temperature.

4. Discard the supernatant and flick the tube to loosen the cell pellet.
5. Carefully resuspend cells in 10 mL prewarmed hypotonic solution and incubate for 10–15 min at 37 °C (*see* **Note 7**).
6. Add 1 mL fresh ice-cold fixative solution and mix by gently inverting the tube to fix the cells and prevent cell clumping.
7. Centrifuge at 400 × *g* for 8 min at room temperature. Discard supernatant and flick the tube to loosen the cell pellet.
8. Carefully add 5 mL fresh ice-cold fixative solution along the wall of the tube.
9. Centrifuge at 400 × *g* for 8 min at room temperature. Discard the supernatant and flick the tube to loosen the cell pellet.
10. Repeat **steps 8** and **9** two more times.
11. Resuspend cells in 1–2 mL ice-cold fixative solution (*see* **Note 8**).
12. Samples can be stored at −20 °C for several years (*see* **Note 9**).

3.2 Preparation of Metaphase Chromosome Spread Slides

1. Prewarm water bath to 80 °C.
2. Resuspend metaphase cell preparation by flicking the tube (*see* **Note 10**).
3. Drop 10–20 μL of cell suspension onto a precleaned microscope slide.
4. After the slide surface becomes "grainy", move the slide briefly through the vapor of the water bath for 1–2 s and dry the slides at room temperature (*see* **Note 11**).
5. Check the quality of the metaphase spreads using a phase-contrast microscope. Proceed with the best preparations or adjust conditions to improve slide quality (*see* **Note 12**).
6. Dehydrate slide in 100 % ethanol for 5 min.
7. Dry at room temperature.
8. Incubate slide in pepsin solution for 2–5 min (*see* **Note 13**).
9. Wash slide three times in fresh 2× SSC buffer for 1 min each.
10. Briefly immerse slide in ddH_2O.
11. Dehydrate slides twice in 70 % ethanol for 2 min each, then twice in 90 % ethanol for 2 min each, and finally in 100 % ethanol for 4 min at room temperature.
12. Dry the slide at room temperature.
13. Age the slide at 65 °C on a hot plate for 1 h (*see* **Note 14**).

3.3 Preparation of FISH Probes (Random Priming)

1. Digest 2 μg of purified herpesvirus BAC, cosmid or viral DNA with the appropriate restriction enzyme (e.g., *HaeIII* or *DpnI*) for 2 h (*see* **Note 15**).
2. Purify digested DNA with a DNA purification kit.

3. Analyze aliquot of digested DNA on a 1 % agarose gel (*see* **Note 16**).
4. Dilute 1–2 μg of the DNA in 16 μL ddH_2O and denature for 10 min at 95 °C. Immediately cool the denatured DNA in an ice bath.
5. Mix DNA solution with 4 μL Biotin or DIG-High Prime and incubate over night at 37 °C (*see* **Note 17**).
6. Purify digested DNA with a DNA purification kit to remove unincorporated nucleotides (*see* **Note 18**).

3.4 Hybridization

1. Prepare staining solution by mixing 20–40 ng of DIG-labeled probe and 20 μg of sheared salmon sperm or Cot-1 DNA in 12 μL hybridization buffer (*see* **Note 19**).
2. Denature staining solution for 5 min at 75 °C and immediately cool the DNA in an ice bath.
3. Add 12 μL of the staining solution onto a coverslip and lower it onto the metaphase slide (*see* **Note 20**).
4. Seal coverslip with rubber cement and let dry at room temperature (*see* **Note 21**).
5. Place slides on a metal surface in the 80 °C incubator and incubate for 2 min (*see* **Note 22**).
6. Open the incubator door slightly (10 mm) and let it cool to 37 °C over approximately 15 min.
7. Transfer slides into a humidity chamber in a preheated 37 °C incubator and incubate for 24–48 h (*see* **Note 23**).

3.5 Wash and Detection

1. Prewarm two coplin jars containing 50 mL stringency wash buffer and two coplin jars with 2× SSC wash buffer in a 42 °C water bath.
2. Remove the slides from the humidity chamber and place them into a glass bowl containing 2× SSC wash buffer. Carefully remove the rubber cement and coverslip (*see* **Note 24**).
3. Transfer slides into prewarmed coplin jar containing the stringency wash buffer and incubate for 5 min at 42 °C. Transfer slides into the second stringency wash buffer jar and repeat wash step (*see* **Note 25**).
4. Wash slides twice for 5 min each in prewarmed 2× SSC wash buffer.
5. Incubate slides for 5 min in detergent wash buffer. If directly labeled probes were used, proceed to **step 8**.
6. Drain slides, apply 75 μL of antibody solution and cover immediately with a piece of parafilm. Incubate the slide in the humidity chamber for 15–20 min at 37 °C (*see* **Note 26**).

7. Remove parafilm from the slide and wash three times with detergent wash buffer in a coplin jar for 4 min each at room temperature.
8. Briefly immerse slide in ddH_2O and let dry at room temperature.
9. Mount slide with 25 µL Vectashield DAPI using a coverslip.
10. Examine FISH slides using a fluorescence microscope, record images and determine the number and locations of the herpesvirus integration sites. Individual chromosomes can be distinguished by their size or by labeling each chromosome with a specific probe. Reactivation of herpesvirus genomes can be readily detected in interphase nuclei (Fig. 1e, f). If no or very weak signal is detectable or if background is present on the slide, follow the troubleshooting guide in Table 1.

4 Notes

1. All buffers and reagents should be prepared using ultrapure ddH_2O with a resistance of 18.2 MΩ cm at 25 °C.
2. The purity of the viral DNA is crucial for the generation of FISH probes. Contamination with cellular DNA can cause unspecific background staining. Optimally, BACs or cosmid clones containing the herpesvirus genome should be used as a template for the generation of FISH probes. DNA of BAC or cosmid clones should be purified using silica column-based DNA extraction kits to ensure the purity and quality of the DNA.
3. Use high-quality deionized formamide for the preparation of the denaturation buffer to ensure an efficient denaturation of the target DNA. Low quality formamide can result in the distortion or thickening of the metaphase chromosomes [17]. Handle solutions containing formamide only in the fume hood as it can cause respiratory tract, eye and skin irritation.
4. Fluorescently labeled avidin and anti-digoxigenin antibodies are commercially (e.g., Sigma-Aldrich, Roche Applied Science) available with a variety of fluorophores including FITC, Cy3 and Cy5. Antibody precipitates can result in unspecific dotted background on FISH slide and should be removed by centrifugation of the antibody solution for 5 min at $>10,000 \times g$ prior to use.
5. Propagate HHV-6 infected cells and LCLs in RPMI medium containing 10 % FBS at 37 °C under a 5 % CO_2 atmosphere. MDV infected LCLs should be propagated in LM Hahn media at 41 °C. HHV-6 and MDV infected lymphocytes can be stimulated using 0.5–5 µg/mL phytohemagglutinin (PHA) over night to increase the number of cells in metaphase.

As shown in Fig. 1d, e, the virus genome can also be detected in the nucleus of interphase cells; however, it is impossible to determine if it is associated with cellular chromosomes.

6. If only few metaphase spreads are detectable, extend colcemid treatment to 12 or 24 h in order to increase the number of cells in metaphase. However, prolonged treatment can lead to chromosome breaks [18] resulting in very short chromosomes in metaphase spread preparations. If adherent cells are being used, trypsinize the cells, wash them two times with PBS, and proceed with the next step.
7. Duration and temperature of the hypotonic treatment have a strong effect on the quality of metaphase preparations. Insufficient hypotonic treatment results in metaphase chromosomes that do not spread well and are surrounded by a large amount of cytoplasmic debris. The proteinaceous debris can prevent the FISH probe from gaining access to target sequences. If the hypotonic treatment is too long, chromosomes may appear thicker and swollen. Adjust hypotonic treatment accordingly.
8. If very few metaphase chromosomes and interphase nuclei are present upon spreading on microscope slides, centrifuge the cells and resuspend in a smaller volume of fixative solution. Add more fixative solution if the cell suspension is too dense.
9. If samples were stored for more than 2 weeks, centrifuge cells at 400 × *g*, discard supernatant and resuspend cells in fresh fixative solution. This ensures the optimal methanol:acetic acid ratio required for high-quality metaphase spreads. Cells can be transferred into 2 mL tubes for long-term storage at −20 °C.
10. Ensure that cell clumps are completely resuspended. Cell clumping can drastically reduce the number of metaphase spreads on the slides.
11. Evaporation of the fixative solution facilitates spreading of the metaphase chromosomes on the microscope slides. Humidity plays a crucial role during this process, which can be optimally controlled using the 80 °C water bath. The drying temperature can be modified to optimize spread of metaphase chromosomes as described previously [17].
12. Metaphase chromosomes should be clearly visible on the microscope slide using a phase-contrast microscope. If metaphase chromosomes do not spread well and remain as a clump on the slide, you may either increase hypotonic treatment (*see* **Note 7**), modify the length of vapor application or drying temperature for the slides (*see* Subheading 3.2, **step 4** and **Note 11**). If cytoplasmic debris is visible as a halo covering the metaphase chromosomes, increase the time of the hypotonic treatment to 15–20 min (*see* **Note 7**) or extend pepsin treatment (*see* **Note 13**).

13. Pepsin treatment allows removal of protein debris covering the chromosomes that can prevent the FISH probe from gaining access to target sequences. Pepsin treatment of metaphase spread slides with a large amount of cytoplasmic debris should be extended to 5–20 min. Excessive pepsin treatment can result in loss of chromosomes or chromosomal structure.
14. *Aging* of the slides is another crucial step that facilitates fixation of the chromosomes and interphase nuclei onto the microscope slide. It is also important for the preservation of the chromosome architecture. Excessive *aging* enhances the rigidity of the chromosomes, which can decrease the hybridization efficiency of the FISH probe to the chromosomes.
15. Large FISH probes usually bind less specifically and less efficiently to chromosomes than shorter probes. Digestion of purified herpesvirus BAC, cosmid or viral DNA with frequent cutters ensures an optimal probe size of <2,000 bp.
16. Analyze the DNA by gel electrophoresis to ensure that the quality is sufficient and that the DNA is completely digested.
17. Incubation can be extended to 24 h to increase the yield of the labeled probe. Alternatively, probes can be also directly labeled using commercially available nucleotides such as Fluorescein-12-dUTP, Cy3-6-dUTP or Cy5-dUTP. Signal intensities of directly labeled probes are usually lower compared to hapten-labeled probes, as every avidin- or hapten-specific antibody carries several fluorophores. Various FISH labeling methods and fluorophore selection have been reviewed by Morrison and colleagues [19].
18. Presence of unincorporated labeled nucleotides results in background staining on FISH slides. Alternatively, unincorporated nucleotides can be removed using Sephadex G-50 columns or by ethanol precipitation [19].
19. Denaturation and hybridization can be performed separately as described previously [20]. This sequential staining procedure provides comparable results; however, it is more labor-intensive than the simultaneous staining described here.
20. Air bubbles trapped under the coverslip will result in uneven staining of the slide. If air bubbles are present, gently apply pressure to the coverslip with a pipette tip to remove the bubbles.
21. Samples should not dry during the staining procedure. Make sure that the coverslip is completely sealed with rubber cement to avoid drying of the slides in the following steps.
22. Complete denaturation is essential for efficient hybridization of the FISH probe to the chromosomes. Measure the temperature adjacent to the slides with a thermometer to ensure proper denaturation of the samples.

23. Usage of a humidity chamber ensures that the slides do not dry out. Hybridization temperature can be increased if unspecific chromosomal staining is detectable. Control the temperature of the incubator to ensure the optimal hybridization temperature.
24. The slides should be covered with 2× SSC wash buffer to ensure that they do not dry out. Remove the coverslips very carefully as the samples can be damaged during this step.
25. Stringency wash steps facilitate removal of unspecific-bound probes. Stringency should be increased if unspecific chromosomal staining is observed by raising the stringency wash temperature up to 65 °C, reducing the salt concentration during the SSC wash steps (*see* Subheading 3.3, **step 4**) by using 0.5× or 1× SSC wash buffers, or by extending the duration of wash steps. Control the temperature of the wash solutions to ensure optimal wash conditions.
26. Ensure that the slides do not dry out during the staining procedure as this would result in unspecific background staining on the slides.

Acknowledgments

The author thanks Dr. Nikolaus Osterrieder for editing the manuscript. This work was supported by the DFG grant KA3492.1-1 and funding from the Freie Universität Berlin to B.B.K.

References

1. Trask BJ (1991) Fluorescence in situ hybridization: applications in cytogenetics and gene mapping. Trends Genet 7(5):149–154
2. Wiegant J, Ried T, Nederlof PM et al (1991) In situ hybridization with fluoresceinated DNA. Nucleic Acids Res 19(12):3237–3241
3. Hackstein H, Jahn G, Kirchner H et al (1996) Fluorescence in situ hybridization with cosmid clones for the detection of human cytomegalovirus DNA in peripheral blood leukocytes. Histochem Cell Biol 106(2):229–234
4. Lawrence JB, Marselle LM, Byron KS et al (1990) Subcellular localization of low-abundance human immunodeficiency virus nucleic acid sequences visualized by fluorescence in situ hybridization. Proc Natl Acad Sci USA 87(14):5420–5424
5. Reisinger J, Rumpler S, Lion T et al (2006) Visualization of episomal and integrated Epstein-Barr virus DNA by fiber fluorescence in situ hybridization. Int J Cancer 118(7):1603–1608
6. Brabec-Zaruba M, Pfanzagl B, Blaas D et al (2009) Site of human rhinovirus RNA uncoating revealed by fluorescent in situ hybridization. J Virol 83(8):3770–3777
7. Robertson KL, Verhoeven AB, Thach DC et al (2010) Monitoring viral RNA in infected cells with LNA flow-FISH. RNA 16(8):1679–1685
8. Hall CB, Caserta MT, Schnabel K et al (2008) Chromosomal integration of human herpesvirus 6 is the major mode of congenital human herpesvirus 6 infection. Pediatrics 122(3):513–520
9. Li M, Mizuuchi M, Burke TR Jr et al (2006) Retroviral DNA integration: reaction pathway and critical intermediates. EMBO J 25(6):1295–1304
10. Delecluse HJ, Hammerschmidt W (1993) Status of Marek's disease virus in established lymphoma cell lines: herpesvirus integration is common. J Virol 67(1):82–92

11. Luppi M, Barozzi P, Marasca R et al (1994) Integration of human herpesvirus-6 (HHV-6) genome in chromosome 17 in two lymphoma patients. Leukemia 8(Suppl 1):S41–S45
12. Luppi M, Marasca R, Barozzi P et al (1993) Three cases of human herpesvirus-6 latent infection: integration of viral genome in peripheral blood mononuclear cell DNA. J Med Virol 40(1):44–52
13. Kaufer BB, Jarosinski KW, Osterrieder N (2011) Herpesvirus telomeric repeats facilitate genomic integration into host telomeres and mobilization of viral DNA during reactivation. J Exp Med 208(3):605–615
14. Arbuckle JH, Medveczky MM, Luka J et al (2010) The latent human herpesvirus-6A genome specifically integrates in telomeres of human chromosomes in vivo and in vitro. Proc Natl Acad Sci USA 107(12): 5563–5568
15. Delecluse HJ, Schuller S, Hammerschmidt W (1993) Latent Marek's disease virus can be activated from its chromosomally integrated state in herpesvirus-transformed lymphoma cells. EMBO J 12(8):3277–3286
16. Calnek BW, Shek WR, Schat KA (1981) Spontaneous and induced herpesvirus genome expression in Marek's disease tumor cell lines. Infect Immun 34(2):483–491
17. Henegariu O, Heerema NA, Lowe WL et al (2001) Improvements in cytogenetic slide preparation: controlled chromosome spreading, chemical aging and gradual denaturing. Cytometry 43(2):101–109
18. Satya-Prakash KL, Liang JC, Hsu TC et al (1986) Chromosome aberrations in mouse bone marrow cells following treatment in vivo with vinblastine and Colcemid. Environ Mutagen 8(2):273–282
19. Morrison LE, Ramakrishnan R, Ruffalo TM et al (2002) Labeling fluorescence in situ hybridization probes for genomic targets. Methods Mol Biol 204:21–40
20. Rens W, Fu B, O'Brien PC et al (2006) Cross-species chromosome painting. Nat Protoc 1(2):783–790

Chapter 11

Fast Generation of Stable Cell Lines Expressing Fluorescent Marker Molecules to Study Pathogen Induced Processes

Jens Bernhard Bosse, Jessica Ragues, and Harald Wodrich

Abstract

Virology has greatly benefited from the introduction of fluorescent proteins (FP's) as tags to viral as well as cellular structures. With advanced imaging technologies it is now possible to observe host–pathogen interactions in living cell systems in real-time. The generation of high-quality genetic tools to study host–pathogen interactions therefore becomes imperative for the further development of this type of analysis. In this chapter we describe a universal and reliable method to rapidly generate stable cell lines expressing FP-tagged proteins to be used for the analysis of host–pathogen interactions. The protocol is exemplified for two cellular structures recognized for their importance in the host–pathogen interplay: autophagosomes and the actin cytoskeleton, but can be applied to virtually any transgene or FP. It is based on the commercial Flp-In™ and Gateway™ systems (Life Technologies) and allows the rapid generation of FP-tagged transgenes by Gateway™ technology followed by recombination into a cell line containing a single transcriptionally active genomic recombination locus.

Key words Stable cell line, Gateway™, Flp-In™, FRT, Fluorescent protein, Live cell imaging

1 Introduction

Advanced microscopy techniques developed over the past decades have shifted the analysis of host–pathogen interactions from indirect methods to direct visualization in living cells. The introduction of fluorescent proteins (FPs) to label organelles, cellular substructures or proteins [1, 2] and the development of ever better and faster optical recording systems made it possible to record the dynamic nature of host–pathogen interactions by live cell microscopy at high temporal and spatial resolution [3]. Most commonly, fluorescent marker proteins derived from either the jellyfish *Aequorea victoria* (e.g., green fluorescent protein, GFP) or the coral genus *Discosoma* (e.g., DsRed) are utilized as N- or C-terminal fusions to target domains or whole proteins. Most commonly the expression of such fusions is achieved by transient transfections or

Susanne M. Bailer and Diana Lieber (eds.), *Virus-Host Interactions: Methods and Protocols*, Methods in Molecular Biology, vol. 1064, DOI 10.1007/978-1-62703-601-6_11, © Springer Science+Business Media, LLC 2013

by stable integration into the cellular genome. Although fast and often very efficient, transient transfection has several drawbacks. First, expression levels are usually non-physiologically high, which might lead to overexpression artifacts. For instance if the tagged protein in question has limited cellular binding partners the overexpression will certainly exhaust the endogenous pool and subsequent mislocalization can provide "false" effects. Moreover, expression levels can vary strongly between transfected cells due to varying amounts of expression units (plasmids) an individual cell incorporates. This may result in a wide variety of phenotypes that impede the analysis. Another less well-known side effect of transient transfection is that the massive introduction of genetic material (plasmid DNA) into a cell can trigger several intracellular pattern recognition pathways, which could specifically surpass host–pathogen interactions, e.g., those subject to the actual study [4].

Stable expression systems for tagged transgenes are more favorable than transient expression systems as they can ensure lower cell-to-cell variance and more physiological expression levels. Moreover, stable transgene expression does not suffer from polar effects like activation of pattern recognition pathways. However, methods to generate stable cell lines expressing a transgene also suffer from drawbacks as they often rely on the random insertion of the transgene expression cassette into the host genome, which could lead to activation/silencing of adjacent genes or insertional mutagenesis [5]. Also, some inserted transgenes undergo transcriptional silencing due to methylation of the insertion locus [6]. This makes tedious clonal isolation and characterization mandatory to find long-term, well expressing cell clones.

In this protocol we provide a detailed reference on how to rapidly generate stable cell lines expressing fusions between fluorescent proteins (FPs) and various cDNAs that do not suffer from most of the above mentioned drawbacks. Although the protocol is of general use, we will detail in this chapter how to generate stable cell lines based on U-2 OS human osteosarcoma cells expressing the autophagy marker LC3 or the F-actin probe Lifeact [12]. We chose U-2 OS cells as they are particularly well suited for live cell imaging due to their morphology, size and cytoplasm to nucleus ratio. A third example is provided in Martinez et al. (*see* Chapter 15 of this series) where we applied this method to study and visualize endosomal membrane penetration and escape of adenoviruses.

Our first example for a cellular process involved in host–pathogen interaction is autophagy. Autophagy is an intracellular degradation pathway that balances cellular homeostasis [7]. During viral infections it also serves as clearance mechanism and feeds the intracellular immune recognition machinery by engulfing viral components into a membraneous organelle termed autophagosome and targeting its content via fusion to the lysosomal compartments, which is enriched in immune sensors [8]. Consequently viruses have evolved multiple mechanisms to manipulate autophagy. Following induction of autophagy the formation of autophagosomes is initiated by the formation of LC3 positive membranes, which serves as classical marker for autophagosomes [7, 9]. LC3 is an ubiquitin-like molecule that is inserted into the autophagosomal membrane upon conjugation with phosphatidylethanolamin thereby expanding the size of the organelle [10]. Thus, studies involving autophagy in host–pathogen interactions benefit from fluorescently labeled LC3, which allows the monitoring of autophagic processes in living cells (Fig. 1).

As second example we choose the actin cytoskeleton (Fig. 2). Various viruses use and modulate the host actin cytoskeleton during their lifecycle to facilitate entry, intracellular transport and egress (for a review *see* ref. 11). Still, it is often not known how the host actin cytoskeleton is utilized in realtime by a given pathogen. Recently, a new peptide-based F-actin probe called Lifeact has been described [12]. It consists of the first 17 N-terminal amino acids of the yeast actin-binding protein Abp140. In comparison to actin-GFP fusions, Lifeact-GFP does not seem to influence F-actin dynamics [12]. Moreover, as it only consists of 17 amino acids, it can easily be fused to any FP by incorporating its coding sequence into the primers used for PCR amplification (see below).

The following protocol is an adaptation that utilizes a combination of the commercial Gateway™ and the Flp-In™ system (both Life Technologies) to generate virtually any cell line suitable for long-term expression of GFP or mCherry (or any other compatible FP) tagged transgenes (overview in Fig. 3).

The key concept is to first search for a genomic site that provides high and stable expression levels of a transgene by random insertion of a reporter gene into the host genome. The system is based on the plasmid pFRT/lacZeo (Life Technologies) (*see* **Note 1**). It codes for a fusion between lacZ and a Zeocin resistance (ZeoR) gene with an in-frame FRT site inserted after the start codon. To generate a target cell line (*see* **Note 2**), this cassette is randomly integrated into the host genome by transfection of linearized vector DNA. Zeocin resistant cell clones are selected and characterized for their ability to sustain long-term, stable and sufficiently high expression of the lacZ-ZeoR fusion gene. Only after clonal purification, the resulting cell line is then used as a hub to insert genes of interest into the very same integration site by Flippase (Flp) mediated recombination. The subsequent use of Gateway™ and Flip-In technology provides convenient and rapid state-of-the-art technology.

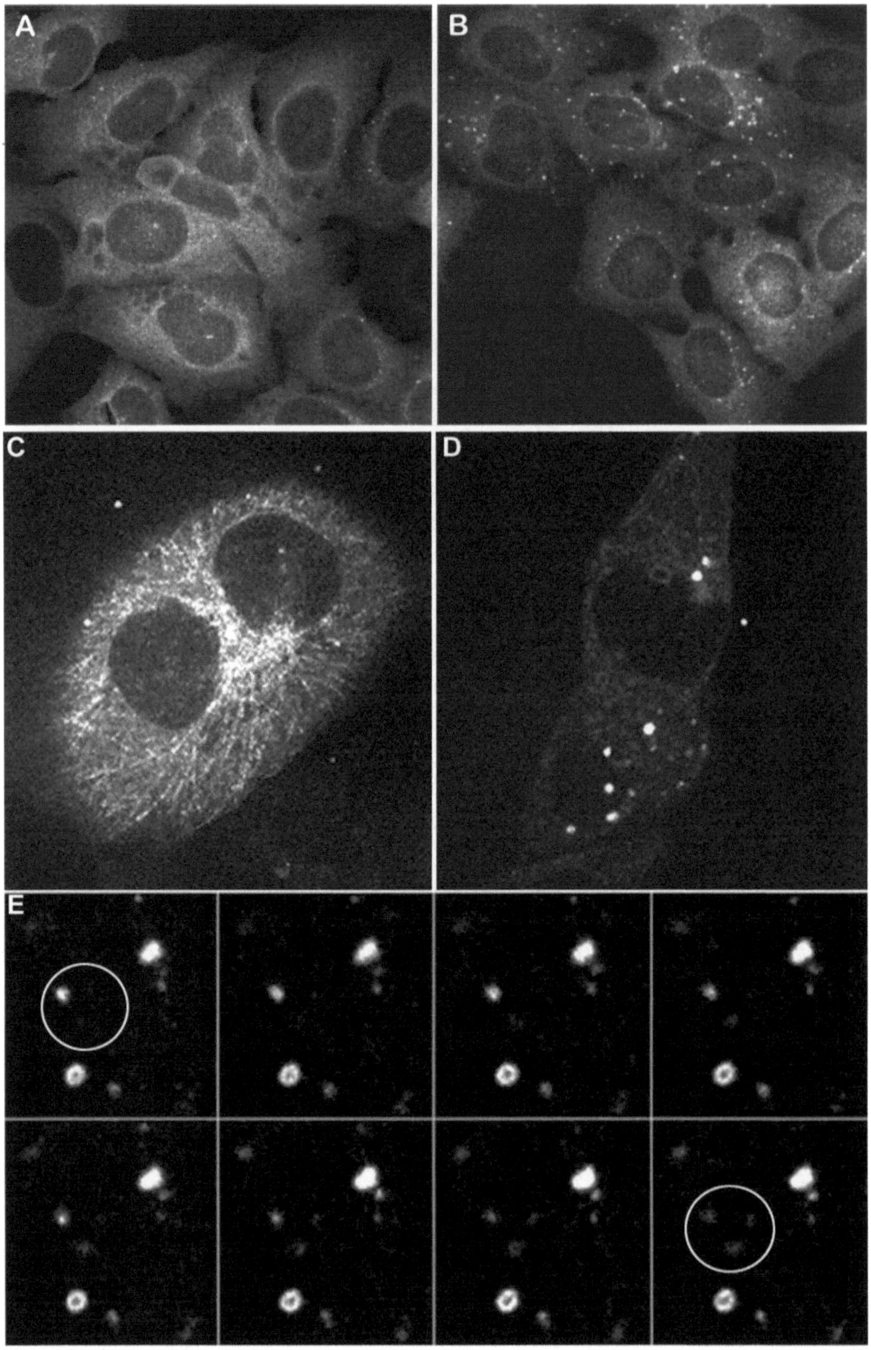

Fig. 1 Generation of a stable cell line expressing FP-tagged LC-3 using FRT recombination. *Top row*: Fixed U-2 OS cells stably expressing GFP tagged LC-3 without (**a**) or following induction of autophagy after 2 h of treatment with HBSS (**b**). *Middle row*: Living cell expressing GFP tagged LC-3 and imaged using spinning disk confocal microscopy without (**c**) or after (**d**) induction of autophagy. Note that the induction of autophagy changes the LC-3 localization from microtubule associated to vesicular. *Bottom rows*: Series of individual frames imaged at ten frames per second from the region with formed autophagosomes in the cell shown in **d**. Note the donut shaped autophagosomal structures

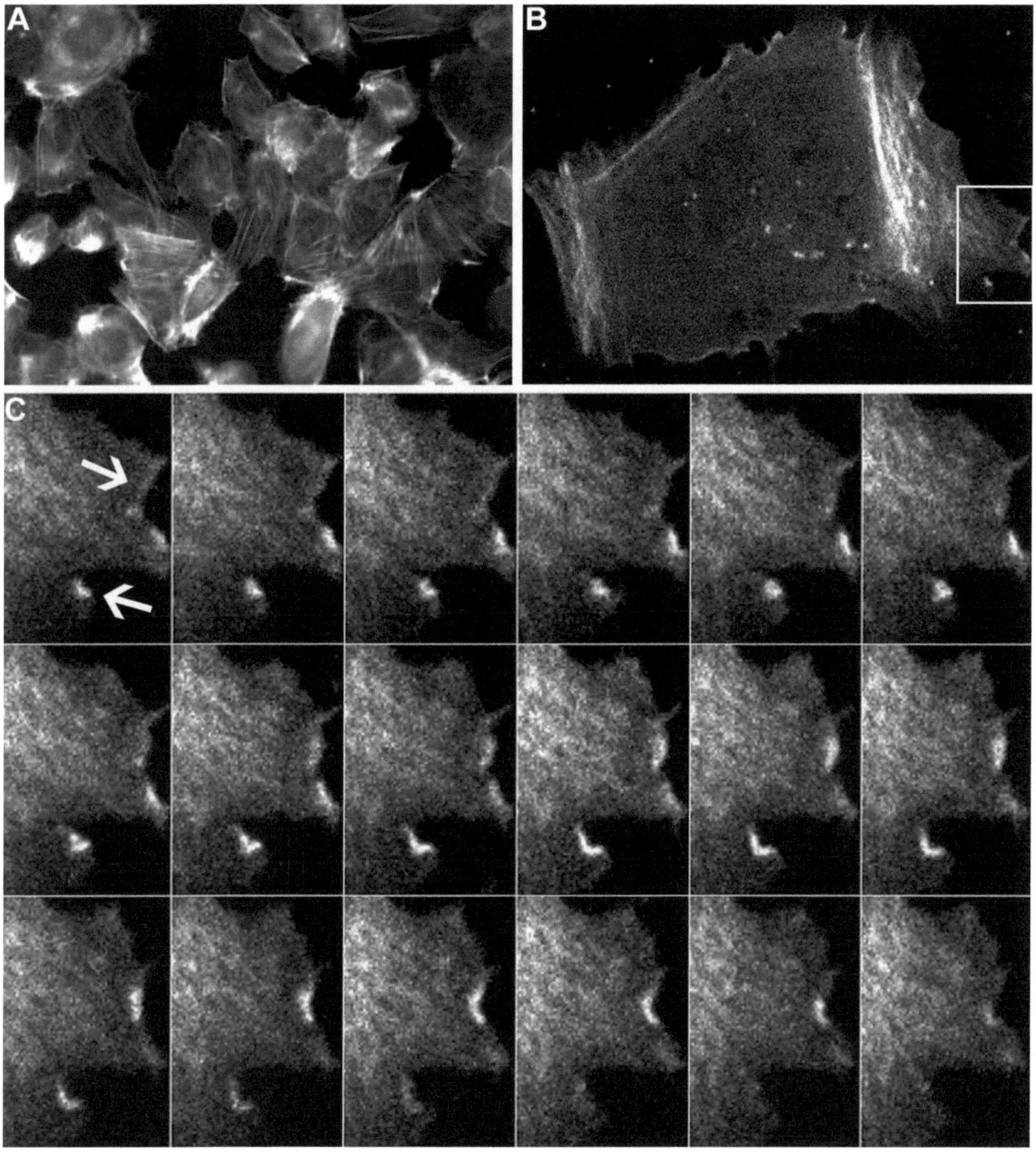

Fig. 2 Generation of a stable cell line expressing FP-tagged Lifeact staining the actin cytoskeleton using FRT recombination. (**a**) Overview of the generated U-2 OS Lifeact cells following fixation with paraformaldehyde. (**b**) Living cell expressing FP-tagged Lifeact imaged using spinning disk confocal microscopy. (**c**) Series of individual frames imaged at ten frames per second from the cell edge of the cell shown in **b** (*white box*). Ruffling membranes are indicated by *white arrows*

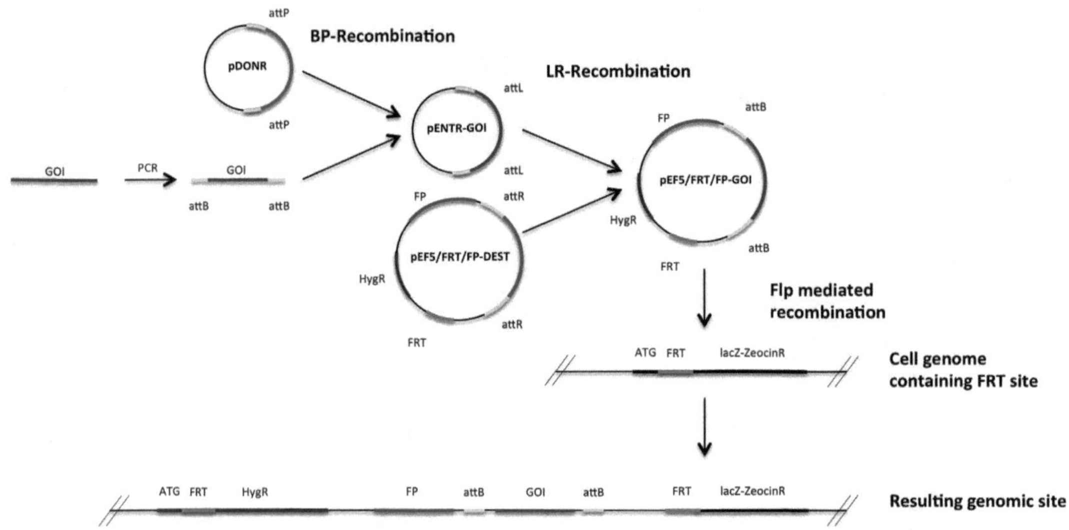

Fig. 3 Gateway™ assisted Flp-In™ mediated generation of stable cell lines: A gene of interest (GOI) is amplified by PCR and Gateway™ recombination sites (attB sites) are introduced through the amplification primers. The resulting product is subjected to BP recombination into a pDONR™ vector containing attP sites, resulting in pENTR™-GOI. This construct is sequence verified and the GOI is transferred into the Flp-In™ compatible plasmid pEF/FRT/FP-DEST described in this chapter by LR recombination between their attL and attR sites. Next, the resulting construct pEF/FRT/FP-GOI is transfected into a cell line detailed in the text that harbors a lacZ-ZeocinR fusion with an FRT site positioned at a well-expressed genomic locus. Expression of cotransfected Flp-recombinase plasmid permits integration of the pEF/FRT/FP-GOI construct into the genomic FRT locus. As the integration also removes the ATG from the lacZ-ZeocinR cassette, resulting cell clones are Zeocin sensitive but Hygromycin resistant due to the HygR locus on pEF/FRT/FP-GOI allowing for optical (FP) and resistance marker (hygromycin) selection

2 Materials

2.1 Cell Culture

1. U-2 OS cell line (ATCC: HTB-96).
2. Dulbecco's Modified Eagle Medium (DMEM) with GlutaMAX (Life Technologies # 31966016).
3. Fetal bovine serum (FBS) (check batch with specific cell line).
4. Opti-MEM GlutaMAX (Life Technologies # 51985026).
5. Penicillin/Streptomycin (Life Technologies # 15140122).
6. Dulbecco's Phosphate-Buffered Saline (D-PBS) (Life Technologies # 14190235).
7. Trypsin/EDTA (Life Technologies # 25300054).
8. Zeocin (Life Technologies # R250-01).
9. Lipofectamine 2000 (Life Technologies # 11668027).
10. Hygromycin B (InvivoGen # ant-hm-1).

11. β-Gal Staining Kit (Life Technologies # K1465-01).
12. Cloning cylinders (Bel-Art products # 378470100).
13. High vacuum grease (Dow corning).

2.2 Cloning

1. Flp-In™ Core System containing pFRT/lacZeo, pOG44 etc. (Life Technologies # K6010-02).
2. pEF5/FRT/V5-DEST (Life Technologies # V6020-20).
3. pDONR™-221 (Life Technologies # 12536017).
4. Gateway™ BP Clonase II Enzyme mix (Life Technologies # 11789020).
5. Gateway™ LR Clonase II Enzyme mix (Life Technologies # 11791020).
6. *E. coli* DH5α (Life Technologies # 18258012).
7. *E. coli* One Shot ccdB Survival 2 T1R (Life Technologies # A10460).
8. Kanamycin (Sigma-Aldrich # K4000).
9. *SwaI* (NEB # R0604).
10. *KpnI*-HF (NEB # R3142L).
11. *SpeI*-HF (NEB # R3133L).
12. *AgeI*-HF (NEB # R3552S).
13. *PvuII*-HF (NEB # R3151L).
14. Expand High Fidelity Taq (Roche # 11732641001).
15. Deoxynucleotide Solution Mix (NEB # N0447S).
16. QIAquick PCR Purification Kit (Qiagen # 28104).
17. UltraPure Agarose (Life Technologies # 16500500).
18. T4 DNA Ligase (NEB # M0202M).

3 Methods

For general information about the Flp-In™ system as well as the Gateway™ system and protocols we refer to the homepage of Life Technologies (www.lifetechnologies.com).

We will outline here the key steps and give advice on crucial procedures.

3.1 Generation of a Stable U-2 OS FRT Cell Line

Successful selection of stable cell clones depends on an efficient, antibiotic-mediated selection. Therefore, cells have to be tested for their level of sensitivity to Zeocin before transfecting them with the pFRT/lacZeo plasmid. Culture the U-2 OS (ATCC: HTB-96) in full medium (DMEM GlutaMAX, 10 % FBS, 1 % Penicillin/Streptomycin).

1. To test for sensitivity of cells against Zeocin, seed a 6-well plate with approximately 2.5×10^5 cells (about 25–50 % confluence the next day) per well in complete medium and incubate over night (*see* **Note 3**).
2. The day after seeding, replenish medium with full medium containing increasing amounts of Zeocin (*see* **Notes 3** and **4**).
3. On days 5–14 of cell culture, replenish medium containing the same amounts of Zeocin every 3 days and observe cell viability. For further experiments, choose the lowest concentration of Zeocin that killed all cells in a period of approximately 7–14 days.

3.2 Transfection of pFRT/lacZeo

1. On day 1, linearize pFRT-LacZeo. We use SwaI, which is a blunt-cutter and cuts in the backbone of pFRT-LacZeo. Use 4 U of enzyme per μg of DNA. Incubate at 25 °C for 1 h. Afterwards, deactivate the enzyme by heating to 65 °C for 20 min.
2. Freeze sample at −20 °C (*see* **Note 5**).
3. Trypsinize U-2 OS cells and seed a 6-well plate (5×10^5/well) in full medium. Incubate over night.
4. On day 2, transfect U-2 OS cells in a 6-well plate with increasing amounts of DNA per well. It is recommended to try 1, 2, 5 and 10 μg of linearized pFRT-LacZeo per well first. Also include a negative control to control the efficiency of the Zeocin concentration used. We use Lipofectamine 2000 (Life Technologies) but the optimal reagent might vary for other cell lines (*see* **Note 6**).
5. Follow the transfection protocol of the reagent of your choice.
6. On day 3, inspect the cells on an inverted microscope for toxic effects. Replenish medium in case the transfection influenced cell viability.
7. On day 4, wash the cells with D-PBS. Replenish the medium with full medium containing the appropriate concentration of Zeocin as determined before.
8. On days 5–14, wash the cells every 3–4 days with D-PBS and apply fresh full medium containing Zeocin.

3.3 Selection of Stable, FRT Containing Cell Clones

Check cell viability regularly under an inverted microscope. Most cells should die in the first few days and resistant cell foci should become visible after 7–10 days. Cells in the mock-transfected sample should all die. After about 2 weeks, but before colonies merge, select about 5–20 Zeocin-resistant cell clones (*see* **Note 7**).

1. On day 1, prepare cloning cylinders by applying a small amount of vacuum grease to one side. Then remove the medium from

the cells and isolate cell foci by tightening a cloning cylinder around it.

2. Apply 15 μL of Trysin-EDTA onto the cell foci and incubate for ~5 min in the cell incubator until cells become detached. In the meantime prepare a 24-well plate and add 0.5 mL of full medium into wells.
3. Pipette the cell-solution of one focus (the interior of the cloning cylinder) into one well each and incubate for 2 days in full medium without Zeocin.
4. On day 4, exchange the cell culture medium with full medium containing Zeocin.
5. The following week and as soon as wells become confluent, expand your clones by passaging them onto 12-well plates and then 6-well plates. Always use selective full medium containing Zeocin.

3.4 Selection of Best Expressing Clones

1. On day 1, seed approximately 2×10^5 cells per well and cell clone in 6-well plates and incubate overnight.
2. On day 2, discard medium and wash cells with D-PBS once. Fix and stain according to an X-Gal staining protocol for cell cultures (*see* **Note 8**). Assess and compare the X-Gal staining intensity of several clones. Choose three wells with expressing clones and make sure that clones are pure by checking that all cells of a given clone are expressing lacZ.
3. In the following days expand clones and freeze at least 20 tubes using a standard cell freezing protocol. Use good cell banking procedures to ensure reproducibility.
4. Thaw one tube of each clone and assess cell viability. Culture clones further. Always use full medium containing Zeocin and recheck purity and stability of β-galactosidase expression over extended periods by X-Gal staining. For best reproducibility choose only one clone for downstream procedures. This clone is your basic FRT-cell line optimized for stable expression.

The generated cell line is called U-2 OS-FRT-LacZ-Zeo further on.

3.5 Utilizing the Gateway™ System to Generate FP Tagged Fusion Proteins

The Gateway™ system (Life Technologies) is based on a two-step recombination protocol. The first step generates a "storage plasmid" for the gene of interest (GOI), termed pDONR™ or pENTR™ when the GOI has been inserted. This clone serves as a hub, from which expression clones with a verified GOI can be easily obtained due to recombination with a Destination vector (often termed pDEST™). This second "functional plasmid" contains expression elements and useful additions to the GOI, e.g., in form of a fluorescent fusion protein tag like GFP, mRFP or derivatives thereof.

Typically, the pENTR™ clone is generated by adding attB1 and attB2 recombination sites to PCR primers. The resulting PCR product is subsequently *in vitro* recombined with a pDONR™ vector that contains attP1 and attP2 sites. Recombination between attB and attP sites results in attL sites replacing the Gateway™-cassette containing the toxic ccdB gene (*see* **Note 9**). The recombination reaction is then transformed into a ccdA negative *E. coli* strain (e.g., DH5α) (*see* **Note 10**). Here again, ccdB mediated negative selection is used to ensure that only recombination products can grow. Moreover, most Destination plasmids encode an Ampicillin resistance gene while pDONR™ and resulting pENTR™ plasmids generally encode for a Kanamycin or other resistance gene, ensuring selection of recombined Expression clones.

As major cDNA collections like the Mammalian Gene Collection (http://mgc.nci.nih.gov/) supply their cDNAs in a Gateway™ compatible format, the Gateway™ system allows the rapid adoption of a GOI for virtually any expression need, provided a suitable Destination vector is available.

In the following part, we describe the modification of the Destination plasmid pEF5/FRT/V5-DEST for the custom generation of stable cell lines expressing the fusion of proteins of interest to various FPs.

Exemplary, we will describe the generation of a GFP-LC3 for the detection of autophagy as well as a Lifeact-GFP fusion to label the actin cytoskeleton, and we will outline all essential steps to modify the pEF5/FRT/V5-DEST with the Gateway™ system.

3.6 Modification of pEF5/FRT/V5-DEST to Express FP-Fusions

To allow the N-terminal fusion of FPs to GOIs, FPs can be inserted between the EF promoter and the Gateway™ (GW) cassette of pEF5/FRT/V5-DEST. To this end, the plasmid has to be cut in between the EF-1α promoter and the attB1 site and the FPs in question have to be made compatible by amplifying them with universal FP primers (*see* **Note 11**) adding restriction sites to the PCR product. This strategy can be modified to insert any desired tag into pEF5/FRT/V5-DEST to generate N-terminal fusion proteins (*see* **Note 12**).

1. Primers:

   KpnI-FP-5′: (GTGACT*GGTACC*TAATGGTGAGCAAGGGCGAGGAG).
   SpeI-FP-3′: (CACAGT*ACTAGT*TTGTACAGCTCGTCCATGCC).

   (italic: RE recognition sites; underlined: FP universal sequence) (*see* **Note 11**).

   The primers add KpnI and SpeI sites to the 5′ and 3′ end respectively to the product. The product also lacks the stop codon as well as the last base pair, keeping the reading frame to genes inserted into the GW cassette.

2. PCR Mix:

39.5 μL PCR-grade water	
5.0 μL Buffer 2 (Expand High Fidelity PCR System)	(10×)
1.0 μL dNTPs	(10 mM each)
1.5 μL 5′ Primer	(10 pmol/μL)
1.5 μL 3′ Primer	(10 pmol/μL)
1.0 μL Template	(50 ng/μL)
0.5 μL Hi-Fidelity Taq Mix (Expand High Fidelity PCR System)	(3.5 U/μL)

3. Thaw all reagents shortly before use. Keep on ice. Add Taq-Mix at last and proceed directly with thermocycling ("hot-start").
4. Thermocycle protocol for "Touchdown" PCR (modified from [13]):

   95 °C hold ("hot-start")

   95 °C for 2 min

   95 °C/15 s

   65–48 °C/30 s, cycle 18 times, with 1 °C decreased annealing temperature per cycle

   72 °C/1 min per kb of product length

   95 °C/15 s

   48 °C/30 s, cycle 17 times

   72 °C/1 min per kb of product length

   72 °C/10 min

   4 °C hold
5. Run analytical 1 % agarose gel with 5 μL of PCR product and check for product as well as purity. The expected product size is 733 bp. Then purify PCR product with a PCR spin column cleanup kit (we use the Qiagen PCR purification kit), elute in 30 μL of 10 mM Tris-Cl, pH 8.5 and measure DNA concentration on a spectrophotometer.
6. Digest PCR product with 4 U of KpnI and SpeI per μg of DNA for 2 h, purify the PCR product with a PCR spin column cleanup kit and elute in 30 μL 10 mM Tris-Cl, pH 8.5.
7. Cut 10 μg of pEF5/FRT/V5-DEST with 40 U of KpnI and SpeI for 4 h. Run the cut plasmid DNA in four lanes on a 1 % agarose gel and let it run long enough to make sure that the digested DNA is well separated from possibly non-cut DNA. Purify the digested DNA fragment (7,514 bp) from the gel and elute in 30 μL 10 mM Tris-Cl, pH 8.5. Measure DNA concentration on a spectrophotometer (*see* **Note 13**).

8. Ligate the pEF5/FRT/V5-DEST backbone with the FP insert.
9. Ligation-Mix:

2 μL T4-ligase Mix (NEB)	(10×)
100 ng Plasmid DNA	
30 ng PCR product (*see* **Note 14**)	
1 μL T4 ligase (NEB)	(2,000 U/μL)

fill to 20 μL with PCR grade water

10. Pipette on ice, mix well and incubate over night at 16 °C. Transform into competent ccdB-resistant *E. coli* by either electroporation or heat shock (*see* **Note 15**).
11. Select clones on Ampicillin containing LB-agar plates (100 μg/mL) over night (*see* **Note 16**) and pick clones. Then make DNA preparations and measure DNA concentration.
12. Digest about 500 ng isolated clone DNA with AgeI and run on an analytical agarose gel. The expected pattern for the parental plasmid is 4,512 bp + 3,016 bp. The addition of an FP shifts the second fragment to about 3,700 bp.
13. Select two right clones, prepare glycerol-stocks for long-term storage and prepare a bigger DNA preparation of one clone. Measure DNA concentration and dilute to 150 ng/μL with 10 mM Tris-Cl, pH 8.5. The resulting clone is named pEF5/FRT/FP-DEST.

3.7 Transferring a cDNA into the Gateway™ System

Full length cDNA clones in a Gateway™ compatible format can be obtained from distributors associated with the mammalian gene collection (http://mgc.nci.nih.gov/). Here, we will detail the generation of an Entry clone coding for the autophagic marker LC3 by amplification from a rat cDNA library. However, virtually any other gene of interest (GOI) can be cloned this way.

1. Primer:

   5′-attB1-ratLC3: GGGGACAAGTTTGTACAAAAAAGCAGGCTCCATGCCGTCCGAGAAGACC
   3′-attB2-ratLC3: GGGGACCACTTTGTACAAGAAAGCTGGGTCTTACACAGCCAGTGCTGTCCCG

   (italic: Gateway™ oligos containing attB1 and attB2 sites; underlined: sequence homologies to ratLC3 cDNA).

   For PCR and purification *see* Subheading 3.6.
2. To generate your Entry clone follow the BP recombinase protocol from Life Technologies.
3. Pick three to six clones, prepare DNA preparations and measure DNA content. Then digest 500 ng DNA for each clone

and run an analytical agarose gel to confirm clone identity (*see* **Note 17**).

4. Select two seemingly right clones and verify by sequencing. Utilize the M13 forward and reverse priming sites, e.g., when using pDONR™221. Prepare glycerol-stocks for long-term storage. Dilute stock to 150 ng/μL with 10 mM Tris-Cl, pH 8.5.

3.8 Generation of a pEF5/FRT-FP-GOI Expression Vector

1. The LR reaction which transfers the verified GOI into the Destination vector is performed analogously to the BP reaction and downstream steps to verify clone integrity are the same as described for the BP reaction (refer to Life Technologies for protocols, www.lifetechnologies.com).
2. Transform and select clones on Ampicillin-containing LB-agar plates. Check three clones by analytical digest. Sequence-verification is not needed if the pENTR™-GOI clone was sequenced. The resulting construct is generally named pEF5/FRT/FP-GOI as in pEF5/FRT/GFP-LC3.

3.9 Generation of Flp-In™ Compatible Expression Vectors Coding for Peptide-Tagged FPs

Instead of cloning FPs 5′ to the Gateway™ cassette resulting in N-terminal fusions of FPs to GOIs as described in Subheading 3.2, it is also possible to target an FP to a specific cellular structure by fusing short targeting peptides directly onto the FP and expressing this fusion from the Gateway™ cassette. Here we utilize Lifeact, a 17 amino acid long peptide that binds reversibly to F-actin.

In the following, the universal FP primers outlined in Subheading 3.2, **step 1** are used as a basis. However, to make the primers Gateway™ compatible, attB sites are added. The Lifeact coding sequence is added to the 5′ end of the sense primer followed by a short GGSG linker sequence. Moreover, the start ATG is deleted from the FP binding part in the sense primer.

1. Primers:

   5′-attB1-Lifeact-FP: *GGGGACAAGTTTGTACAAAAAAGCAGGCT*TA**ATGGGTGTCGCAGATTTGATCAAGAAATTCGAAAGCATCTCAAAGGAAGAA**GGCGGCAGCGGCGTGAGCAAGGGCGAGGAG

   3′-attB1-FP: *GGGGACCACTTTGTACAAGAAAGCTGG*GTATTACTTGTACAGCTCGTCCAT

   (italic: Gateway™ oligos containing attB1 and attB2 sites; bold: Lifeact coding sequence; non-formatted: GGSG linker; underlined: FP universal sequences).

2. For PCR, the BP reaction and clone verification *see* Subheadings 3.6–3.8.

   The resulting construct is called pENTR™-Lifeact-FP.

3. In a next step, the Lifeact-FP construct is transferred into pEF5/FRT/V5-DEST by a LR reaction (*see* Subheading 3.8). The resulting construct is called pEF5/FRT-Lifeact-FP.

3.10 Cell Culture

Culture the U-2 OS-FRT-LacZ-Zeo in complete DMEM medium (DMEM GlutaMAX, 10 % FBS, 1 % Penicillin/Streptomycin). Add 100 μg/mL of Zeocin shortly before use (*see* **Note 18**). Cells have to be tested for their level of sensitivity to Hygromycin B before initial use as in the case of Zeocin described in Subheading 3.1. Typically, the range of sensitivity for mammalian cells lies between 50 and 200 μg/mL.

3.11 Generation of Stable Cell Lines

1. Follow the Flp-In™ protocol from Life Technologies to insert the pEF5/FRT-plasmid into the FRT-site containing cell line by co-transfection of your FRT-plasmid and the flippase encoding plasmid pOG44 at a ratio of 10:1 using 1–2 μg total DNA. Use the transfection protocol suitable for your preferred transfection reagent.
2. On day 4, wash the cells with D-PBS. Replenish the medium with full medium containing the appropriate concentration of Hygromycin B.
3. During the following days, wash the cells every 3–4 days with D-PBS and apply fresh full medium containing Hygromycin B.
4. Check cell viability regularly under an inverted microscope. Most cells should die in the first few days and resistant, fluorescent cell foci should become visible. Most colonies should be fluorescent.
5. After about 2 weeks, but before colonies merge, select fluorescent clones (*see* **Note 19**).
6. To isolate clones discard the medium and isolate foci by tightening a cloning cylinder around them.
7. Apply 15 μL of Trysin-EDTA onto the cell foci and incubate for 5 min in the cell incubator until all cells are detached.
8. In the meantime prepare a 24-well plate and add 0.5 mL of full medium into wells.
9. Pipette the cell-solution of one focus into one well each and incubate for 4 days. Isolate at least three clones. Add full medium with Hygromycin B.
10. As soon as wells become near confluent, passage clones onto 12-well plates, 6-well plates, T25 flasks and finally T75 flasks or equivalent containers. Always use selective full medium containing Hygromycin B.
11. Check clonal purity regularly by assessing the fluorescence pattern and intensity on an inverted fluorescence microscope (*see* **Note 20**). Check functional features if applicable (*see* Fig. 2 for an example).
12. Freeze clones using a standard cell freezing protocol. After a few days thaw one tube and assess cell viability and fluorescence. Use good cell banking procedures to ensure reproducibility.

4 Notes

1. An optimized version with a truncated SV40 promoter called pFRT/lacZeo2 is available from the same vendor, which might be better suited to select for insertions into genomic sites that are under the control of transcriptional enhancers.
2. A few cell lines are available for purchase from Life Technologies and other vendors. Also services to generate individual FRT-cell lines are available.
3. In our hands, 300 μg Zeocin per mL of full medium was sufficient to kill nonresistant U-2 OS. However, the level of resistance might vary for other cell lines, due to culture conditions or the batch of FBS used.
4. Zeocin is light-sensitive and should always be kept in the dark.
5. Linearized plasmid DNA is stable for extended periods of time. If it is likely that other cell lines will be made, a bigger amount of DNA can be digested and stored for future use.
6. Scale the amount of transfection reagent with the amount of DNA used (i.e., 2.5 μL of Lipofectamine 2000 per μg of DNA). U-2 OS cells are sensitive to high concentrations of Lipofectamine 2000.
7. Try to select clones from wells transfected with the lowest amount of DNA. This will reduce the number of clones having integrated more than one copy of the lacZ-ZeoR cassette. The original protocol by Life Technologies recommends Southern blotting of genomic DNA and probing for the lacZ gene to ensure selection of clones with only one copy. We found this to be not absolutely necessary in our hands. However, it should be noted that this might depend on the transfection efficiency and the concentration of Zeocin used as a too efficient transfection might result in frequent multiple insertions while a too high concentration of Zeocin might select for multiple insertions or gene amplification of the resistance cassette.
8. We use the β-Gal Staining Kit (Life Technologies) but several other commercial kits as well as protocols are available.
9. Amplification of plasmids containing the Gateway™ cassette requires an *E. coli* strain expressing the ccdA gene to overcome the ccdB toxicity like ccdB Survival. Strains containing the F-plasmid like XL1-blue are not recommended as the ccdA expression level might not be sufficient.
10. A wide variety of commercial and noncommercial Destination plasmids is available from Addgene (www.addgene.org).
11. These primers work with all EGFP derived FP variants and also with the mFruit series as the first and last 21 bp of these proteins are identical [14].

12. We routinely use the Expand High Fidelity PCR System (Roche). However, other high-fidelity Taq mixes might also be suitable for the employed Touch-Down PCR protocol but should be tested before.
13. We routinely use HF-enzymes from NEB, which all cut in the same buffer. If others are used, enzyme amounts might have to be adjusted due to incompatible buffer conditions.
14. The molar ratio of 1:3 (Plasmid/Insert) is calculated by the formula (size of insert [bp]/size of plasmid [bp]) × 300 ng if 100 ng plasmid DNA is used.
15. We routinely use the *E. coli* strain ccdB Survival (Life Technologies) for that purpose. However, other ccdA expressing strains might also work.
16. To make sure, that the Gateway™ cassette is present, clones can be plated on Ampicillin/Chloramphenicol-containing LB plates.
17. PvuII is a good enzyme to check pDONR™221 derived plasmids. The parental plasmid generates the following pattern: 2,218 bp, 1,942 bp, 602 bp. After recombination, the 1,942 bp band remains unchanged while additional bands indicate the insert.
18. Zeocin is light sensitive! Keep stock as well as cultures in the dark.
19. Check if clones exhibit stable fluorescence during the selection process using an inverted microscope with corresponding filter sets for your desired fluorescence. Most clones should display a homogeneous fluorescence within all cells of a clone. The Flp-In™ system does not explicitly require clonal selection as most resulting Hygromycin B resistant cells should have integrated the FRT-containing vector at the same genomic position. In our hands this seems to be true as for most constructs the variability between individual clones is indeed low. However, we have observed construct-specific differences (depending on your gene of interest), so clonal selection can be beneficial as it further increases the homogeneity of the cell pool.
20. Some populations will show nonfluorescent but Hygromycin B resistant cells. Do not use these clones. We also noted that the efficiency of the protocol and purity of isolated cell clones strongly depends on the GOI inserted. Some GOIs have a negative effect on the host cells and therefore have a selection disadvantage. In our hands this was the case for some transcriptional repressors. Also some small GTPases and especially their dominant negative variants can have detrimental effects on cell viability. In such cases we recommend to use inducible systems.

Acknowledgments

Part of this work was supported by a Franco-Bavarian bilateral grant BFHZ-CCUFB FK22-10 (H.W.). J.B.B. received support via a PhD studentship from the German National Academic Foundation (Studienstiftung des Deutschen Volkes). We acknowledge the Bordeaux imaging centre (BIC) for help in setting up the live cell imaging acquisition and Lynn Enquist for support given to J.B.B. H.W. is an INSERM fellow.

The protocols in this chapter are based on technologies that are the intellectual property of Life Technologies. General information as well as protocols regarding the Flp-In™ system as well as the Gateway™ technology can be found at: www.lifetechnologies.com.

References

1. Tsien RY (1998) The green fluorescent protein. Annu Rev Biochem 67:509–544
2. Giepmans BN, Adams SR, Ellisman MH et al (2006) The fluorescent toolbox for assessing protein location and function. Science 312(5771):217–224
3. Wiedenmann J, Oswald F, Nienhaus GU (2009) Fluorescent proteins for live cell imaging: opportunities, limitations, and challenges. IUBMB Life 61(11):1029–1042
4. Tamassia N, Bazzoni F, Le Moigne V et al (2012) IFN-beta expression is directly activated in human neutrophils transfected with plasmid DNA and is further increased via TLR-4-mediated signaling. J Immunol. doi:10.4049/jimmunol.1102985
5. Baum C (2007) Insertional mutagenesis in gene therapy and stem cell biology. Curr Opin Hematol 14(4):337–342
6. Mutskov V, Felsenfeld G (2004) Silencing of transgene transcription precedes methylation of promoter DNA and histone H3 lysine 9. EMBO J 23(1):138–149. doi:10.1038/sj.emboj.7600013
7. Mizushima N (2009) Physiological functions of autophagy. Curr Top Microbiol Immunol 335:71–84
8. Jordan TX, Randall G (2012) Manipulation or capitulation: virus interactions with autophagy. Microbes Infect 14(2):126–139
9. Mizushima N, Yoshimori T, Levine B (2010) Methods in mammalian autophagy research. Cell 140(3):313–326
10. Moreau K, Ravikumar B, Renna M et al (2011) Autophagosome precursor maturation requires homotypic fusion. Cell 146(2):303–317
11. Taylor MP, Koyuncu OO, Enquist LW (2011) Subversion of the actin cytoskeleton during viral infection. Nat Rev Microbiol 9(6):427–439. doi:10.1038/nrmicro2574
12. Riedl J, Crevenna AH, Kessenbrock K et al (2008) Lifeact: a versatile marker to visualize F-actin. Nat Methods 5(7):605–607. doi:10.1038/nmeth.1220
13. Don RH, Cox PT, Wainwright BJ et al (1991) 'Touchdown' PCR to circumvent spurious priming during gene amplification. Nucleic Acids Res 19(14):4008
14. Shaner NC, Campbell RE, Steinbach PA et al (2004) Improved monomeric red, orange and yellow fluorescent proteins derived from Discosoma sp. red fluorescent protein. Nat Biotechnol 22(12):1567–1572. doi:10.1038/nbt1037

Chapter 12

Determination of HSV-1 Infectivity by Plaque Assay and a Luciferase Reporter Cell Line

Diana Lieber and Susanne M. Bailer

Abstract

Quantification of infectious virus is crucial to many experimental approaches in virological research. A broadly used and facile technique is the so-called "plaque assay" which provides precise information on the absolute quantity of infectivity in a given volume. Due to advances in the understanding of viral gene expression, transactivator-promoter pairs have been identified which can be used in transgenic cell lines as reporters of viral infection. Even though such "cellular reporter assay" systems are mostly restricted to relative quantification, they are attractive tools which can complement or replace the conventional plaque assay. Cellular reporter assays become especially interesting in state-of-the-art high-throughput screening approaches, as for instance RNAi and compound library screens, since they are often compatible with small-scale and automated experimentation. In this chapter, a regular plaque assay as well as a cellular reporter assay employing a luciferase reporter gene are described. As an example, HSV-1 infectivity is assessed with both methods yielding complementary information. Advantages and disadvantages of the two techniques and possible applications are discussed.

Key words Herpes Simplex Virus type 1, HSV-1, Varicella Zoster Virus, VZV, Infectivity, MV9G, Plaque assay, Luciferase assay, Antiviral screen, RNAi

1 Introduction

Accurate virus research depends on methods for the quantification of infectivity. Virus stocks are routinely titrated to determine the amount of infectious virus and to allow standardization of experiments and dosage control. Titration by *plaque assay* is a standard technique which yields an absolute estimate of infectious units in a given volume [1, 2]. In this assay, cells are infected with serial dilutions of virus. After plaques have formed, the cells are fixed and stained and the number of plaques counted. This approach is easy to perform and inexpensive and does not require sophisticated equipment. However, large amounts of cells and inoculum virus are needed, and depending on the replication time of the analyzed virus, it may take several days for plaques to appear. Therefore, the

Susanne M. Bailer and Diana Lieber (eds.), *Virus-Host Interactions: Methods and Protocols*, Methods in Molecular Biology, vol. 1064, DOI 10.1007/978-1-62703-601-6_12, © Springer Science+Business Media, LLC 2013

plaque assay tends to be relatively time consuming. Finally, plaque assays are laborious thus hampering high-throughput approaches required for antiviral screening.

A *cellular reporter assay* where a reporter gene encoding a fluorescent or enzymatic protein responds to virus infection represents an attractive alternative. If the reporter gene is introduced into a broadly permissive cell line and responds to various virus species, different viruses including recombinant as well as wild type strains can be analyzed. Similarly, the impact of various treatments on infection can be compared, e.g., the administration of antiviral substances like small compounds, siRNAs, miRNAs, or neutralizing antibodies. These options make the approach highly flexible in application. The assay can be performed on a small scale using 96-well microtiter plates which reduces the required amount of reagents, cells and inoculum virus. Furthermore, the approach is suitable for automation and therefore compatible with high-throughput experiments. By using a fast-responding reporter gene, e.g., under control of a strong promoter, the assay is very sensitive and requires much less time for readout than the plaque assay. If a promoter reactive to an immediate early transactivator drives the reporter gene, virus entry and initiation of viral gene expression can be analyzed. A disadvantage of such a reporter cell line is that absolute quantification of infectivity is difficult or even impossible. Thus, the approach is restricted to comparative analyses with relative quantification of virus infection.

In this chapter, we describe and compare two procedures for exemplary quantification of Herpes Simplex Virus type 1 (HSV-1) infectivity. First, a regular plaque assay is detailed for titration of a viral stock solution. Second, a firefly luciferase reporter cell line is employed for quantitative analysis of infection with the same viral stock [3]. The reporter cell line used has originally been generated for the detection of Varicella Zoster Virus (VZV; [3, 4]), like HSV-1 a member of the alphaherpesvirinae. The luciferase reporter gene is under control of a VZV ORF9 minimal promoter (orthologous to HSV-1 UL49) which is responsive to VZV infection, ectopic expression of the major transactivator of VZV (the immediate early antigen 62 (IE62) which is orthologous to HSV-1 RS1), as well as to HSV-1 infection. In both parts of this chapter, serial dilutions of the HSV-1 stock solution are prepared. Subsequently, cells are infected in triplicates and infection quantified by counting plaques or by measuring luciferase activity, respectively. The number of plaques is used to calculate the actual titer of the stock solution while the reporter cell assay reveals the dilution range of the virus stock within which a good resolution can be achieved in subsequent experiments. An intrinsic difference of the two assays is that the plaque assay readout relies on removal of infected cells from the culture whereas the reporter cell assay measures the amount of infected cells present in the culture. The two approaches can either complement each other or stand alone depending on the underlying scientific question. Finally, the MV9G cell line represents a broadly applicable reporter system comparable to another infection-responsive reporter line described in this book series (*see* Scheibe et al., Chapter 25). Such reporter cell lines are ideally suited for the generation of stable expression lines of infection modulators, either proteins or miRNAs (*see* Bosse et al., Chapter 11 and Lieber, Chapter 13 of this series).

2 Materials

2.1 Cell Culture

1. Vero cells: African green monkey kidney cells (ATCC *CCL-81*™) (*see* **Note 1**).
2. MV9G cell line: MeWo cells transfected with the plasmid pGL-T9G and a plasmid encoding the G418-resistance gene [3].
3. Vero cell culture media: DMEM containing 10 % FCS, 1 % Pen/Strep solution.
4. MV9G cell culture medium: DMEM containing 10 % FCS, 1 % Pen/Strep solution. Complete medium contains Geneticin G-418 at a concentration of 50 μg/ml. Antibiotics are excluded from the medium during the reporter assay.
5. Trypsin-EDTA (0.05 %).
6. Plastic dishes.
7. Neubauer chamber for cell counting or similar device.

2.2 Plaque Assay

1. Virus strain: Herpes Simplex Virus type 1 strain F, kindly provided by B. Roizman, University of Chicago.
2. Methyl cellulose media: 50 mL 10× MEM, 5 % FCS, 1 % Glutamin, 2.5 % Nonessential aminoacids (NEAA), 1 % Pen/Strep solution containing 3.75 g Carboxymethyl cellulose, 24.7 mL $NaHCO_3$, adjust with deionized water to 500 mL (*see* **Note 2**).
3. Plaque staining solution: 0.2 % (w/v) Crystal violet (*see* **Note 3**), 11 % (v/v) Formaldehyde, 2 % (v/v) Ethanol, 2 % (w/v) Paraformaldehyde in water.
4. Dulbecco's phosphate buffered saline PBS.
5. Plastic ware: 12- and 24-well dishes (*see* **Note 4**).

2.3 Seeding and Infection of MV9G Cells

1. Virus strain: Herpes Simplex Virus Type 1 strain F, kindly provided by B. Roizman, University of Chicago.
2. Sterile cell culture supplies, plastic ware for single use.
3. 1.5 mL reaction tubes, sterile.
4. T75 cell culture flask (75 cm^2).
5. 96-well cell culture plates (with lid).

6. Phosphate-buffered saline (PBS), sterile.
7. 0.05 % Trypsin-EDTA.
8. Incubator for cell culture, 37 °C, humidified, 5 % CO_2.

2.4 Luciferase Assay

1. 5× Passive Lysis Buffer (PLB) (Promega).
2. Luciferase Assay Reagent (Promega).
3. Phosphate-buffered saline (PBS).
4. Multichannel pipette (10–100 μL) with suitable pipet tips.
5. 96-well assay plate, white.
6. Rotary shaker.
7. Incubator for cell culture, 37 °C, humidified, 5 % CO_2.
8. Plate-reading luminescence reader with injector.
9. Washing solution: 70 % ethanol in water.
10. Distilled water.

3 Methods

Work with infectious and genetically modified virus is a potential biohazard. Adhere to the general guidelines and local regulations for work with BL-2 organisms.

3.1 Plaque Assay to Determine Titer of Virus Stock

1. Maintain Vero cells in cell culture media at 37 °C and 5 % CO_2. Once the cells reach confluence, split them at a ratio of 1:3 to 1:5.
2. The day before the plaque assay (day 1), seed the Vero cells in 24-well plates at a density of 8×10^4 per well: remove media, wash once with PBS, add 2 mL Trypsin-EDTA (0.05 %) and wait until cells are rounding, remove Trypsin-EDTA, and harvest cells by washing the plate with 5 mL of media. Count cells in a Neubauer chamber, dilute them to a density of 8×10^4 per 500 μL of media, and add 500 μL to each well (Fig. 1).
3. Prepare as many wells as required for a particular experiment (to determine the plaque forming units of 1 virus stock), perform experiments in triplicates, if testing five dilutions, 3×5 wells are required which accounts for 15×500 μL of Vero cells (1.2×10^6 cells in media).
4. Let cells grow for 1 day when they should be subconfluent (*see* **Note 5**).
5. On day 2, prepare a tenfold serial dilution of the virus stock solution (*see* **Note 6**) in triplicates: add 450 μL of media into each well of a 48-well tissue culture plate (Fig. 1). Add 50 μL of the virus stock solution to be tested into the first well and

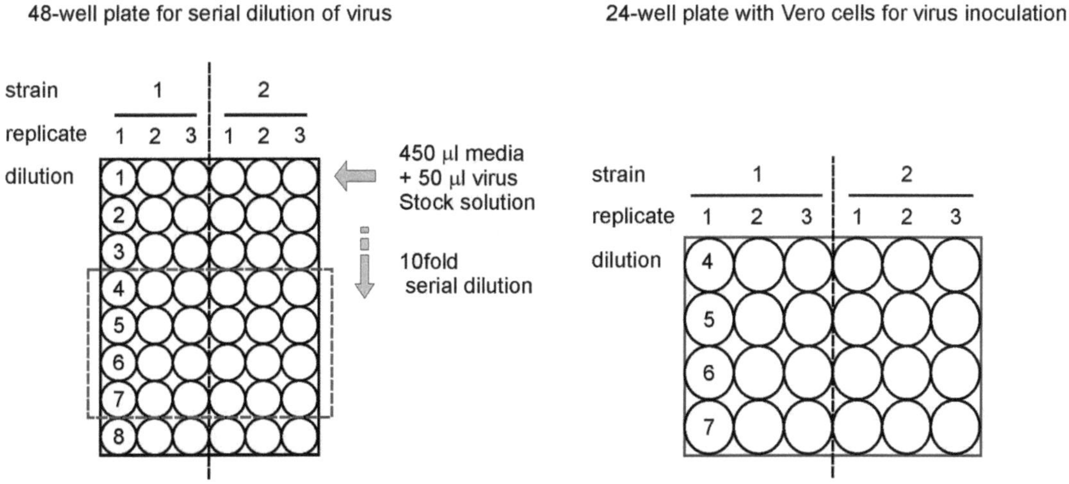

Fig. 1 Plaque assay to determine titer of virus stock solution. Day 1 of the plaque assay Vero cells are seeded in 24-well plates (*right panel*). The next day, a serial dilution of the virus stock is prepared in triplicates in 48-well plates containing 450 μL of medium (*left panel*). 400 μL of the dilution steps 4–7 are plated on 24-well plates for inoculation of Vero cells (*right panel*)

carefully pipet up and down (*see* **Note 7**). Then transfer 50 μL of this first dilution into the second well and continue as described for the first dilution to generate a serial dilution. In total prepare five dilutions dependent on the expected virus concentration at the given MOI and time after infection of a culture.

6. Collect 24-well plate containing Vero cells out of the incubator, remove media.
7. Add 400 μL of virus dilutions prepared in 48-well plates and transfer them to the 24-well plate for inoculation of Vero cells (Fig. 1). Include negative control by providing cells with medium lacking virus inoculum. A virus stock with known titer should be included as positive control.
8. Incubate the 24-well plates for 1 h in a cell culture incubator while gently rocking the plates (*see* **Note 8**).
9. Collect the 24-well plate from the incubator and suck off the inoculum.
10. Wash the Vero cells once with PBS.
11. Add 400 μL of Methylcellulose media to each well and transfer the plates to the incubator.
12. On day 3–5, remove 24-well plates from the incubator and suck off the Methylcellulose media (*see* **Note 9**).
13. Add staining solution containing Crystal violet to fix and stain the cells, incubate for 10 min (*see* **Note 10**). Remove Crystal

violet solution and collect it in a waste container (*see* **Note 11**). Wash plates with tap water until the water is clear (*see* **Note 12**), let the plates dry at room temperature. Seal the plates with tape and keep them in the dark (*see* **Note 13**).

14. Count plaques in wells where the plaques can clearly be distinguished but still are frequent, average the counts of the triplicates and calculate the plaque forming units (pfu). Determine the dilution of the virus stock as following, assuming that 100 plaques are present in dilution 5 (dilution 10^{-5}) (Fig. 1; *see* **Note 14**):

$$100 \times 2.5 \times 10^5 = 2.5 \times 10^7 \text{ pfu/mL}$$

3.2 Luciferase Assay to Determine Infectivity of Virus

1. Culture MV9G cells with complete culture medium containing geneticin in a T75 cell culture flask until they are 70–80 % confluent.
2. Wash cells once with PBS.
3. Detach cells with 2 mL Trypsin and add 8 mL medium without antibiotic.
4. Count cells.
5. Adjust cell density with medium without antibiotic to 5×10^5 cells/mL at a final volume of at least 3 mL per virus (*see* **Note 15**).
6. Seed cells into a 96-well plate: 100 μL cell suspension per well corresponding to 5×10^4 cells per well. Prepare 27 wells for each virus to be tested (*see* **Note 16**).
7. Allow cells to settle and attach overnight.
8. Inspect cells to confirm viability and even distribution.
9. Prepare a fivefold dilution series of each virus in culture medium (Fig. 2): Add 100 μL of virus to 400 μL of medium. Mix well. Use a fresh pipet tip to transfer 100 μL of this dilution to 400 μL fresh medium and mix well. Repeat another nine times. This will yield 11 sequential virus dilutions (*see* **Note 17**).
10. Infect MV9G cells by replacing medium with the sequential virus dilutions (Fig. 2). Prepare triplicates. To three wells, add fresh medium without virus as background control (mock).
11. Incubate plate for 24 h at 37 °C and 5 % CO_2.
12. Equilibrate 5× PLB and luciferase assay reagents to room temperature (*see* **Note 18**).
13. Inspect plate for cytopathic effects as indicator of successful infection.
14. Prepare 1× PLB cell lysis buffer by diluting the 5× PLB 1:4 in distilled water. Mix well.
15. Carefully remove the medium from the wells with the MV9G cells and replace with 20 μL 1× PLB per well (*see* **Note 19**). Do this one by one and avoid drying of the cells. Place the lid back on the plate.

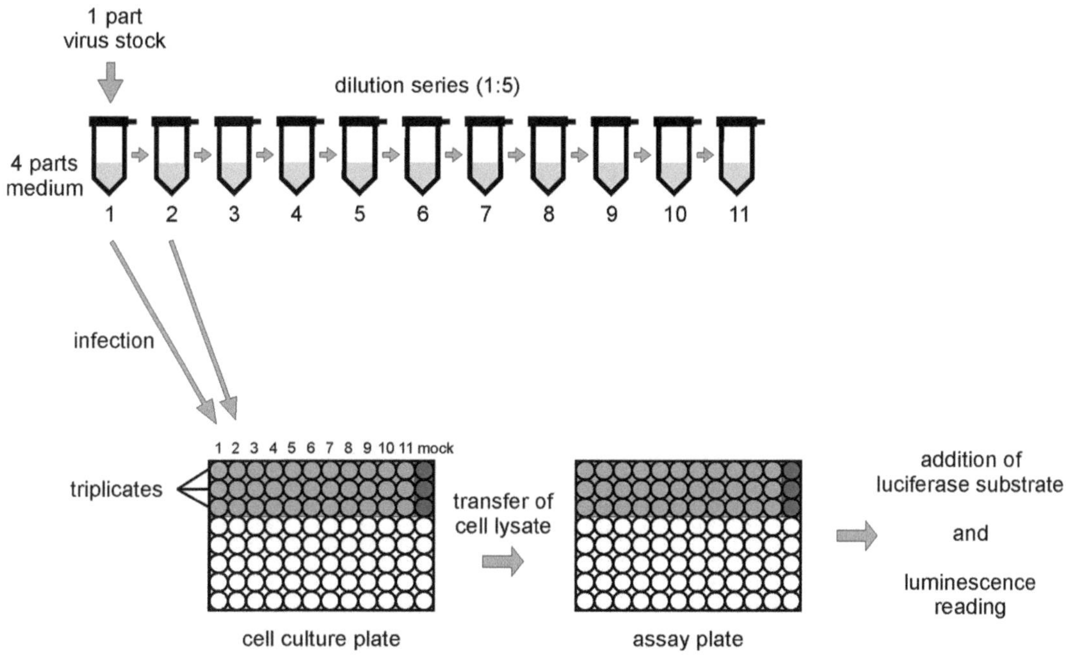

Fig. 2 Luciferase assay: serial dilutions and infection. For preparation of the fivefold HSV-1 dilution series, 11 tubes are each charged with 400 μL of cell culture medium. Then, 1 part (100 μL) of HSV-1 virus stock is added to 4 parts (400 μL) of fresh cell culture medium in tube 1. After thorough mixing, 100 μL of this initial dilution is transferred to tube 2 and mixed. This is repeated for the remaining tubes until all tubes have received virus. For infection, the culture medium is removed from the MV9G cells in the 96-well cell culture plate and replaced by 100 μL of the respective virus dilution. Each dilution is added to three replicate wells. 24 h after infection, the cells are lysed. An aliquot of the cell lysate is transferred to the assay plate for luminescence reading

16. Gently rock the plate on a rotary shaker for 15 min at room temperature.
17. Add 80 μL of PBS to each well to dilute the cell lysate 1:5 and mix by pipetting. This can be done with a multichannel pipette.
18. Transfer 10 μL of each cell lysate with the multichannel pipet to a white 96-well assay plate.
19. Set up the luciferase substrate (*see* **Note 20**).
20. Prime the luminescence reader with the luciferase substrate (*see* **Note 21**).
21. Program the luminometer to dispense 50 μL substrate per well followed by a 2 seconds measurement delay and a 10 seconds measurement read.
22. Load the assay plate into the luminescence reader.
23. Assay luciferase activity and record the luminescence reads.
24. Save unused luciferase substrate and store, protected from light, at −80 °C for future use.
25. Wash injector of the luminometer with 5 mL 70 % ethanol and rinse with distilled water according to the manufacturer's instructions.

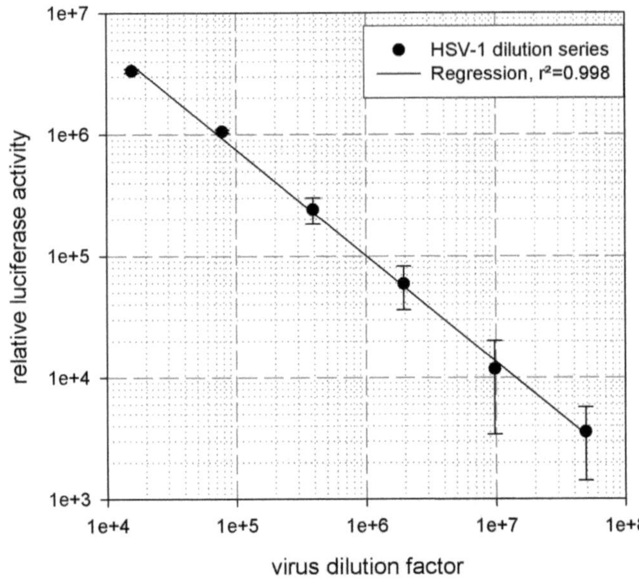

Fig. 3 Response of the MV9G-encoded luciferase reporter to HSV-1 infection. This figure shows the linear range of the exemplary HSV-1 luciferase assay as described in Subheading 3.2. Background luminescence has been subtracted from the raw data and means of triplicates have been plotted. Error bars indicate standard error of the mean. The linear regression curve nicely fits the data as indicated by the coefficient of determination (r^2). Within the shown range of virus dilution, the resolution of the assay is optimal for relative quantification of infection. Plot of data and determination of r^2 was done using SigmaPlot11.0 by Systat Software

26. Discard all liquid in the assay plate and wash the plate with 70 % ethanol and then with distilled water for later reuse.
27. For data analysis, calculate the mean luciferase activity of the triplicate mock infected samples (background) (*see* **Note 22**).
28. Subtract the background from all raw data.
29. Calculate the mean of the triplicate values for each virus dilution.
30. In order to determine the linear range, plot mean luciferase activities versus the dilution factor with linear scales (*see* **Note 23**).
31. Plot the values of the linear range with logarithmic scales for both axes in order to receive a linear curve (*see* **Note 24**) (Fig. 3).

4 Notes

1. For HSV-1 plaque assays, Vero cells can be replaced by BHK cells.
2. Methylcellulose medium is viscous and retains viral particles at the site of release and thus in vicinity of the developing viral

plaque. This is important to avoid the formation of plaques by progeny virus.

3. Crystal violet is toxic. To prepare the staining solution, wear coat, protective gloves and masks. Crystal violet waste has to be deposited properly.
4. Plaque assays require high quality cell culture dishes. Otherwise cell layers are prone to detach making analysis unreliable and tedious.
5. Cells should be subconfluent and just about to touch each other. Cells that are too tightly grown upon infection will produce very small plaques due to "overgrowth" by uninfected neighboring cells.
6. Thaw the virus stock prior to preparing the serial dilutions. Prepare the serial virus dilution immediately before inoculating the cells.
7. For accuracy and reproducibility, perform plaque assays in triplicates. To prepare serial dilutions, carefully pipet up and down the virus dilution. Exchange the pipet tip after each dilution step to avoid carryover of adherent virus.
8. For inoculation, keep the cells on a rocking platform in an incubator at 37 °C and 5 % CO_2. In case a rocking platform is not available, manually rock the plates regularly.
9. Infected cells can be discriminated from uninfected ones by rounding and detaching from the plate surface (cytopathic effect). Often, however, in particular in wells with few plaques, the plaques go unnoticed until the cells were stained thereby detaching the infected cells and revealing the cell-free areas which are the plaques.
10. Crystal violet is toxic, solutions containing crystal violet have to be deposited according to safety guidelines.
11. Once the infected cells and the virus released from them have been exposed to Crystal violet solution for 10 min, the infected cells and viral particles are inactivated. Thus, neither the Crystal violet solution nor the water used for rinsing the plates should be infectious.
12. After staining and fixing the cells, the plates can be processed outside of the sterile hood.
13. Crystal violet is light sensitive, and therefore, the dried plates should be sealed and kept in the dark.
14. Virus plaques will be determined as plaque forming units pfu/mL. The calculation takes into account that just 400 μL of virus inoculum are added (*see* Subheading 3.1, **step** 7).

15. If more than one virus is analyzed simultaneously increase the cell number and volume accordingly.
16. The luciferase assay described here as an example includes eight virus dilutions and a mock infection and is carried out in triplicates. If more replicates or more virus dilutions are intended, more wells need to be prepared according to the individual setup.
17. If virus is scarce and a high virus titer is expected, the dilution series can be started with a higher dilution factor, e.g., 25-fold (20 μL virus in 480 μL medium) or 125-fold (5 μL virus and 620 μL medium), and then continued with 5-fold serial dilutions. This limits virus consumption to the bare minimum.
18. The assay can be adapted from the endpoint-measurement mode described here to a kinetics measurement mode with consecutive readings by choice of a different substrate (see Chapter 25 of this series). Nontoxic substrates are available which can penetrate living cells. This allows long-term monitoring of luciferase activity in the same culture. If no live-assay substrate is available, kinetics measurements can still be performed by preparing triplicate infected cultures separately for each time point to be recorded.
19. Infected cells are rounded and easily detach from the surface of the cell culture dish. The luciferase assay protocol (Promega) asks to rinse the cells with PBS before adding the lysis buffer. In our hands, this led to massive loss of infected cells. Therefore, we routinely omit the washing step.

    In addition, we found that sucking the medium away with a vacuum pump was too harsh and led to loss of cells in wells with strong infection. Therefore, we recommend to remove the medium carefully with a pipet.

    Another possibility to avoid excessive loss of infected cells would be to lyse cells earlier (<24 h) when the cytopathic effects are less pronounced. This would probably lead to lower luciferase signals and increase the detection limit. Therefore, a pilot experiment to identify ideal parameters for a given virus strain should be carried out.
20. Follow the manufacturer's instructions. For the Promega Luciferase Assay System reconstitute one aliquot of the lyophilized "Luciferase Assay Substrate" with 10 mL "Luciferase Assay Buffer" at room temperature. Mix well and transfer to a 15 mL tube.
21. Using an automated luminometer reduces the pipetting work and assay time and renders the assay compatible to high-throughput applications. However, manual sample processing is also possible if such a reader is not accessible. Please refer to the manuals of the available instrument and to the luciferase substrate guidelines.

22. Modern plate readers mostly are equipped with an analysis software to do the following calculations. We prefer spreadsheet programs such as Excel (Microsoft).
23. Expect an S-shaped curve for the complete range with high virus doses showing saturation of the luminescence signal and the luciferase activities of low virus doses being indiscernible from the background.
24. Subsequent comparative analysis of virus infection, e.g., under different treatments can be carried out with essentially the same protocol. Serial dilutions are generally not required. Use a virus dose within the predetermined range to achieve optimal resolution and perform the assay in triplicates. Keep in mind to include mock infections for background estimation and compare the relative luciferase activities after background subtraction.

References

1. Schmidt T, Striebinger H, Haas J et al (2010) The heterogeneous nuclear ribonucleoprotein K is important for Herpes simplex virus-1 propagation. FEBS Lett 584(20):4361–4365. doi:10.1016/j.febslet.2010.09.038
2. Ott M, Tascher G, Hassdenteufel S et al (2011) Functional characterization of the essential tail anchor of the herpes simplex virus type 1 nuclear egress protein pUL34. J Gen Virol 92(Pt 12): 2734–2745
3. Wang GQ, Suzutani T, Yamamoto Y et al (2006) Generation of a reporter cell line for detection of infectious varicella-zoster virus and its application to antiviral studies. Antimicrob Agents Chemother 50(9):3142–3145. doi:10.1128/AAC.00342-06
4. Inoue N, Matsushita M, Fukui Y, Yamada S, Tsuda M, Higashi C et al (2012) Identification of a varicella-zoster virus replication inhibitor that blocks capsid assembly by interacting with the floor domain of the major capsid protein. J Virol 86(22):12198–12207

Chapter 13

Generation of a Stable Cell Line for Constitutive miRNA Expression

Diana Lieber

Abstract

miRNAs have in recent years emerged as novel players in virus–host interactions. While individual miRNAs are capable of regulating many targets simultaneously, not much is known about the role of distinct host or viral miRNAs in the context of infection. Analysis of the function of a miRNA is often hampered by the complexity of virus–host interactions and the enormous changes in the host cell during infection. Many viral miRNAs as for example from Kaposi sarcoma-associated Herpesvirus (KSHV) are probably exclusively expressed in latent infection. This might lead to a steady-state situation with offense and defense mechanisms counteracting each other. Cellular miRNAs involved in defense against pathogens on the other hand might be suppressed in infection. A cell culture system allowing for constitutive expression of individual miRNAs at high levels is a useful tool to enhance miRNA-specific functions and to uncouple viral miRNA function from other infection-related mechanisms.

Here, a protocol is described to generate stable cell lines for constitutive expression of single cellular or viral miRNA precursors in absence of infection. The procedure comprises cloning of the precursor sequence, generation of the lentiviral expression vector, transduction of the cells of interest, selection for polyclonal cell lines, and isolation of monoclonal cell lines by limiting dilution.

Key words miRNA, Host miRNA, Viral miRNA, miRNA precursor, miRNA expression, Constitutive miRNA expression, Stable cell line, Gateway cloning, Lentiviral transduction, Limiting dilution

1 Introduction

microRNAs (miRNAs) are short nucleic acid molecules of approximately 22 nucleotides in length which are excised from hairpin-shaped precursors in a multistep process (Fig. 1) [1, 2]. Based on sequence complementarity, mature miRNAs repress gene expression of specific targets mainly on the posttranscriptional level. As interaction with a specific target mRNA requires only partial sequence complementarity, a single miRNA might regulate up to several hundred different genes [3, 4]. Initially, RNA-induced gene silencing has been described in *C. elegans*, plants and fungi [5–11]. To date, small regulatory RNAs such as miRNAs and related small-interfering RNAs (siRNAs) have been shown to be encoded by

Susanne M. Bailer and Diana Lieber (eds.), *Virus-Host Interactions: Methods and Protocols*, Methods in Molecular Biology, vol. 1064, DOI 10.1007/978-1-62703-601-6_13, © Springer Science+Business Media, LLC 2013

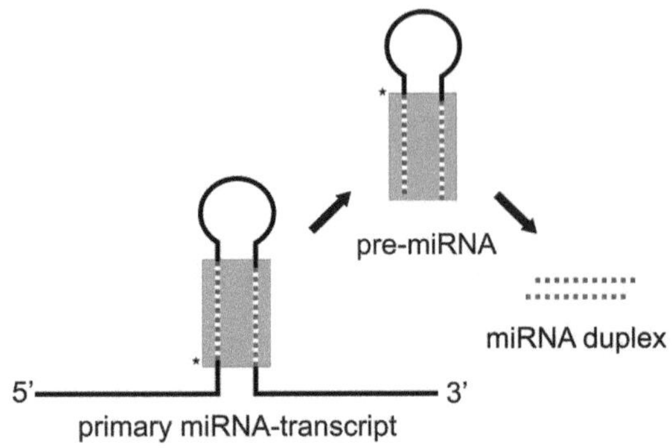

Fig. 1 miRNA precursor structure and processing. Mature miRNAs are excised from primary single stranded transcripts forming characteristic hairpin structures of approximately 70–80 nucleotides in length and with a terminal loop. Sequential cleavage of the hairpin results in an intermediate product, the miRNA precursor or pre-miRNA, followed by the mature miRNA duplex. Note that both strands of the duplex can function as miRNA but in most cases one strand is more active than the other due to more efficient incorporation into the RNA-induced silencing complex (RISC). Cleavage sites are indicated by *asterisks*. *Broken lines* highlight mature miRNA sequences. *Grey box* shows the position of the mature miRNA sequences in the hairpin structure

diverse eukaryotes. Currently known miRNAs are listed in Web-based repositories such as miRBase [12]. Amongst other functions, a role in defense against pathogens has been associated with host-encoded miRNAs. The discovery of miRNA genes in members of the herpesvirus, polyomavirus, and adenovirus families revealed a novel mode of viral influence on the host cell [13, 14]. A range of putative host target genes of viral miRNAs has since been predicted. However, only few targets have actually been validated. An overview of currently validated cellular targets of miRNAs encoded by Kaposi Sarcoma-associated herpesvirus (KSHV) is provided in a recently published review [15]. To a large extent, the function of viral miRNAs in infection remains elusive.

Systematic or individual analysis of miRNA function in the context of viral infection can be done by transient transfection of miRNA mimics and inhibitors similar to the RNAi screen described by S. Griffiths (*see* Chapter 5 of this series). Transfection of a miRNA mimic leads to increased abundance and enhanced function of a host miRNA within the cell. In contrast, an inhibitor oligonucleotide with complementarity to the endogenously expressed miRNA interferes with and relieves target regulation. Although convenient, this strategy has certain disadvantages. In long-lasting experiments, abundance of transfected mimics and inhibitors obviously decreases with ongoing cell division due to dilution effects. But maintenance of an antiviral or proviral state might require long-term constitutive delivery, depending on expression kinetics and the molecular function of target genes. Moreover, several miRNAs can share the same target and target abundance and function are often affected only mildly by single miRNAs [16]. Thus, disruption of a miRNA mostly does not cause clear phenotypic alterations. Functional characterization of virally encoded miRNAs is additionally hampered by the complexity of virus–host interactions resulting from the presence of viral proteins and other viral miRNAs. Thus, a cell line constitutively over-expressing an individual miRNA precursor at high abundance is a valuable tool. This approach allows studying the impact of isolated heterologous miRNAs on the host cell in absence of infection. Equally well, the function of host miRNAs can be analyzed in the context of viral infection, even if the virus represses the endogenous miRNA gene. Priming of cells with high levels of a distinct miRNA prior infection may identify otherwise masked antiviral or proviral functions.

A protocol is described in this chapter for the generation of a stable cell line for constitutive expression of a selected miRNA precursor. Stable genome integration of the expression cassette is achieved by using a lentiviral vector. The method includes cloning of the desired miRNA precursor sequence, generation of the lentiviral vector, transduction of the target cells and selection of polyclonal and monoclonal cell lines.

2 Materials

2.1 Cloning

1. Template DNA for PCR (*see* **Note 1**).
2. PCR reagents: 40 mM dNTP mix (10 mM of each dATP, dTTP, dCTP, and dGTP), 5 U/μL GoTaq-Polymerase and 10× GoTaq reaction buffer (Promega), nuclease-free water.
3. 10 μM miRNA precursor-specific primers: pre-miRNA-forward primer 5′-AAAAAGCAGGCT-template-specific sequence-3′, pre-miRNA-reverse primer 5′-AGAAAGCTGGGT-template-specific sequence-3′ (*see* **Note 2**).
4. 10 μM attB-adapter primers: attB1-adapter 5′-GGGGACAAGTTTGTACAAAAAAGCAGGCT-3′, attB2-adapter 5′-GGGGACCACTTTGTACAAGAAAGCTGGGT-3′.
5. Agarose, electrophoresis grade.
6. 10 mg/mL ethidium bromide.
7. 1× TAE-buffer: 4.84 g Tris, 1.14 mL glacial acetic acid (17.4 M), 2 mL EDTA-solution (0.5 M, pH 8), make up to 1 L with distilled water.
8. DNA marker: e.g., 100 bp DNA ladder, N3231, New England Biolabs.
9. 6× DNA-loading buffer (Fermentas).
10. Gel electrophoresis chamber and power supply.

11. Gel documentation system with UV-light.
12. PCR product and gel band extraction kit (e.g., illustra GFX PCR DNA and Gel Band Purification Kit, GE Healthcare).
13. 75 ng/μL pDONR207 vector (Life Technologies) (*see* **Note 3**).
14. 75 ng/μL pLENTI6/V5-DEST destination vector (Life Technologies).
15. Gateway® BP Clonase™ II Enzyme Mix (Life Technologies).
16. Gateway® LR Clonase™ II Enzyme Mix (Life Technologies).
17. TE-Buffer: 10 mM Tris–HCl pH 8.0, 1 mM EDTA.
18. *E. coli* DH5alpha or similar: heat shock transformation-competent bacteria.
19. LB-medium: 10 g NaCl, 5 g yeast extract, 10 g Tryptone, make up to 1 L with distilled water, autoclave.
20. LB-agar: 10 g NaCl, 5 g yeast extract, 10 g Tryptone, 15 g/L agar, make up 1 L with distilled water, autoclave.
21. Gentamicin, working concentration 15 μg/mL.
22. Ampicillin, working concentration 50 μg/mL.
23. Plasmid purification kit (e.g., Plasmid Mini Kit, Qiagen).
24. Restriction enzymes: BanII, EcoRV (New England Biolabs).
25. Incubator for bacteria, 37 °C.
26. Shaker for bacterial cultures, 37 °C.
27. Glycerol: 75 % in distilled water, autoclaved.
28. Cryovials.
29. Photometer.

2.2 Lentivirus Production

1. ViraPower™ Lentiviral Packaging Mix, contains three packaging plasmids pLP1, pLP2, and pLP/VSVG (Life Technologies).
2. pLENTI control vector with GFP reporter gene expression cassette (e.g., Life Technologies).
3. 293FT cell line (Life Technologies).
4. OptiMEM I reduced serum medium (Life Technologies).
5. FuGene 6 transfection reagent (Roche Applied Science).
6. 293FT culture medium: Dulbecco's Modified Eagle Medium (DMEM), 10 % fetal bovine serum, 6 mM L-glutamine, 0.1 mM MEM nonessential amino acids, 500 μg/mL Geneticin (G418).
7. Lentivirus producer medium: Dulbecco's Modified Eagle Medium (DMEM), 10 % fetal bovine serum, 6 mM L-glutamine, 0.1 mM MEM nonessential amino acids.
8. Trypsin.
9. Incubator for cell culture, 37 °C, humidified, 5 % CO_2.

10. Sterile cell culture dishes, 10 cm in diameter.
11. Sterile cell culture supplies, plastic ware for single use.
12. Syringe and appropriate syringe filters with 0.45 μm pore size.
13. Cryovials.

2.3 Determination of Antibiotic Sensitivity of Target Cells

1. Sterile cell culture supplies, plastic ware for single use.
2. Target cell line for transduction.
3. Cell culture medium according to the target cell line.
4. Trypsin.
5. T75 cell culture flask.
6. 6-well cell culture plates.
7. Blasticidin S HCl: stock at 10 mg/mL in sterile water, filter sterilize, store aliquots at −20 °C.
8. Incubator for cell culture, 37 °C, humidified, 5 % CO_2.

2.4 Transduction

1. Sterile cell culture supplies, plastic ware for single use.
2. Target cell line for transduction.
3. Cell culture medium according to the target cell line.
4. Trypsin.
5. T75 cell culture flask.
6. 12-well cell culture plates.
7. Blasticidin S HCl: stock at 10 mg/mL in sterile water, filter sterilized, store in aliquots at −20 °C.
8. Incubator for cell culture, 37 °C, humidified, 5 % CO_2.
9. Fluorescence microscope for observing GFP-expression.

2.5 Limiting Dilution

1. Cell line of interest.
2. Cell culture medium, suitable for the cell line of interest.
3. Sterile cell culture supplies, plastic ware for single use.
4. 96-well cell culture plates.
5. Neubauer Chamber for cell counting or similar device.
6. Cell culture medium according to the cell line.
7. Trypsin.
8. Incubator for cell culture, 37 °C, humidified, 5 % CO_2.

3 Methods

3.1 Cloning of a miRNA Precursor into the Lentiviral Vector

This procedure describes the cloning of a single miRNA precursor from cellular DNA by PCR-amplification and site-specific recombination. Cloning of a viral miRNA precursor follows the same protocol but obviously might require a different template (*see* **Note 1**).

Here, the Gateway® recombinational cloning strategy (Life Technologies) is used to insert the PCR product into a cloning vector (donor vector) first and then to transfer the sequence-verified insert into the lentiviral destination vector. Briefly, short attB recombination sites (~25 nucleotides) are added with adapter primers to both flanks of the sequence of interest. The recombination enzyme mixture BP Clonase™ recombines attB sites with attP sites in the vector, thereby inserting the sequence of interest into the cloning vector. At the same time, attL recombination sites are generated (~100 nucleotides) which are in a second step recognized and recombined by LR Clonase™ to transfer the insert into the final destination vector. As detailed by Bosse et al. (*see* Chapter 11 of this series), the cloning vector can be used as resource for subcloning of the miRNA-precursor into a multitude of other Gateway®-compatible destination vectors such as transient expression constructs.

In this section, a miRNA precursor is amplified by PCR and Gateway® recombination sequences are added with the primers. As attB-primers spanning the whole recombination site are quite large, a two-step Adapter-PCR is performed with the first primer pair tagging the miRNA-specific sequence with partial recombination sites and the second primer pair completing the recombination sites.

1. Prepare PCR reaction mix on ice:

   33.75 μL nuclease-free water.

   10.0 μL 10× GoTaq Reaction Buffer.

   1.0 μL nucleotide mix.

   2.0 μL pre-miRNA-forward primer.

   2.0 μL pre-miRNA-reverse primer.

   1.0 μL template at <10 ng/μL end concentration.

   0.25 μL GoTaq DNA Polymerase.
2. Perform a "Touchdown" PCR according to the indicated cycling conditions (Table 1) (*see* **Note 4**).
3. Analyze 10 % of the PCR product on a 1 % agarose gel with ethidium bromide at a final concentration of 0.25 μg/mL in 1× TAE-buffer (*see* **Note 5**). Use a 100 bp DNA ladder to control for the correct product size.
4. Inspect the gel in UV light. If the product size is as expected and only one band is visible, purify the remaining PCR product with a spin column based purification kit (*see* **Note 6**). Elute in 30–50 μL nuclease-free water.
5. Dilute the PCR product to obtain a template for the second PCR. A dilution factor between 100 and 1,000 worked well but this might depend on the template and PCR conditions.
6. Repeat **steps 1** through **4**. Use attB1- and attB2-adapter primers in the second PCR. Measure concentration of the PCR

Table 1
Thermocycler protocol for "Touchdown" PCR (*see* Note 4)

Step	Temperature	Duration	Number of cycles
Initial denaturation	95 °C	2 min	1
Denaturation	95 °C	30 s	10
Primer annealing	65 °C, reduce by 1 °C per cycle	30 s	
Extension	72 °C	30 s	
Denaturation	95 °C	30 s	20–25
Primer annealing	56 °C	30 s	
Extension	72 °C	30 s	
Final extension	72 °C	5 min	
Cooling	4 °C	Indefinite	1

product with a photometer and adjust DNA concentration to 75 ng/μL.

7. In order to clone the PCR product into the donor plasmid, set up a BP recombination reaction:

   1.0 μL PCR product (75 ng/μL).

   1.0 μL pDONR207 vector (75 ng/μL).

   2.0 μL TE-Buffer, pH 8.0.

   1.0 μL BP Clonase™ II Enzyme Mix.

   Mix well by pipetting and incubate for 1 h at 25 °C (*see* **Note 7**).

8. Terminate the reaction by adding 1 μL Proteinase K solution, mix and incubate at 37 °C for 10 min. Proteinase K is provided with the BP Clonase™.
9. Transform 2–5 μL of the reaction into heat-transformation competent *E. coli* DH5alpha. Keep cells on ice. Mix carefully with the DNA and leave on ice for 10 min. Transfer to 42 °C for 2 min and place immediately back on ice for 5 min.
10. Add ice-cold LB-medium without antibiotic and incubate at 37 °C for 30 min.
11. Streak out on LB-agar containing the appropriate antibiotic (e.g., gentamicin for pDONR207) and allow colonies to form overnight at 37 °C.
12. Inoculate four clones in 4–5 mL LB-medium with gentamicin and allow to grow overnight in a shaker at 37 °C.
13. Secure the clones for long-term storage by preparing glycerol stocks in cryovials (e.g., mix 1 part bacterial culture with 2 parts 75 % glycerol giving a final glycerol concentration of 50 %). Store at −80 °C.

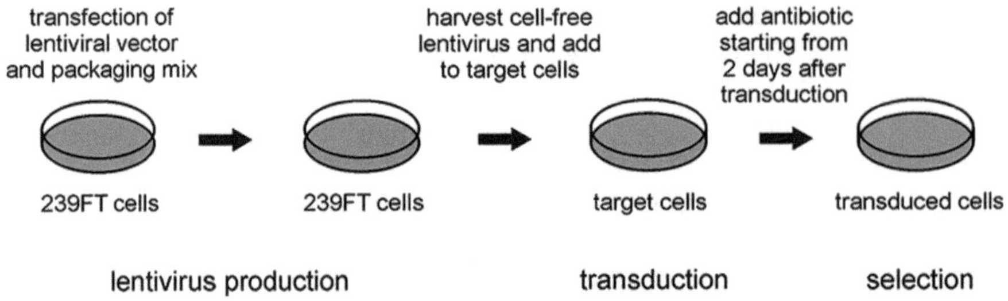

Fig. 2 Lentivirus production, transduction, and selection of transduced cells. Lentivirus is produced by transfection of the lentiviral vector and plasmids encoding structural proteins (packaging mix) into 293FT cells. Infectious supernatant is harvested followed by removal of cellular debris. Target cells are then inoculated and transduced with cell-free virus. After onset of expression of marker and resistance genes, an antibiotic selection regimen is initiated to kill non-transduced cells

14. Isolate plasmid DNA from 3 mL culture using a spin column plasmid purification kit (e.g., Plasmid Mini Kit, Qiagen), elute the plasmid in 30 μL 10 mM Tris–HCl pH 8.
15. Perform a restriction digest with plasmid DNA and a suitable restriction enzyme, e.g., BanII, according to the manufacturer's instructions and analyze digested DNA on an analytical agarose gel (*see* **Note 8**).
16. Choose two clones with correct restriction pattern for verification by sequencing (*see* **Note 9**).
17. Choose one correct clone for generating the lentiviral expression vector. Adjust DNA concentration to 75 ng/μL.
18. Set up an LR recombination reaction:

    1.0 μL sequence-verified vector (75 ng/μL).

    1.0 μL pLENTI6/V5-DEST destination vector (75 ng/μL).

    2.0 μL TE-Buffer, pH 8.0.

    1.0 μL LR Clonase™ II Enzyme Mix.

    Mix well by pipetting and incubate for 1 h at 25 °C (*see* **Note 7**).
19. Terminate the reaction by adding 1 μL Proteinase K solution, mix and incubate at 37 °C for 10 min.
20. Transform into *E. coli* DH5alpha, isolate and verify resulting clones essentially as described in **steps 9** through **15** except that selection must be performed with ampicillin and restriction digest carried out with EcoRV (*see* **Note 10**).
21. Choose one correct clone, from now on referred to as pLENTI expression clone, for packaging of the lentivirus.

3.2 Production of Lentivirus

This section describes the packaging of the lentiviral expression vector into infectious viral particles which will be used to transduce the target cells (Fig. 2). Lentivirus produced in this section can be stored at −80 °C or used immediately for transduction of target cells. If transduction is intended to be carried out with freshly produced virus, sensitivity to Blasticidin has to be tested first and target cells should be seeded 2 days after transfection of the producer 293FT cells. *See* Subheadings 3.3 and 3.4 for procedures.

1. Culture 293FT cells in complete culture medium with geneticin.
2. Detach cells with Trypsin, add medium and seed 1.5×10^6 293FT cells in one 10 cm cell culture dish for each pLENTI vector (*see* **Notes 11** and **12**). Important: Use medium without antibiotics for transfection. Allow cells to settle overnight.
3. Inspect cells to confirm viability.
4. Dilute transfection reagent. Pipet 200 μL Opti-MEM I and 16 μL FuGene 6 into a sterile tube (*see* **Note 13**).
5. Mix by vortexing for 1 s. Incubate for 5 min at room temperature.
6. Prepare plasmid mixture containing packaging mix and pLENTI expression plasmid:

   3 μg packaging mix, containing 1 μg of each pLP1, pLP2 and pLP/VSVG

   1 μg pLENTI expression plasmid.
7. Add the plasmid mixture to the diluted transfection reagent and mix by vortexing for 1 s. Allow transfection complexes to form for at least 15 and up to 45 min.
8. While complexes form, replace 293FT culture medium by 15 mL Opti-MEM I reduced serum medium in the 10 cm dish.
9. Pipet transfection complexes to the cells in a drop-wise manner and swirl the dish gently to distribute evenly.
10. Place the dish back into the incubator and allow transfection to occur for 24 h.
11. Replace OptiMEM I with 15 mL producer medium (without antibiotic since the target cells in transduction might be sensitive to the antibiotic!) and allow cells to release packaged lentivirus into the medium (*see* **Notes 14** and **15**).
12. Harvest supernatant 72 h after transfection. Remove cell debris by centrifugation at $3{,}000 \times g$ for 15 min.
13. Filtrate cell-free supernatant with a syringe and filter with 0.45 μm pore size.
14. Prepare 1 mL aliquots and store at −80 °C or use immediately to transduce target cells (*see* **Notes 14** and **16**).

3.3 Determination of Antibiotic Sensitivity of Target Cells

Different cell lines may differ in sensitivity to the antibiotic used for selection of transduced cells. In order to allow for efficient selection while avoiding preferential selection of cells with multiple insertions, the optimal antibiotic concentration needs to be determined for each cell line before transduction. A kill curve experiment serves this purpose (*see* **Note 17**). In short, a range of antibiotic concentrations

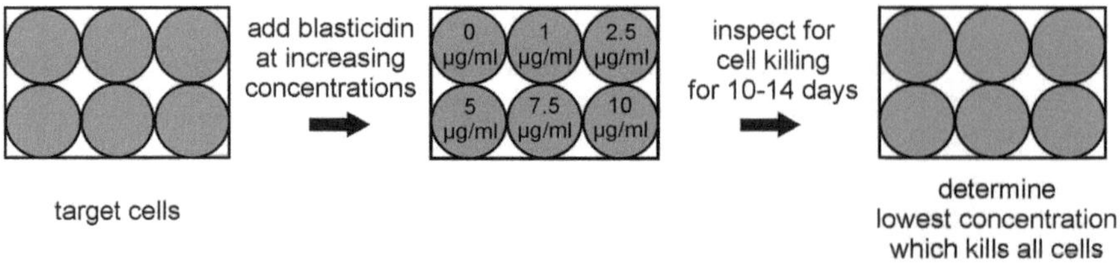

Fig. 3 Antibiotic sensitivity testing of target cells. In order to determine the optimal antibiotic concentration for selection of transduced cells, a kill curve experiment is carried out. Cells of interest are subjected to a range of antibiotic concentrations, in this case Blasticidin, and then monitored for the lowest concentration that just kills all cells within 10–14 days

is tested to identify the lowest concentration which kills the non-transduced cells of interest within a given time (Fig. 3).

1. Culture target cells with complete culture medium in a T75 cell culture flask (*see* **Note 18**).
2. Detach cells with Trypsin when they are approximately 75 % confluent and add medium without antibiotic to a total volume of 22.5 mL.
3. Pipet 1 mL of the cell suspension in each well of a 6-well plate (*see* **Note 19**). Adjust volume to 2.5 mL with medium. This should result in about 25 % confluence.
4. Allow cells to settle and adhere overnight.
5. Replace medium with fresh medium containing Blasticidin at different concentrations: 0, 1, 2.5, 5, 7.5 and 10 μg/mL (*see* **Note 20**).
6. From now on, inspect the cells regularly in order to observe antibiotic-induced cell killing.
7. Replace the medium with fresh medium containing Blasticidin twice per week, meaning every 3–4 days.
8. Determine the lowest concentration that kills all cells in the well within 10–14 days. This is the Blasticidin concentration you will use later in Subheading 3.4 for selection of transduced cells.

3.4 Transduction of the Cell Line of Interest

This protocol describes the stable transduction of the target cell line with the lentivirus produced in Subheading 3.2 (Fig. 2).

1. Culture target cells with complete culture medium in a T75 cell culture flask (*see* **Note 18**).
2. Detach cells with Trypsin when they are approximately 75 % confluent and add medium to a total volume of 14 mL.
3. Pipet 0.5 mL of the cell suspension into one well of a 12-well plate for each lentivirus. Prepare one well as control for antibiotic-induced cell killing in absence of transduction. Adjust volume to 2 mL with medium. This should result in about 50 % confluence.

4. Allow cells to settle and adhere overnight.
5. Inspect cells to confirm viability.
6. Thaw cell-free lentivirus stock from Subheading 3.2, if stored at −80 °C.
7. Add 1 mL of cell-free lentivirus to one well with cells and mix carefully by pipetting. Add 1 mL of medium to one well with cells for the mock control. Total volume per well is now 3 mL. Incubate overnight.
8. One day after addition of virus, replace medium in all wells with 3 mL fresh medium in order to remove residual lentivirus in the supernatant.
9. If appropriate, confirm expression of the GFP reporter gene in cells transduced with the control virus by fluorescence microscopy (*see* **Note 21**).
10. Two days after transduction, replace medium in all wells including mock-transduced cells with 3 mL fresh medium containing the appropriate concentration of Blasticidin as determined in Subheading 3.3 (*see* **Note 22**).
11. Replace medium with fresh Blasticidin-containing medium every 3–4 days for 10–12 days until antibiotic-resistant colonies can be distinguished (*see* **Note 23**).
12. Detach resistant colonies with Trypsin, add medium and allow the cells to spread in order to avoid cells overgrowing each other (*see* **Note 24**).
13. Expand cells as required and determine abundance of the respective miRNA (*see* **Notes 25** and **26**).

3.5 Isolation of Individual Clones by Limiting Dilution

Insertion of the lentiviral vector is largely random. Polyclonal cell lines do have advantages over monoclonal cell lines since positional effects from the site of integration are leveled out. In some applications, however, a monoclonal cell line could be favorable where particularly high expression levels are aspired, for instance. The following protocol provides instructions for generating monoclonal cell lines starting from a polyclonal situation.

1. Culture cells with complete culture medium.
2. Detach cells with Trypsin, add culture medium and count cells in a Neubauer Chamber.
3. Adjust cell number to 10^4 cells per mL in 20 mL total volume. Mix well.
4. Dilute 5, 30 and 100 μL of the cell suspension in 10 mL medium each (*see* **Note 27**) (Fig. 4).
5. Pipet 100 μL of the diluted cell suspensions in each well of a 96-well plate (*see* **Note 28**). Use one full plate for each dilution (*see* **Note 29**).

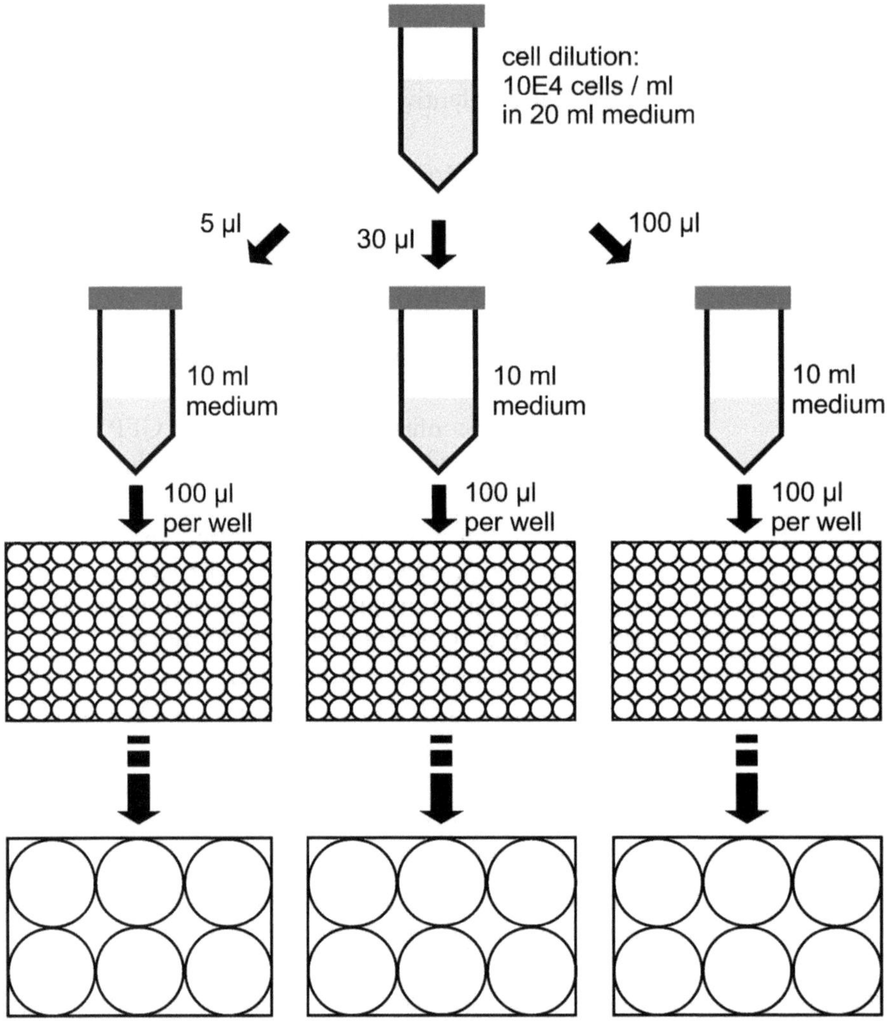

Fig. 4 Limiting dilution cloning. The pipetting scheme for singularization of cells from a given population is shown. The protocol starts with a cell density of 10^4 cells/mL from which dilutions are prepared that yield densities of 5, 30, and 100 cells/mL. Thus, on the three 96-well plates, 0.5, 3, or 10 cells per well are theoretically expected, if evenly distributed. Wells where single cells/colonies grow up are selected and the respective clones are successively expanded

6. Allow the cells to sediment for 2 h at 37 °C in the incubator.
7. Inspect the plates and mark wells that have received only a single cell (*see* **Note 30**).
8. Allow cells to grow for 5–7 days. Do not change the medium in this time.
9. Inspect plates and mark wells where only a single colony has grown up.
10. If required, detach selected colonies when cells are overgrowing each other and allow to reattach until the well is 80–90 % confluent (*see* **Note 31**).

11. Expand the colonies by successively transferring the cells into 48-, 24-, 12- and then 6-well plates and then into 10 cm cell culture dishes or small flasks.
12. Assay clones for optimal miRNA expression (*see* **Notes 25** and **32**).

4 Notes

1. Cellular DNA to be used as PCR template can be purified from cultured cells employing the FlexiGene DNA Kit from Qiagen according to the manufacturer's instructions. Any other DNA purification method that delivers DNA in a quality suitable for PCR can be applied. Infected cells can serve as source for viral templates from DNA viruses such as herpesviruses.
2. miRNA precursor-specific primers are designed based on the natural miRNA-encoding transcript sequence. In principle, a cluster of miRNA precursors can be cloned equally well as a single precursor. Be aware that a precursor contains two miRNA strands which can both possess regulatory activity. The sequence of a miRNA precursor can be obtained e.g., from the miRBase repository (www.miRBase.org) [12]. Precursors have a length of approximately 70–80 nucleotides. In order to maintain the physiological excision pattern of the miRNA precursor and mature miRNA in the cell and to ensure efficient precursor processing, the cloned sequence should be flanked by a single-stranded stretch of the natural transcript sequence at either end of the precursor. The flanking sequences should be comprised of at least 40 nucleotides on either side of the precursor [17, 18]. 80–100 nucleotides on either side work well. This sequence information is available from Genbank (http://www.ncbi.nlm.nih.gov/genbank). Special care has to be taken to select the correct miRNA-encoding strand. For generating the miRNA-specific primers, miRNA-specific sequence elements are fused to partial attB1 and attB2 sequences at their 3′ ends. attB-sites are the recombination sites used by the Gateway® BP Clonase™ enzyme mixture. Primers should ideally have a GC content of 40–60 % and one or two G or C at the 3′ end. Check sequences of primer pairs for absence of extensive dimer and hairpin formation and ensure low differences (max. 3 °C) in melting temperatures (approximately 70 °C). This can be done with any primer design and analysis tool such as e.g., the free primer analysis software NetPrimer (http://www.premierbiosoft.com/netprimer).
3. There are different pDONR vectors available from Life Technologies. In this protocol pDONR207 is used. An alternative is pDONR221 which is supposed to yield higher copy numbers. When choosing a pDONR vector pay attention to the respective antibiotic resistance gene encoded as well as the

required sequencing primers. For instance, pDONR207 confers gentamicin resistance, while pDONR221 carries a kanamycin resistance gene and universal M13 primer sites.

4. These cycling parameters might need to be optimized for your equipment, reagents and primers.
5. Maximum voltage and duration of the gel run are dependent on your electrophoresis system. Adhere to the operating manual of your system.
6. If there is more than one band per lane, separate the products in the remaining reaction volume by gel electrophoresis, cut out the relevant band and purify, e.g., with the gel extraction protocol of the above mentioned kit.
7. The number of resulting colonies can be increased by over night incubation at 8 °C. The DNA amount of the cloning vector should not exceed 75 ng in a 5 μL LR Clonase™ reaction (15 ng/μL) as this can lead to colonies with multiple DNA molecules, often with a so-called "small colony"-phenotype.
8. BanII is suitable for analyzing pDONR207-derived plasmids. The parental plasmid yields two fragments: 2,966 base pairs (bp) backbone sequence and 2,619 bp encompassing the Gateway® cassette. After the BP reaction, the backbone fragment remains unaltered while the second fragment composed of the PCR-amplified sequence, recombination sites and a short stretch of vector sequence will be significantly reduced in length. Therefore, the actual expected size of the non-backbone fragment is the amplified template sequence (240–300 bp) plus approximately 400 bp.
9. Inserts in pDONR207 can be accessed by universal sequencing primers such as forward primer DONR-F 5′-TCGCGTTA ACGCTAGCATG-3′ and reverse primer SeqL-E 5′GTTGA ATATGGCTCATAACAC-3′, e.g., from AGOWA GmbH, Berlin, Germany.
10. Restriction digest of the parental pLENTI6/V5-DEST vector with EcoRV yields two fragments: 6,975 bp backbone and 1,713 bp encompassing the Gateway® cassette. Upon LR recombination, the backbone remains unaltered and the second fragment decreases in size. The expected size of the non-backbone fragment is the amplified template sequence plus approximately 50 bp which are contributed by the attB1 and attB2 recombination sites.
11. Include a control if required such as a pLENTI expression vector encoding a green fluorescent protein (GFP) reporter under control of the CMV immediate early promoter for constitutive expression. This control facilitates estimation of the transduction efficiency in the target cells and monitoring of the selection progress. A control vector can be prepared by cloning a suitable marker gene with Kozak sequence into the pLENTI6/

V5-DEST vector similar to the cloning procedure of miRNA precursors. Control vectors are also commercially available, e.g., from Life Technologies. An ideal control comprises the same resistance gene and identical promoters for expression of the sequence of interest and the resistance gene as compared to the pLENTI expression clone.

12. The following protocol indicates reagent amounts used for one pLENTI vector or one 10 cm cell culture dish, respectively. Adjust depending on the number of constructs you want to package.
13. Undiluted FuGene 6 must not get into contact with the plastic tube wall. Thus, pipet OptiMEM first and then add FuGene 6 into the center of the tube without touching the tube wall.
14. Be aware that the lentiviruses produced are classified as biohazards and subject to biological safety regulations. The viruses are replication-incompetent. However, they have a broad mammalian host range and therefore can primarily transduce human cells. Strictly adhere to Biosafety Level 2 (BL2) regulations when handling the virus.
15. Inspect cells to observe the appearance of large syncytia due to expression of the glycoprotein VSV-G. This is expected and indicates successful transfection.
16. The ViraPower™ manual recommends to titer the lentivirus stock before proceeding to transduction. This is useful if the number of integrated copies in the target cells or reproducible expression levels in independent experiments need to be controlled. As the aim of this protocol is to generate polyclonal cell lines and the cell lines to be compared are produced in parallel, this step is omitted. However, if required for your purpose, a protocol for titering the lentivirus stock can be obtained for free from the manufacturer's Web site (included in the User Manual no. 25-0501 on ViraPower™ Lentiviral Expression Systems www.invitrogen.com).
17. The antibiotic used for selection naturally depends on the lentiviral vectors used. This protocol employs pLENTI6 vectors with a Blasticidin resistance gene. However, vectors conferring Zeocin resistance are also in use. Choose the correct antibiotic for your constructs. The kill curve protocol described here also works for alternative antibiotics although the concentration range to be tested may differ, e.g., 50–1,000 μg/mL Zeocin is recommended.
18. The third-generation lentiviral system exemplarily used in this protocol relies on a pseudotype virus. This virus exhibits a broadened host range due to glycoprotein G of Vesicular Stomatitis Virus (VSV-G). Thus, any mammalian cell line you wish to use in subsequent experiments should be susceptible to

transduction. Also suspension cells such as human B-cell lines are successfully transduced [19].

19. You will need one well per Blasticidin concentration and one well for the mock control. This makes six wells in total in this protocol. If you want to test more concentrations or several antibiotics adjust the number of wells.
20. Blasticidin-containing medium may be stored at 4 °C for up to 2 weeks. You may wish to prepare larger volumes of medium with different antibiotic concentrations initially rather than sacrifice a fresh aliquot of Blasticidin with each medium change.
21. If a GFP-encoding control lentivirus is included in the experiment, reporter expression will be detectable approximately 2 days after transduction. Keep in mind that not all cells will be GFP-positive at this stage.
22. The original manual of the ViraPower™ System asks to go straight to the full antibiotic concentration. However, if the lowest concentration that just killed all cells within 10–14 days immediately causes extensive cell death, a step-wise increase of the concentration every 2–3 days can be considered, starting from 1 μg/mL and going up in two steps to the final concentration. Keep in mind that this can theoretically lead to development of resistance in non-transduced cells.
23. In the meantime, non-transduced cells are dying and floating in the medium. These are repeatedly removed when the medium is changed.
24. This protocol favors the generation of polyclonal cell lines. Therefore, colonies are not harvested individually. If monoclonal cell lines are preferred, a protocol for harvesting single colonies is given by Bosse et al. (*see* Chapter 11 of this series). In case it appears only later that monoclonal cell lines are required, a limiting dilution protocol can be employed to grow up cell lines from individual cells (See Subheading 3.5).
25. miRNA over-expression expressed as relative miRNA abundance can be quantified by qRT-PCR. This method is routinely used by miRNA researchers and exceeds the scope of this chapter [20]. In short, total RNA is isolated from the cells of interest as well as from control cells. For cellular miRNAs, either cells transduced with the GFP control lentivirus or non-transduced cells are used. For viral miRNA-expressing cell lines, virus-infected cells in the appropriate infection phase are needed as control. The isolated RNA is subjected to a DNase I treatment to remove residual genomic DNA. The RNA is *in vitro* polyadenylated, purified by a phenol-chloroform procedure, and then subjected to a first-strand cDNA synthesis with an oligo d(T) primer. Individual miRNAs are analyzed by quantitative Real Time-PCR with a miRNA strand-specific primer and oligo d(T).

26. Estimation of the efficiency of the selection procedure can be obtained by analyzing the rate of GFP-positive cells in the control transduction sample, e.g., by flow cytometry.
27. Self-conditioned medium can significantly increase cloning efficiency. Culture the parental cell line used in the transduction experiment in a T75 cell culture flask at about 75 % confluence. Exchange medium to fresh culture medium and place the cells back in the incubator. After 24 h, harvest the conditioned medium and pass through a 0.45 μm filter to avoid accidental carryover of cells. Mix conditioned medium with fresh medium. Undiluted self-conditioned medium worked well to support the growth of single MeWo cells within the first week.
28. Swirl cell dilutions repeatedly to avoid sedimentation and unequal distribution of the cells.
29. This will result in 0.5, 3 and 10 cells per well on average in the three plates, respectively.
30. This is tedious, especially when several polyclonal cell lines are processed in parallel. Selecting and marking wells with single cells can be omitted at this stage but equal distribution of the cells over the plates should be controlled immediately.
31. Harvest and analyze at least four to five clones from one polyclonal cell line. If more clones are selected and expanded, the supernumerary ones can be harvested during the expansion phase and stored at −80 °C as backup.
32. Ideally, at least two cycles of the cloning procedure should be carried out in order to ensure clonality. However, this might not be necessary for all applications.

Remark

The Gateway® recombinational cloning technology and the ViraPower™ Lentiviral Expression System are the intellectual property of Life Technologies, formerly Invitrogen. General information as well as manuals are available from www.lifetechnologies.com.

References

1. Bartel DP (2004) MicroRNAs: genomics, biogenesis, mechanism, and function. Cell 116:281–297
2. He L, Hannon GJ (2004) MicroRNAs: small RNAs with a big role in gene regulation. Nat Rev Genet 5:522–531
3. Lewis BP, Shih IH, Jones-Rhoades MW et al (2003) Prediction of mammalian microRNA targets. Cell 115:787–798
4. Lim LP, Lau NC, Garrett-Engele P et al (2005) Microarray analysis shows that some microRNAs downregulate large numbers of target mRNAs. Nature 433:769–773
5. Napoli C, Lemieux C, Jorgensen R (1990) Introduction of a chimeric chalcone synthase gene into petunia results in reversible co-suppression of homologous genes in trans. Plant Cell 2:279–289

6. Van Der Krol AR, Mur LA, Beld M et al (1990) Flavonoid genes in petunia: addition of a limited number of gene copies may lead to a suppression of gene expression. Plant Cell 2:291–299
7. Lindbo JA, Dougherty WG (1992) Untranslatable transcripts of the tobacco etch virus coat protein gene sequence can interfere with tobacco etch virus replication in transgenic plants and protoplasts. Virology 189:725–733
8. Lee RC, Feinbaum RL, Ambros V (1993) The C. elegans heterochronic gene lin-4 encodes small RNAs with antisense complementarity to lin-14. Cell 75:843–854
9. Cogoni C, Irelan JT, Schumacher M et al (1996) Transgene silencing of the al-1 gene in vegetative cells of Neurospora is mediated by a cytoplasmic effector and does not depend on DNA-DNA interactions or DNA methylation. EMBO J 15:3153–3163
10. Fire A, Xu S, Montgomery MK et al (1998) Potent and specific genetic interference by double-stranded RNA in Caenorhabditis elegans. Nature 391:806–811
11. Hamilton AJ, Baulcombe DC (1999) A species of small antisense RNA in posttranscriptional gene silencing in plants. Science 286:950–952
12. Griffiths-Jones S (2004) The microRNA registry. Nucleic Acids Res 32:D109–D111
13. Pfeffer S, Zavolan M, Grasser FA et al (2004) Identification of virus-encoded microRNAs. Science 304:734–736
14. Grundhoff A, Sullivan CS (2011) Virus-encoded microRNAs. Virology 411:325–343
15. Lieber D, Haas J (2011) Viruses and microRNAs: a toolbox for systematic analysis. Wiley Interdiscip Rev RNA 2:787–801
16. Bartel DP (2009) MicroRNAs: target recognition and regulatory functions. Cell 136:215–233
17. Chen CZ, Li L, Lodish HF et al (2004) MicroRNAs modulate hematopoietic lineage differentiation. Science 303:83–86
18. Zeng Y, Cullen BR (2005) Efficient processing of primary microRNA hairpins by Drosha requires flanking nonstructured RNA sequences. J Biol Chem 280:27595–27603
19. Dolken L, Malterer G, Erhard F et al (2010) Systematic analysis of viral and cellular microRNA targets in cells latently infected with human gamma-herpesviruses by RISC immunoprecipitation assay. Cell Host Microbe 7:324–334
20. Shi R, Chiang VL (2005) Facile means for quantifying microRNA expression by real-time PCR. Biotechniques 39:519–525

Chapter 14

Applications for a Dual Fluorescent Human Cytomegalovirus in the Analysis of Viral Entry

Kerstin Laib Sampaio, Gerhard Jahn, and Christian Sinzger

Abstract

The existence of cell type-specific entry pathways of human cytomegalovirus is an unresolved question as the course of viral entry in different cell types is still not fully understood. To gain more insight into these processes, we generated a dual fluorescent HCMV, where the capsid-associated tegument protein pp150 is labelled with EGFP and the envelope glycoprotein gM with mCherry. This dual labelled virus allows for the separate tracking of the viral envelope fusing with a cellular membrane and the viral capsid during its movement from the cellular membrane to the nucleus. We describe two applications for this virus in the analysis of viral entry: (a) Dynamic live-cell imaging allows for the visualization of viral de-envelopment and transport processes within the living cell. (b) Imaging of cell cultures fixed at different time points after infection enables a more comprehensive statistical analysis of the kinetics of viral entry events such as adsorption, fusion, and nuclear translocation. The techniques are described on the example of fibroblasts and endothelial cells, but can be adapted to other cell types as well. Furthermore, these protocols could provide suggestions for the establishment of live cell applications to other viruses.

Key words Dual fluorescent virus, Human cytomegalovirus, HCMV, Live-cell imaging, Entry

1 Introduction

Human cytomegalovirus (HCMV) belongs to the betaherpesvirus subfamily and usually causes an asymptomatic, lifelong persistent infection within the immunocompetent human host. In contrast, infections of immunocompromised patients as well as intrauterine infection of fetuses can result in life-threatening disease. Under such conditions, HCMV can infect virtually any organ due to its broad cell tropism [1]. Entry into fibroblasts and endothelial cells is mediated by different glycoprotein complexes, and antibodies against the respective glycoprotein complexes were shown to inhibit virus infection in a cell-type-specific way [2–7]. Hence, a more detailed knowledge of cell-type-specific entry pathways of HCMV is desirable for the development of novel therapy or vaccine strategies. The application of fluorescent viruses for visualization of

Susanne M. Bailer and Diana Lieber (eds.), *Virus-Host Interactions: Methods and Protocols*, Methods in Molecular Biology, vol. 1064, DOI 10.1007/978-1-62703-601-6_14, © Springer Science+Business Media, LLC 2013

certain steps in HCMV entry offers new possibilities in understanding the dynamics during the viral life cycle and will help to overcome limitations inherent in established approaches.

Here we describe the application of a dual fluorescent HCMV that was designed for simultaneous tracking of viral capsid and viral envelope. The capsid-bound tegument protein pp150 was labelled by fusing its C-terminus with EGFP [8], whereas the viral envelope was labelled by insertion of the red fluorescent protein mCherry into glycoprotein M directly after the first transmembrane region. The abundance of both tagged proteins [9] within the virion facilitates their detection in living cells.

On the example of two different target cells (fibroblasts and endothelial cells), we provide a basic protocol to generate time lapse movies that enable the visualization of attachment of virions, viral de-envelopment events, and transport of complete virions or naked capsids. While time lapse movies are ideal for the detection of fast dynamic processes in individual cells, the collection of statistically significant amounts of data is very laborious due to the need to focus on a small number of cells for each movie.

To enable kinetic analyses of viral entry steps in large numbers of infected cells, we provide a second protocol for preparation of fixed cultures at different time points after infection. The fixation procedure leaves the native fluorescence of both proteins intact and is compatible with the staining of the cytoskeletal marker tubulin, which facilitates subsequent data analysis.

2 Materials

1. HCMV-TB40-BAC$_{KL7}$-UL32EGFP-UL100mCherry (Laib Sampaio et al., unpublished).
2. Human foreskin fibroblasts (HFFs) cultured in minimal essential medium (MEM5) supplemented with 5 % fetal calf serum, 2.4 mmol/L glutamine, 100 μg/mL gentamicin (*see* **Note 1**).
3. Human umbilical vein endothelial cells (HUVECs) cultured in RPMI1640 with 50 μg/mL ECGS (endothelial cell growth supplement), 10 % human serum (HCMV seronegative), 5 IU/mL heparin, and 100 μg/mL gentamicin.
4. 0.1 % gelatine in water, autoclaved.

2.1 Live Cell Imaging

1. 35 mm Petri dishes with a #1.5 coverslip like bottom (μ-dish, Ibidi) (*see* **Note 2**).
2. Imaging medium: MEM without phenol red supplemented with 2 % fetal calf serum, 2.4 mmol/L glutamine, 100 μg/mL gentamicin, and 0.5 mM Trolox (Trolox should be added shortly before usage).

3. 100 mM Trolox stock solution (200×): dissolve 50 mg Trolox in 230 μL methanol, add 1.6 mL ddH_2O and redissolve the precipitate by adding up to 190 μL 1 M NaOH. The stock solution can be stored at −20 °C for up to 3 months; however, we preferred to prepare it freshly the day before imaging (*see* **Note 3**).
4. Inverted fluorescence microscope suitable for live cell imaging (*see* **Note 4**).

2.2 Fixed Images

1. 8-well chamber slides (μ-slide, Ibidi) (*see* **Note 2**).
2. PHEM buffer: 25 mM HEPES, 10 mM EGTA, 60 mM PIPES, 2 mM $MgCl_2$.

   Add in the given order and adjust pH to 6.9 with 10 M KOH. Filter-sterilize and store at 4 °C.
3. 4 % paraformaldehyde (PFA): Dissolve 4 g PFA in 20 mL 0.1 M NaOH by heating it in a 80 °C water bath for 2–3 min. Cool down to room temperature and add 80 mL PHEM buffer (*see* **Note 5**).
4. 100 % methanol, stored at −20 °C.
5. Permeabilization solution: 10 % Sucrose, 1 % FCS in PHEM buffer. Filter-sterilize and store at 4 °C. Before usage remove the desired volume and add Nonidet P40 to a final concentration of 0.5 %.
6. Antibodies: anti-α-Tubulin mouse IgG, Alexa Fluor 350 $F(ab')_2$ goat anti-mouse IgG (Invitrogen).

3 Methods

3.1 Preparation of Virus Suspension

The inoculation of the producer culture should be scheduled 5 days prior to the experiment (*see* **Note 6**).

1. Seed 7×10^5 HFF in a 25 cm^2 culture flask together with infectious virus at a multiplicity of infection (MOI) of about 5.
2. Exchange medium the next day to normal growth medium (MEM5).
3. Repeat medium exchange in the evening of day 4 either with MEM5 for fixed images or with imaging medium for live cell analysis.
4. Harvest the virus preparation on day 5 post infection and remove cell debris by centrifugation at $3{,}220 \times g$ for 10 min. Remove supernatant and place into a new tube.
5. Prepare dilutions if necessary (*see* **Note 7**). Store the supernatant in an incubator at 37 °C and 5 % CO_2 with the lid of the tube slightly opened (to ensure gas exchange) and remove it only shortly before adding it to the cells.

3.2 Analysis of Viral Entry by Live Cell Imaging

1. Coat the dishes by adding 2 mL of 0.1 % gelatine to the 35 mm culture dish and incubate at 37 °C for 20 min. Aspirate the solution thoroughly before seeding the cells (*see* **Note 8**).
2. Seed the cells in the respective growth medium supplemented with 0.5 mM Trolox to the 35 mm dishes (culture volume: 2 mL/well) at a density of 2×10^5 cells/well the day before the experiment.
3. Arrange the microscope: equilibrate the heating chamber with humidity and CO_2, switch on the fluorescent light source.
4. Prepare virus suspension as described, pipet 2 mL into a self-standing tube, and place in the heating chamber of the microscope.
5. Wash cells twice with prewarmed imaging medium, reduce culture volume to 1 mL and place the dish in the heating chamber. Incubate for 30 min.
6. Pour the virus suspension into the dish, focus on an individual cell with the aid of the brightfield channel and immediately start imaging (*see* **Note 9**) (Fig. 1).

3.3 Generation of Fixed-Images at Different Time Points of Infection

1. Coat the 8-well slides by adding 0.3 mL of 0.1 % gelatine per well and incubate at 37 °C for 20 min. Aspirate the solution thoroughly before seeding the cells (*see* **Note 8**).
2. Seed the cells in their respective growth medium into the 8-well slide (culture volume: 0.3 mL/well) at a density of 4×10^4 cells/well 1 day prior to the experiment. Incubate the slide at ambient temperature for 5–10 min to allow for homogenous cell attachment before placing it into the incubator [10].
3. Prepare virus suspension as described.
4. Exchange medium for prewarmed MEM5 and incubate for 30 min at 37 °C.
5. Replace medium with prewarmed virus preparation and incubate for 1 h at 37 °C.
6. Replace virus inoculum with prewarmed MEM5 and incubate further until the desired time point.
7. Fix the cells with 4 % PFA for 30 min at ambient temperature with gentle shaking. Wash three times with PHEM buffer.
8. Remove buffer and gently add precooled 100 % methanol. Incubate the slide for 5 min in a −20 °C freezer, remove and wash three times with PHEM buffer (*see* **Note 10**).
9. Permeabilize with freshly prepared permeabilization solution for 10 min at ambient temperature and wash three times with PHEM buffer.
10. Incubate with primary antibody dilution (anti-tubulin, 1:200 in PHEM buffer) for 1 h at 37 °C. Wash three times with PHEM buffer.

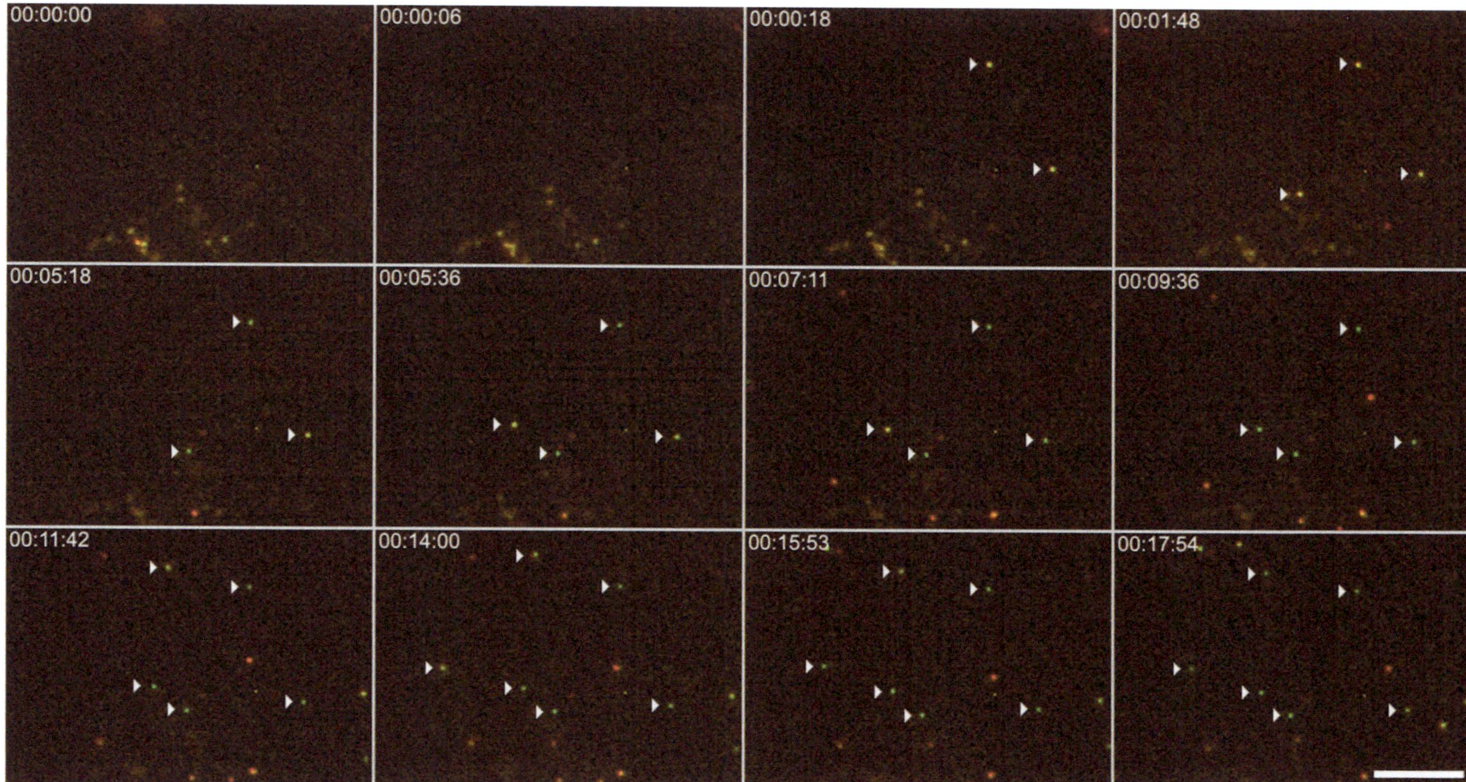

Fig. 1 Live imaging of fibroblasts infected with the dual fluorescent virus. Fibroblasts were infected with freshly prepared infectious supernatant and immediately processed for capturing pictures of *red*, *green* and brightfield channels at a frame rate of 10 images/min. Selected pictures of one movie with a duration time of 30 min are shown. *Arrowheads* point to particles that first attach to the cell (appearing *yellow*) and subsequently lose their envelope (color change from *yellow* to *green*). The time specification (shown as *h:min:sec*) indicates time post infection and the scale bar represents 10 μm. The average de-envelopment time calculated from the first particle appearance to the loss of *red* signal is 3 min ($n = 60$). The complete movie can be accessed on the video companion site "Springer Videos" (video.springer.com)

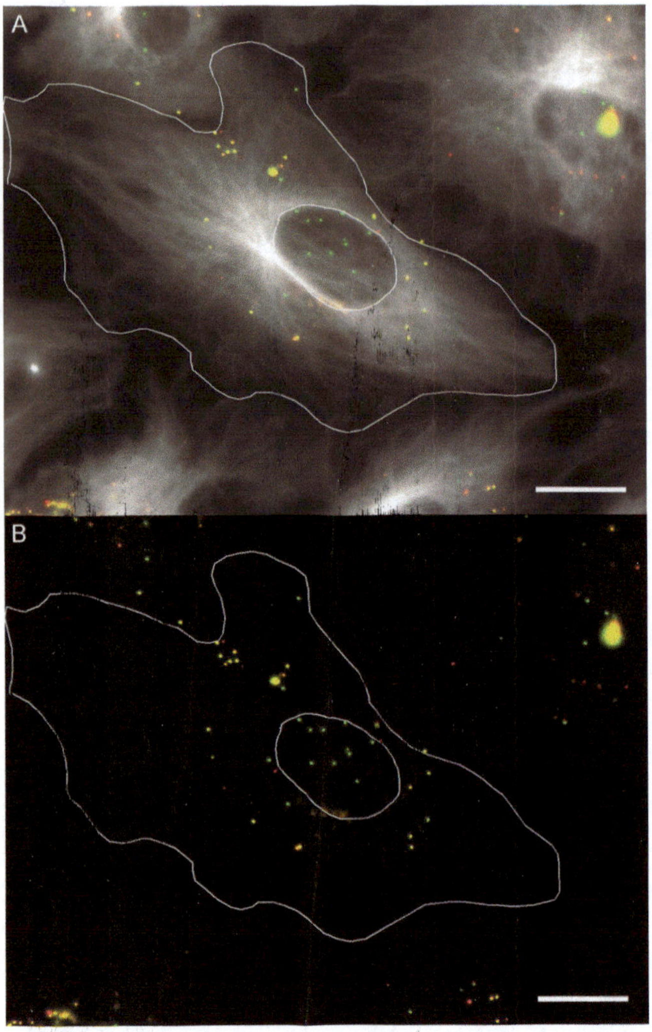

Fig. 2 (**a**) Representative fluorescence micrographs of an endothelial cell fixed at 5 h post infection showing overlays of tubulin (*white*), pp150-EGFP (*green*), and mCherry-gM (*red*). On the basis of the tubulin staining, the periphery of the cell and the margin of the nucleus were traced. (**b**) The delineation of the cell supports the subsequent analysis when the tubulin staining channel is hidden for a more sensitive detection of the *green* and *red* fluorescence. The scale bar represents 20 μm

11. Incubate with secondary antibody dilution (Alexa350-anti-mouse-Ig, 1:100 in PHEM buffer) for 2 h at 37 °C. Wash three times with PHEM buffer.
12. Proceed with picture grabbing: use the blue tubulin staining for an unbiased selection of individual cells and for focussing. Take images of each fluorescent channel (Alexa350, EGFP and mCherry) in two focal planes and overlay (*see* **Note 11**).
13. Mark the contour of the cell with the aid of the tubulin staining and count the numbers of enveloped virions (yellow) or naked capsids (green) (Fig. 2) (*see* **Note 12**).

4 Notes

1. HFFs are commonly used at cell culture passages up to 30; this is only suitable for routine virus propagation. When HFFs are utilized as target cells with the methods described here, passage numbers should be kept as low as possible (<passage 15). Older cells tend to form punctate autofluorescence signals [11] that are visible under any excitation wavelength and might thereby be misinterpreted as virion signals. A negative control has to be included in each experiment to verify the absence of autofluorescence.
2. In case of working with cell types that poorly adhere on glass surfaces (e.g., endothelial cells), the usage of tissue culture dishes with thin bottoms (like the Ibidi µ-dish/slide system) is essential. Attempts to treat glass surfaces to culture endothelial cells never yielded satisfying results in our hands. However, as long as working with cell types tolerating glass surfaces, dishes with cover glass #1.5 and chamber slides with a removable chamber that allows for coverslipping may be used.
3. The addition of an antioxidant such as Trolox is helpful in avoiding phototoxicity [12] during live cell imaging, but it is not absolutely required. The major factor in reducing phototoxicity is to minimize exposure times and to set the level of fluorescent light illumination to a minimum [13]. If the light source does not allow for a direct adjustment, this can be achieved by using neutral density filters. It is advisable to carefully optimize the signal-to-noise ratio for each individual microscope and cell culture system.
4. A large assortment of high-end fluorescence microscopes suitable for live-cell imaging is available. As a guideline, we provide the details of the system used in our studies: Axiovert 200 microscope equipped with an Axiocam MRm CCD camera (Carl Zeiss), a Ludl filter wheel set controlled by the MAC 5000 system (Ludl Electronic Products Ltd.), a heating chamber with CO_2 supply and a HBO 100 mercury lamp that was reduced to 32 % intensity by a neutral density filter (ND 0.5). Movies were processed with the Axiovision 4® software (Carl Zeiss).
5. For consistency of results, it is best to use freshly prepared PFA solutions. However, the PFA solution can be stored in aliquots at −20 °C for up to 2 months. Frozen aliquots should be heated in an 80 °C water bath for 10 min after thawing.
6. In both methods described here, it is obligatory to use freshly prepared virus supernatant to minimize the fraction of biologically nonfunctional virions, as the half life of infectious HCMV virions *in vitro* is about 1 day (unpublished observations). Damaged or inactive virus particles still give a fluorescent

signal and will therefore distort the results. To determine the MOI of each individual experiment, a TCID50 assay [14] can be performed on the deployed supernatant in parallel.

It remains enigmatic, but the double labelled virus seems to exhibit starting difficulties upon freezing (resulting in sporadically lower viral yields in the harvest of the first round of infection). We achieved the best and continuous virus titers in the second round of infection as follows: thaw a virus aliquot, infect HFF as described, and on day 7 post infection use 1 mL of this culture to infect a new 25 cm^2 flask of HFF. Use this culture for the experiment.

7. The adsorption rate of HCMV virions differs between cell types, hence the amount of input virus has to be adjusted. If a comparative study between HFFs and HUVECs is done, the virus preparation for HFFs should be diluted sixfold with the respective medium (MEM5 or imaging medium). For other cell types, the respective adsorption ratio has to be determined, e.g., by adsorption of the virus suspension on ice for 90 min followed by fixation and analysis as described (*see* Subheading 3.3, **step 6** and following).
8. When comparing HFFs and HUVECs we coat the dishes with gelatine for both cell types for sake of comparability.
9. The maximum number of frames in a time lapse series is limited by the fading of the fluorescent signals. The higher the frame rate, the shorter the time of the total imaging time will be. The ideal settings depend on the aim of the experiment.

   For example, to depict particle movements we grabbed 3-channel-movies (EGFP, mCherry, and brightfield) with a frame rate of 30–60 images/min and a total imaging time of 10–15 min. To capture de-envelopment events (the loss of mCherry signal), we sampled images at a lower frame rate of 10–12 images/min with a total imaging time of up to 30 min.

   We do not want to conceal that the generation of a movie dataset, that allows statistical evaluation of a specific topic, is in some ways hard work (especially in cell types, that exhibit a lower adsorption rate like endothelial cells). One should always consider that a single cell within one movie might not reflect the behavior of the whole culture and therefore an extensive set of individual short-takes has to be compiled.
10. The fixed cells can be stored at 4° over night covered with PHEM buffer.
11. If using a standard wide-field fluorescence microscope, two individual "*z*-stacks" were found to be absolutely sufficient to gain the full information of an individual fibroblast or endothelial cell. However, if this protocol is applied to other cell types or microscope systems, the required number of *z*-stacks has to be

explored. The application of a confocal system can lead to a reduced signal-to-noise ratio and the elimination of out-of-focus fluorescence might lead to the requirement of a higher *z*-stack rate. *See* ref. 15 for an explanation on the common misconceptions of the preference for confocal microscopy.

12. In some cases the overlay of an enveloped virion will not yield perfectly yellow signals and switching between the channels might be necessary to detect the dual labelling.

References

1. Sinzger C, Digel M, Jahn G (2008) Cytomegalovirus cell tropism. Curr Top Microbiol Immunol 325:63–83
2. Hahn G, Revello MG, Patrone M et al (2004) Human cytomegalovirus UL131-128 genes are indispensable for virus growth in endothelial cells and virus transfer to leukocytes. J Virol 78(18):10023–10033
3. Wang D, Shenk T (2005) Human cytomegalovirus virion protein complex required for epithelial and endothelial cell tropism. Proc Natl Acad Sci USA 102(50):18153–18158
4. Adler B, Scrivano L, Ruzcics Z et al (2006) Role of human cytomegalovirus UL131A in cell type-specific virus entry and release. J Gen Virol 87(Pt 9):2451–2460
5. Ryckman BJ, Rainish BL, Chase MC et al (2008) Characterization of the human cytomegalovirus gH/gL/UL128-131 complex that mediates entry into epithelial and endothelial cells. J Virol 82(1):60–70
6. Macagno A, Bernasconi NL, Vanzetta F et al (2010) Isolation of human monoclonal antibodies that potently neutralize human cytomegalovirus infection by targeting different epitopes on the gH/gL/UL128-131A complex. J Virol 84(2):1005–1013
7. Saccoccio FM, Sauer AL, Cui X et al (2011) Peptides from cytomegalovirus UL130 and UL131 proteins induce high titer antibodies that block viral entry into mucosal epithelial cells. Vaccine 29(15):2705–2711
8. Sampaio KL, Cavignac Y, Stierhof YD et al (2005) Human cytomegalovirus labeled with green fluorescent protein for live analysis of intracellular particle movements. J Virol 79(5): 2754–2767
9. Varnum SM, Streblow DN, Monroe ME et al (2004) Identification of proteins in human cytomegalovirus (HCMV) particles: the HCMV proteome. J Virol 78(20):10960–10966
10. Schuessler A, Sampaio KL, Sinzger C (2010) Improvement of a quantitative cell tropism assay for rapid and reliable characterization of human cytomegalovirus mutants. J Virol Methods 167(2):218–222
11. Andersson H, Baechi T, Hoechl M et al (1998) Autofluorescence of living cells. J Microsc 191(Pt 1):1–7
12. Cordes T, Vogelsang J, Tinnefeld P (2009) On the mechanism of Trolox as antiblinking and antibleaching reagent. J Am Chem Soc 131(14):5018–5019
13. Stephens DJ, Allan VJ (2003) Light microscopy techniques for live cell imaging. Science 300(5616):82–86
14. Mahy BWJ, Kangro HO (1996) In: virology methods manual. Academic, San Diego, pp 35–37
15. Waters JC (2007) Live-cell fluorescence imaging. Methods Cell Biol 81:115–140

Chapter 15

High Temporal Resolution Imaging Reveals Endosomal Membrane Penetration and Escape of Adenoviruses in Real Time

Ruben Martinez, Andrew M. Burrage, Christopher M. Wiethoff, and Harald Wodrich

Abstract

Imaging host–pathogen interactions in real time can provide significant insight into dynamic processes and provide information about time and space of their occurences. Here, we present detailed experimental instructions on how to image the membrane penetration process of the non-enveloped adenovirus in real time. The system is based on a cell line stably expressing the lectin galectin-3 fused to a fluorophore. Membrane-lytic events during adenovirus cell entry can be monitored by the recruitment of galectin-3 to galactose-containing membrane glycoproteins on the exo-surface of ruptured membranes. The simultaneous use of fluorescently labeled adenoviral capsids allows to image the events in unmatched temporal resolution.

Key words Adenovirus, Galectin-3, Membrane damage, Endosomal escape, Trafficking, Live cell imaging, Virus entry

1 Introduction

While considerable structural and mechanistic detail of enveloped virus fusion with cell membranes has allowed the development of fusion inhibitory antiviral drugs much less is known about analogous events in non-enveloped virus cell entry. Non-enveloped virus capsids often exist as hydrophilic particles in the extracellular environment, retaining a much higher stability and prolonged infectivity than enveloped viruses [1]. In many cases these capsids must physically penetrate and cross hydrophobic cellular membranes to initiate replication. A major limitation to our understanding of non-enveloped virus membrane penetration is in part due to the lack of experimental systems that allow a detailed study of this process.

Adenoviruses (Ads) are prototypic non-enveloped viruses with a large capsid of ~90 nm in diameter that are mild pathogens for

Susanne M. Bailer and Diana Lieber (eds.), *Virus-Host Interactions: Methods and Protocols*, Methods in Molecular Biology, vol. 1064, DOI 10.1007/978-1-62703-601-6_15,

immune competent individuals but can cause more severe infections in immune suppressed subjects. Adenovirus disease pathogenesis is thought to result from the strong proinflammatory response to the virus [2]. This response is thought to provide some advantage in the exploitation of Ads as vaccine vectors but has limited the utility of Ad vectors for gene therapy [3].

Ads enter cells by receptor-mediated endocytosis, but must rapidly exit the endosome to avoid lysosomal degradation. We and colleagues have previously shown that the membrane lytic activity of Ad resides in an internal capsid protein, protein VI, which is exposed upon entry during or shortly after endosomal uptake [4, 5]. As a consequence, the virus is thought to penetrate the endosomal membrane and to gain access to the cytoplasm for subsequent microtubule-directed transport towards the nucleus. Further analysis of protein VI using *in vitro* studies showed that membrane fragmentation is achieved by protein VI through an amphipathic helix and the induction of positive membrane curvature leading to membrane rupture and leakage of endosomal contents [5–9]. This model for protein VI membrane lysis is consistent with a lack of size selectivity in adenovirus rupture of endosomal membrane [4, 6] and observations that leakage of endosomal contents during adenovirus entry is a danger signal [7–9].

In this chapter we describe a new method that permits the visualization of the membrane lysis and escape from endosomal structures of the non-enveloped adenoviruses using live cell fluorescence microscopy [10]. The method relies on the recruitment of the cytosolic galactose-binding lectin galectin-3 (Gal-3) to ruptured membranes. Under physiological conditions galactose-containing polysaccharides are present exclusively on the extracellular or intraluminal domains of membrane glycoproteins. Upon membrane damage cytosolic Gal-3 can access and bind to exposed galactose residues, effectively concentrating on damaged membranes and thus providing a measurable fluorescent signal when fluorescently labeled Gal-3 is used. This technique was originally established to visualize vacuole lysis by invasive bacteria [11, 12]. Here, we adapted this technique to show membrane lysis and penetration by adenoviruses. Using this approach, we recently showed that membrane rupture and endosomal escape are mechanistically distinct in time and space with Ad particles residing and trafficking in Gal-3 positive structures followed by viral egress from these structures occurring later and often at sites distal from the site of membrane rupture [10]. In this detailed protocol we highlight the experimental setup to image the membrane lysis and escape from ruptured membranes by adenoviruses. However, provided that viruses can be fluorescently labeled without influencing infectivity, this approach should be easy to adapt for the use of any viral system whether they are enveloped or non-enveloped.

2 Materials

2.1 Cell Culture

1. U2OS cell line (ATCC: HTB-96).
2. HEK293 cells (ATCC: CRL-1573).
3. DMEM GlutaMAX with 10 % fetal calf serum (FCS).
4. Opti-MEM GlutaMAX.
5. Penicillin/Streptomycin (100×, Gibco).
6. Phosphat-buffered saline (PBS).
7. Trypsin/EDTA, 0.05 % (1×, Life Technologies).
8. Hygromycin B, 100 mg/mL.

2.2 Virus Labeling and Quantification

1. Purified virus (>0.2 μg/μL, *see* **Note 1**).
2. Microscale protein labeling kit (e.g. Molecular Probes® ref. A30006).
3. Zeba Spin Desalting Columns, 7K MWCO.
4. DMSO (1 mg/mL) or milliQ water to dissolve fluorophore.
5. Virus storage buffer: PBS/10 % glycerol.
6. Labelling buffer: PBS/10 % glycerol, 100 mM Na-bicarbonate buffer pH 8.3.
7. 1 M Tris–HCl pH 7.4.
8. SDS.
9. 0.2 M EDTA.
10. 6-well dish.
11. Low melting agarose.
12. Ad-lysis buffer: 10 mM Tris–HCl pH 7.4, 0.1 % SDS, 1 mM EDTA.
13. Low binding centrifuge tube.

2.3 Material Used for Microscopy

1. Ibidi μ-slides $VI^{0.4}$, 6-channel slides (Ibidi, ref. 80606).
2. Live cell imaging solution (Molecular Probes® ref. A14291DJ).

2.4 Microscopy: Hardware Setup

Imaging of adenovirus membrane rupture and endosomal escape events occur within time scales of seconds to minutes, with very fast acceleration events due to microtubule directed transport [10]. For this reason an advanced imaging system capable of rapid dual color image acquisition is recommended. To image membrane rupture and endosomal escape in real time we used a spinning disk microscope based on a Leica DMI6000 (Leica Microsystems, Wetzlar, Germany) equipped with a confocal Scanner Unit CSU-X1 (Yokogawa Electric Corporation, Tokyo, Japan) using objectives HCX PL Apo 100× oil NA 1.4 and an Evolve EMCCD camera (Photometrics, Tucson, USA). The diode lasers used were at 491 and 561 nm. The 37 °C atmosphere was created with an incubator box and an air heating system (Life Imaging Services, Basel, Switzerland) (*see* **Note 5**).

2.5 Microscopy: Software Setup

Acquisitions were done using MetaMorph software (Molecular Devices, Sunnyvale, USA). For maximum time resolution all acquisitions were performed in streaming mode using two wavelengths (491 and 561 nm) and a dual band pass filter (BP502-552/BP615-675). Image acquisition was at 50 ms per channel without binning at 1,024 × 1,024 px at 10 Hz. This results in rates of ten frames per second (fps, dual-color) of total cells with an excellent signal-to-noise ratio.

3 Methods

3.1 Virus Labeling Protocol

Ads are non-enveloped viruses that can be labeled with different fluorophores using primary amine groups on the exposed surface of the capsid. The labeling reaction does not interfere with the infectivity of the virus and allows very good signal-to-noise ratios for live cell imaging.

Virus preparations to start the labeling reactions should be >0.25 mg/mL and extensively dialyzed against virus storage buffer (*see* **Note 1**). To label the virus we routinely use the Alexa Fluor® 488 Microscale Protein Labeling Kit from Invitrogen (*see* **Note 2**).

1. Prepare labeling solution following kit procedure by diluting the Alexa Fluor® 488 dye in 10 μL DMSO. One vial of fluorophore generally allows labeling of up to three virus preparations of 100 μL each.
2. Combine 100 μL virus in labelling buffer and 3 μL dye in a reaction tube (e.g. a Screwcap Eppendorf tube).
3. Mix the reaction slowly and incubate for 30–60 min on a rotating mixer at room temperature. Cover tube using aluminum foil to protect mix from light.
4. Following the labeling reaction, separate the remaining dye from labeled virus/vector using the Zeba desalting column. Desalting columns allow viruses to pass the resin, while the free dye is retained. This procedure will dilute the virus/vector preparation two or threefold but is fast and efficient (*see* **Note 3**).
5. Equilibrate the column by adding 300 μL PBS/10 % glycerol on a Zeba column and spin it at 1,500 × *g* in a microcentrifuge tube for 1 min.
6. Repeat **step 5** at least three times.
7. Add the reaction mix to the top of the resin carefully without destroying the resin pellet and spin again as in **step 5** for 2 min and collect the flow-through. The resin in the column should

be colored by the free dye and the flow-through should also be colored by the labeled virus/vector.

8. Use the flow-through containing the labeled virus/vector and repeat the dye separation process at least one more time with new equilibrated columns. Free dye trapped in the remaining resin should be low and result in a weaker coloration of the resin after the second or third desalting while the virus/vector flow-through should remain slightly colored.
9. After the last dye removal step aliquot the flow-through and prepare 10 μL aliquots. Snap freeze aliquots in liquid nitrogen and store aliquots at −80 °C. Each aliquot can be thawed up to three times without loss of titer.

3.2 Quantification of Virus

The fastest way to quantify CsCl purified virus/vector preparations is by determining the physical particle number (pp/μL) by measuring the absorbance of particle containing DNA at 260 nm [13]. The same method can be applied for the labeled end product.

1. To release the DNA, particles are lysed in Ad-lysis buffer. Mix virus preparation of purified virus (and original stock for comparison) with Ad-lysis buffer at 1:10 (or 1:20 for preps that exceed 0.4 mg/mL). For example use 10 μL virus preparation and complete to 100 μL using Ad-lysis buffer in low binding microcentrifuge tube.
2. Heat the mix 10 min at 56 °C and measure DNA in a spectrophotometer at OD 260 nm. Measured absorbance should have an OD between 0.5 and 0.05. To calculate the physical particle per μL ratio, multiply the measured OD by the dilution factor and divide by the extinction coefficient for adenoviruses (9.09×10^{-13}). An OD of 1 equals 1.1×10^{9} pp/μL. Labeling reactions depend on the protein concentration. Thus, labeling reactions work better and faster if the virus/vector preparation is more concentrated. We do not proceed with labeling reactions if the concentration drops below an OD of 1 (1.1×10^{9} pp/μL).

3.3 Viral Infectivity of Labeled Virus Determined by Plaque-Forming Assays

To retain the biological properties of labeled adenoviruses it is essential to avoid over-labeling of the capsid with dye. We routinely determine the post-labeling infectivity using plaque-forming assays or measure fluorescence protein (FP) expression from Ad derived vectors and compare it to the parental preparation.

1. To determine the infectivity of labeled virus, plaque-forming assays (PFA) are used. On day 1, seed HEK293 E1 transcomplementing cells at a density of 1×10^{6} cells/well in a 6-well plate. Keep cells at 37 °C, 5 % CO_2 and 95 % humidity. Use one 6-well plate per virus tested.
2. On day 2, three different MOIs in duplicate can be tested by plate. To determine the infectivity of the labeled virus assume

that your labeled virus has the same ratio of physical to infectious particles as your parental preparation. For wild-type Ad5, we use MOIs of 0.1–0.001 per cell. For attenuated viruses higher MOIs should be used. Prepare virus dilutions accordingly, adjust to 1 mL final volume and infect cells by adding 1 mL per well final volume.

3. Incubate cells over night at 37 °C, 5 % CO_2 and 95 % humidity in the incubator. Make independently pipetted duplicates for each MOI (*see* **Note 4**).
4. On day 3 the cells are washed with PBS to remove unbound virus and are then covered with an overlay of DMEM agar solution to avoid virus spread. Some training on noninfected cells can be beneficial for this step, because DMEM agar solution preparation and pouring has to be done fast and carefully.
5. Make 2 % stock solution by dissolving 2 g of low melting agarose into 100 mL 1× PBS. Warm solution can be kept soluble at 60 °C (e.g. in a water bath) for later use.
6. Fill a 50 mL tube with 20 mL prewarmed DMEM (37 °C).
7. Cool the 2 % PBS-agar until hand warm and add 10 mL to the DMEM and mix five times by inverting the tube. If PBS-agar is too warm it will degrade thermosensitive antibiotics and kill cells (they will detach immediately or the next day). If too cold, agar will solidify before pouring.
8. Remove PBS from the infected cells and add 2 mL of the mix to each well and keep under the hood until agar solidifies, then move to the 37 °C incubator. From now on the agar overlay should not dry out. If the gel surface appears dry add carefully two or three drops of DMEM with 10 % FCS on top of the well.
9. Move plates as little as possible to allow nice plaque formation. Plaque formation will take around 10 days. If a GFP expressing vector was used, plaque formation can be followed by spread of GFP fluorescence, otherwise plaques become visible as "hole" in the cell layer.
10. After 8–12 days plaques should be ready for counting. Choose an MOI resulting in ~10–100 plaques for highest accuracy and count the number of plaques in each of your duplicates and divide by two.
11. Quantification will lead to Plaque Forming Units (PFU)/physical particle or PFU/μL of virus preparation.

3.4 Viral Infectivity of Labeled Virus Determined by Fluorescence Activated Cell Sorting

In case an Ad is used that expresses a fluorescent protein (FP), fluorescence activated cell sorting (FACS) is a faster way to determine the infectivity of labeled virus. Moreover, it allows for quantification of FP expression following a single round of infection. As in Subheading 3.2. this requires the quantification of the original and the labeled vector preparation in pp/μL. In addition, access to a FACS machine is required. We prefer this assay because for most applications GFP expressing Ad5 derived vector particles are sufficient.

1. On day 1, seed U2OS at 5×10^4 cells/well in a 24-well plate in a final volume of 500 μL DMEM. Prepare enough wells for triplicates and two additional wells for nontransduced control cells. Keep cells at 37 °C, 5 % CO_2 and 95 % humidity.
2. On day 2, prepare dilutions of vector particles for transduction. To compare transduction efficiency of labeled versus nonlabeled vectors dilute them at 1,000:1–100:1–10:1–1:1 pp/cell of e.g. a GFP expressing Ad derived vector in a final volume of 500 μL DMEM/10 % FCS and infect cells for 3 h without pre-binding step.
3. After 3 h discard the infection solution and replace with fresh 500 μL DMEM/10 % FCS.
4. On day 3 discard the medium and wash once with 500 μL PBS.
5. Add two drops trypsin (~50 μL), incubate for ~1 min at 37 °C until cells detach.
6. Neutralize trypsin by addition of 450 μL complete DMEM and transfer 200 μL of the cell suspension into a 96-well plate (for FACS devices compatible with plates, e.g. BD FACS Canto II) or the entire 500 μL in a tube compatible with other FACS devices.
7. Proceed with FACS analysis to count GFP positive cells in your samples. Typically 10,000 events (from cell population) are counted for each sample and GFP transduction levels can be compared for the labeled versus the nonlabeled vector.

3.5 Imaging: Infection Without Pre-binding

1. To monitor Ad membrane lysis and escape we use stably expressing U2OS-FRT-Galectin-3-mCherry cells. This cell line was essentially generated as described in Bosse et al. (*see* Chapter 11 of this series). U2OS-FRT-Galectin-3-mCherry are kept at 200 μg/mL of hygromycin in complete medium and split at regular intervals at 1:4 for maintenance. To use the cells for imaging we seed cells and perform infections in Ibidi μ-slides $VI^{0.4}$ (6-channel slides, Fig. 1, *see* **Note 5**).
2. On the day prior to your experiment prepare cell solution at a density of approximately 1×10^5 cells per mL by trypsinizing and resuspending the cells in complete medium.
3. Position the Ibidi slide in a 15 cm dish at a slight angle by using the cover as indicated in Fig. 1a.
4. Add 150 μL of your cell solution into the bottom reservoir and start filling the connecting channel as indicated in Fig. 1b, c.

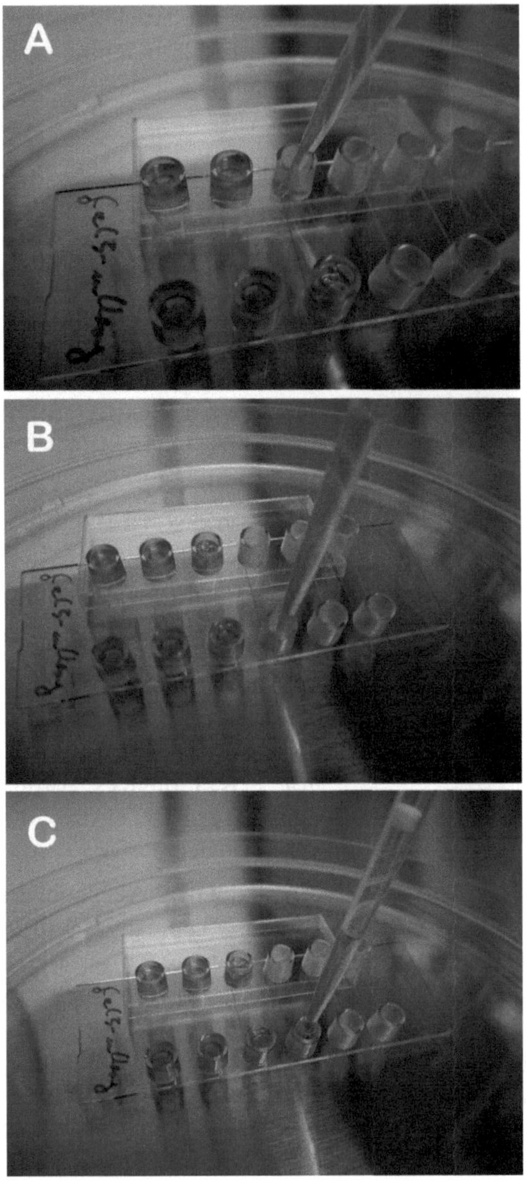

Fig. 1 How to seed cells for the live cell imaging experiment. (**a**) Place the Ibidi slide at an angle by placing one edge of the slide on the cover. (**b**) Start by filling cell solution into the lower reservoir. The channel between the two reservoirs will fill by capillary forces. (**c**) Finish by filling the *lower* reservoir first and then the *upper* reservoir with cell solution. The reservoirs should be filled about 2/3 with DMEM/10 % FCS

The cell solution will fill the channel by capillary forces and provide a connection to the second reservoir.

5. Finish by adding the remaining cells into the second reservoir and put the slide flat to reach a homogenous distribution of cells between both reservoirs so that they are filled approximately by 2/3 and place the ibidi-slide cells back in the incubator.

6. Keep cells at 37 °C, 5 % CO_2 and 95 % humidity. This procedure will result in about 25–50 % confluent cells the next day per channel or ~3,000 cells per channel (*see* **Note 6**).
7. On the day of the experiment exchange the medium in the channels for live cell imaging solution just prior to start your experiment using the same technique as shown in Fig. 1. This clear, HEPES-buffered, physiological solution allows keeping the cells for several hours under ideal imaging conditions.
8. Exchange the liquid two to three times to fully remove any residual medium and cover the slide with the lid.
9. Transfer your slide to the preheated microscope and let the slide equilibrate for 10 min.
10. Before adding the virus/vector to the cells optimize the image acquisition parameters on the microscope using the uninfected cells. Galectin-3 positive cells provide a solid red signal, with few vesicular structures visible.
11. To start the infection dilute your labeled virus/vector at 5×10^7 to 1×10^8 physical particles per 500 μL of live cell imaging solution (you will need ~200 μL virus solution for one channel). At this dilution you have approximately 1,000 pp/cell or 5,000 pp/cell, respectively.
12. To add the virus connect a female Luer-adapter to one of the reservoirs and connect it with a tube to a 1 mL insulin syringe as depicted in Fig. 2a.
13. Next, using the syringe, slowly remove the liquid from the second reservoir and replace it with your virus solution as depicted in Fig. 2b.
14. Slowly aspirate the virus solution through the channel into the tube/syringe making sure that there is always solution covering the cells in the channel as shown in Fig. 2c.
15. Repeat the procedure by refilling the opposite reservoir two to three times (Fig. 2d). This setting allows to quickly and safely fill the channel with virus solution (*see* **Note 7**).
16. Transfer your slide to the preheated microscope and start the acquisition.
17. After infection has commenced severe membrane damage can be seen (*see* Fig. 3, compare left and right panel).

3.6 Imaging: Infection with Pre-binding

The protocol in Subheading 3.5 does not synchronize the steps for infection. An alternative protocol is to pre-bind the virus/vector and perform a synchronized infection. In this case virus solutions and imaging media should be kept on ice.

1. Essentially follow the steps as described in Subheading 3.5. but omitting the temperature equilibration and using cold solutions (virus and imaging solution).

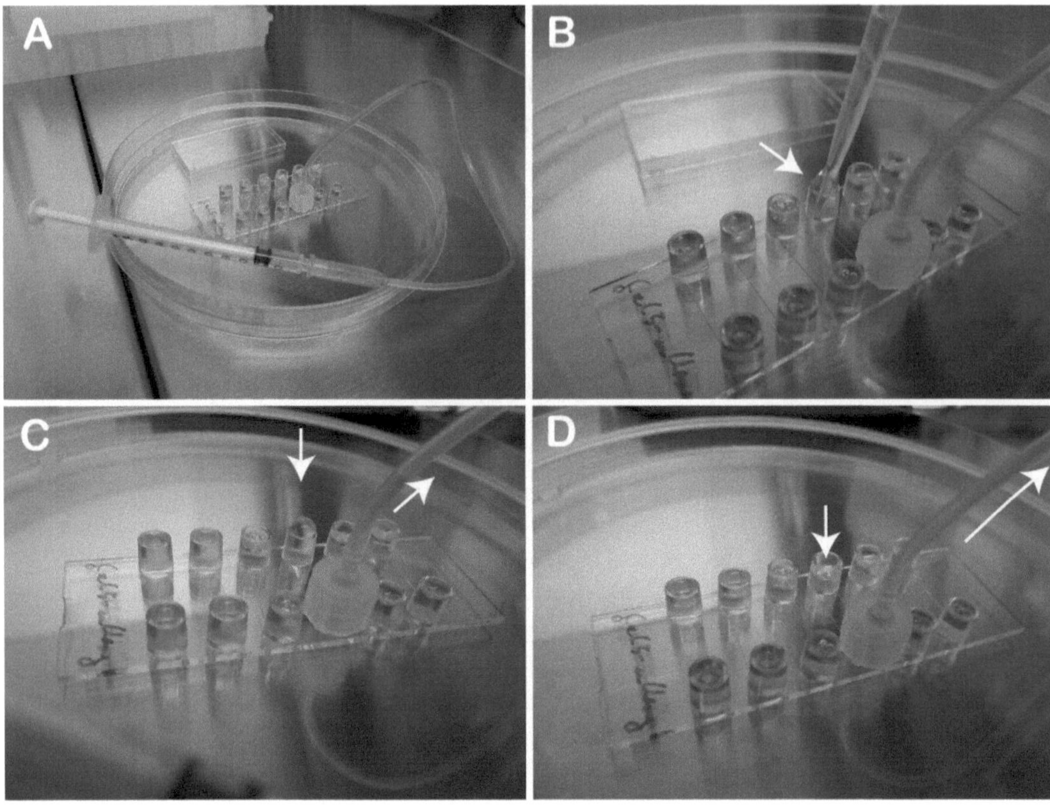

Fig. 2 How to infect cells for the live cell imaging experiment. (**a**) Connect a female Luer adapter with plastic tubing to one reservoir and on the other end to a 1 mL syringe and carefully aspirate the liquid to empty the opposite reservoir. (**b**) and (**c**) Add fresh virus/vector containing solution to reservoir. (**d**) Aspirate the virus/vector containing solution into the channel and repeat the procedure two to three times

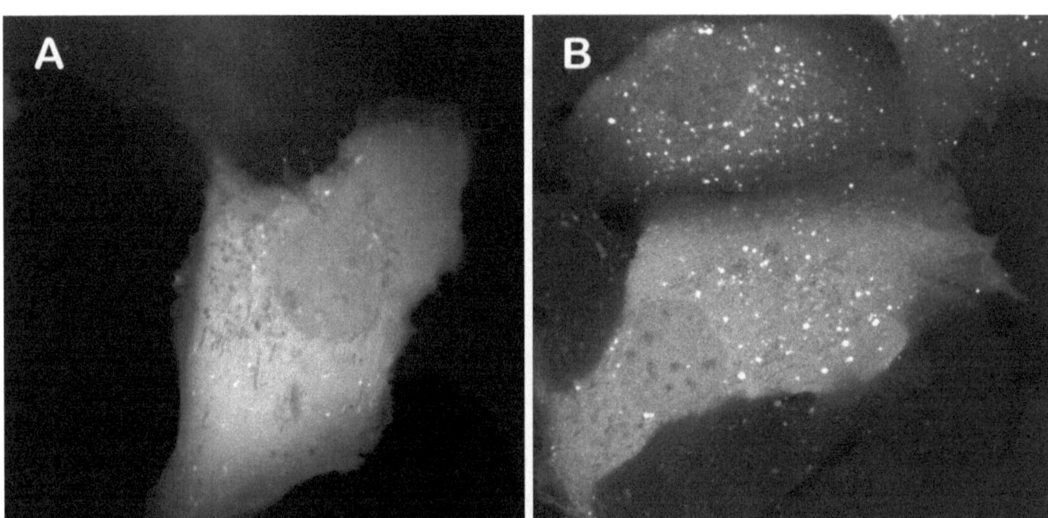

Fig. 3 U2OS-FRT-Gal-3 cell line. (**a**) Midcellular confocal section of a single U2OS-FRT cell stably expressing Gal-3 fused to mCherry. (**b**) As in **a**, following infection with adenovirus. Note the strong increase in Gal-3-labeled vesicular structures

2. Following the addition of the ice-cold virus solution place the slide on a cold metal block (e.g. kept on ice) and incubate for 30 min to allow virus binding.
3. Prior to the transfer to the microscope replace the virus solution with fresh cold imaging medium. The bound virus will remain attached to the cells. However, for interpretation of the results one should be aware that incubating cells at 4 °C also de-polymerizes the microtubule network [14].

3.7 Imaging: General Remarks

1. High temporal imaging resolution, especially when involving two colors, can be achieved by avoiding time consuming steps such as *Z*-stack acquisition, physical filter changes and data transfer. To achieve the best time resolution (in our case 50 ms per frame per color) we restrict imaging to a single confocal slice and acquire individual frames in stream mode using a dual-bandpass filter. This setting reduces the length of the acquisition to the storage capacity of the camera but allows excellent temporal and spatial resolution.
2. We use U2OS cells for most imaging approaches. U2OS cells are flat with a large cytoplasm-to-nucleus ratio and thus a good choice for imaging.
3. For optimal imaging we select cells as shown in Fig. 3, where the cell periphery and the nucleus can be displayed in a single confocal slice and which have a brightness that gives good signal-to-noise ratios at the desired acquisition speed. Selecting cells with flat morphology is optimal for single confocal frame acquisitions and is best achieved by seeding cells to subconfluence as indicated in the protocol under Subheading 3.5.
4. Depending on the events (e.g. for membrane lysis at the plasma membrane by acquisition of Gal-3 stain) we select a more peripheral slice (Fig. 4).
5. The time delay between virus/vector addition and first cell surface binding is approximately 5 min. This is enough time to transfer the cells with the virus to the microscope and adjust focus and laser intensities.
6. The membrane lysis through adenoviruses occurs at or near the cell surface but seldom immediately after binding so bound virus can remain for several minutes before membrane lysis (Gal3-acquisition) occurs.
7. To image membrane lysis as shown in Fig. 4 the focus should be adjusted to an area of the cell surface with several bound viruses. Membrane lysis can be seen by the appearance of Gal-3 positive signal at the site of the virus attachment (Fig. 4, *see* **Note 8**). This recruitment of Gal-3 occurs rapidly, faster than a second. Thus, image acquisition must occur with high temporal resolution (e.g. >5 fps).

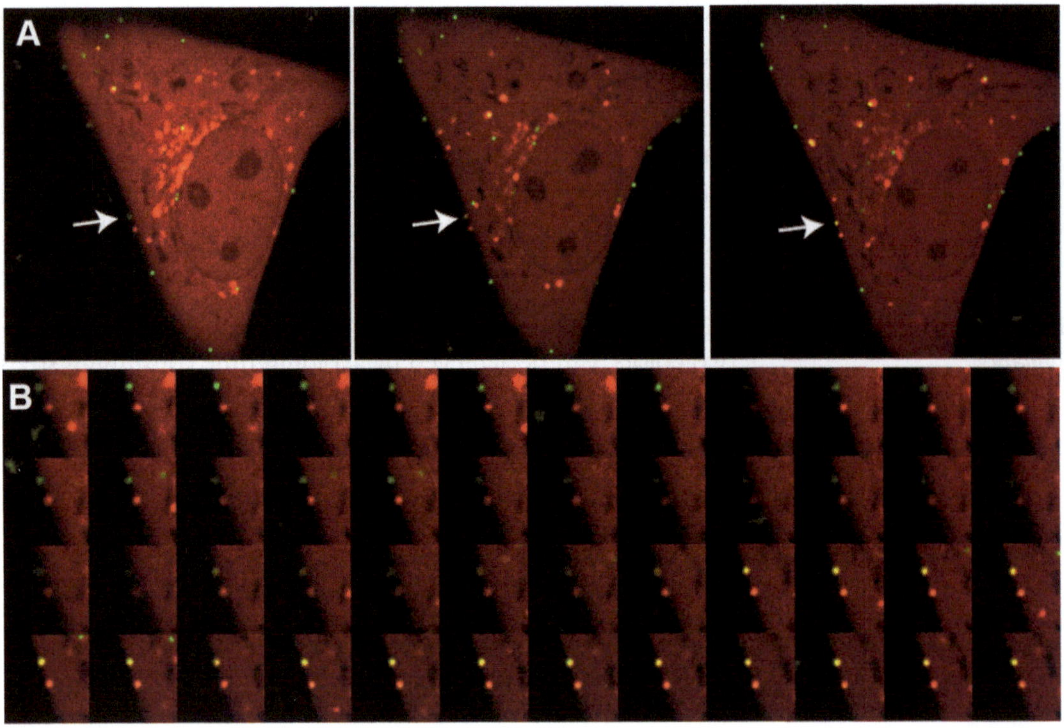

Fig. 4 Adenovirus causing membrane lysis. (**a**) Overview of Alexa488-labeled adenoviruses (*green signal*) bound to the cell surface of a single U2OS-FRT-Gal-3-mCherry cell (*red signal*). From *left* to *right* the *arrow* points at an individual particle that acquires Gal-3 stain. (**b**) Detailed image series of the acqusition of Gal-3 by an Alexa488-labeled particle. Note that the acquisition of the Gal-3 coat becomes visible as *yellow color* from the overlay of both signals

8. Following membrane lysis transport of virus/vector particles in Gal-3 positive vesicular structures from the site of lysis can be observed. In most cases post-lysis transport is initiated by short rapid local movements followed by long range movements and/or separation of the virus/vector from the Gal-3 positive structure (Fig. 5).
9. To image endosomal escape events it is best to focus on the perinuclear region because most escape events occur in nuclear proximity [10].
10. Adjusting the focal plane during acquisition can be helpful because the relative mobility of the Gal-3 positive particle containing structures and the thickness of the cell makes capturing escape events more difficult. Rapid image acquisition at >5 fps is essential to capture the event in detail (*see* **Note 9**).

3.8 Controls

1. The assay can be controlled by using the adenovirus temperature sensitive mutant *Ad2ts (ts1)* [15, 16]. The *ts1* particles can be produced, labeled and used as described in this protocol. When grown at the permissive temperature of 33 °C this virus

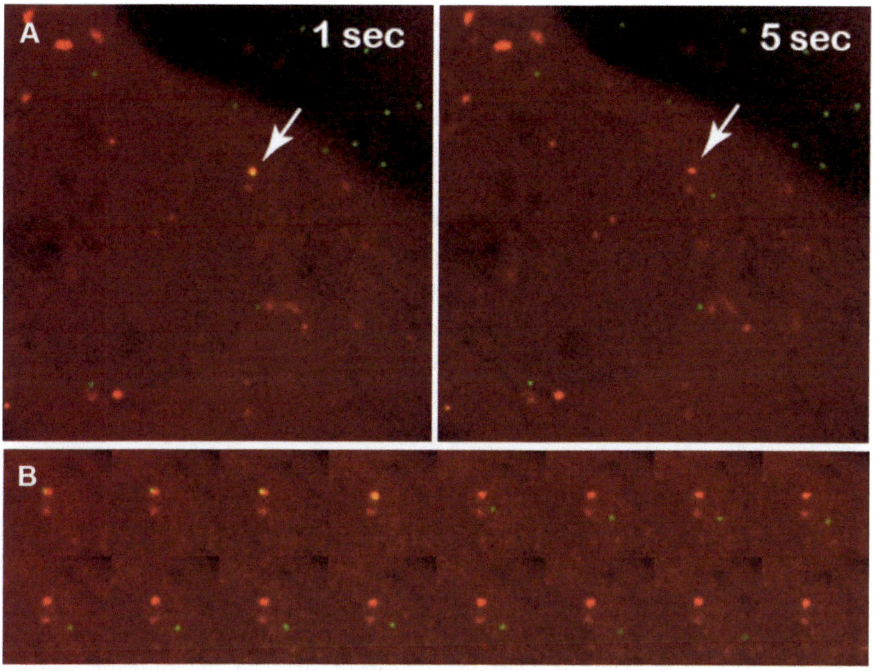

Fig. 5 Adenovirus escaping from Gal-3 positive vesicular structures (**a**) Detail of a single U2OS-FRT-Gal-3-mCherry cell. The *arrow* points to an Alexa488-labeled particle before (*left*) and after (*right*) escaping from the inside of a Gal-3 positive vesicular structure. (**b**) Detailed image series of the endosomal escape process. Note that separation of the combined *yellow* signal into the *green* signal for the escaping particle and the *red* signal for the empty vesicular structure *left* behind

behaves like wild-type viruses/vectors permitting the visualization of membrane lysis and escape events as described above. When grown at the nonpermissive temperature above 38.5 °C the *ts1* mutation located in the viral protease results in the production of hyperstable particles due to unprocessed capsid proteins. When used with the Gal-3 expressing U2OS cells these particles enter cells. Because they lack subsequent disassembly steps and protein VI release they do not acquire Gal-3 and end up in structures presumed to be lysosomes.

4 Notes

1. All work with infectious adenovirus and adenovirus derived vectors should be performed in agreement with the local safety guidelines. We work with adenoviral vectors derived from the human serotype Ad5 deleted for the E1/E3 region. Several excellent methods for CsCl purification of Ad or Ad derived vectors have been published [17–19]. Some of these protocols use TRIS-based buffers for purification. Because most labeling reactions are based on primary amines TRIS has to be eliminated

from purified virus stocks prior to the labeling reaction. We dialyze purified viruses against PBS/10 % glycerol for long-term storage at −80°C. The purity of CsCl purified virus should be confirmed by SDS-PAGE. Since contaminating proteins will likely be labeled as well, these contaminants could influence the background staining and may lead to misinterpretations of results.

2. Because the labeling solution cannot be stored after reconstitution, we label different viruses/vectors in parallel. Although the microscale protein labeling kit is designed to label antibodies and small scale protein preparations we have used this protocol successfully to label Ad and Ad derived vectors with the protein microscale labeling kits for Alexa Fluor®488, Alexa Fluor®555, Alexa Fluor®594 and Alexa Fluor®647. However, in the literature several examples exist where groups have successfully used dyes from other suppliers to label Ad [15, 17]. In such cases, it is advised that the labeling is optimized to maximize fluorescence signal and minimize loss of virus infectivity upon labeling.
3. The kit contains a limited amount of columns and purification matrix. Because we use only a third of the recommended dye solution we often require more column and resin material as the kit provides. Thus we use alternative sources as described in the protocol. However, dye separation should be feasible using several size exclusion columns from many suppliers. The labeling reaction can also be terminated by the addition of TRIS to 40 mM to quench unreacted dye. However, in our hands efficient dye removal is sufficient to terminate the labeling reaction.
4. The physical particles (pp/μL) to infectious particles per μL (multiplicity of infection; MOI) ratio should always be determined for the parental virus preparation. The reason for using different MOIs is to obtain ~10–100 plaques per well at a given MOI for reliable quantification.
5. We prefer an environmental system that surrounds and heats up the whole microscopic setup to avoid thermal aberrations in the optical system. In our experience heated stages alone can result in focus fluctuations during acquisition. Alternatives such as infrared autofocus systems to keep the focus can be used. Likewise we prefer to use ibidi-slides because of their good optical quality and versatility for the addition of the virus/vector. Alternative slide systems from other companies might be used instead.
6. Depending on the application, cell densities can be varied. The conditions provided in the protocol are optimized for single cell imaging. We sometimes also use medium with reduced serum (Opti-MEM®) to prevent cells from fast growing

or to keep the slide for the next day (e.g. when not all channels have been used in one experiment).

7. This step is crucial because of the involved safety issues. Tubing and syringes can easily be obtained from any hospital (e.g. ask for a "butterfly" blood drawing system). To reduce the risk of contaminating the microscope setup we use (a) viral vectors instead of replication competent viruses and (b) fill the channels physically separated from the microscope hardware so that the slide can be covered with the lid before imaging. However, the syringe setup as described in the protocol can also be conveniently fixed to the microscope allowing to image while adding solutions. The long flexible tubing holds the slide in place while changing the solution. Lastly, always make sure to discard the aspirated virus/vector solution as infectious liquid and rinse the tubing/syringe system with bleach, plenty of water and ethanol before the next use.
8. We observed membrane lysis almost exclusively at the cell surface. Viruses/vectors can sit on the cell surface for prolonged times (>30 min) without significant movement or any sign of membrane lysis. Even following membrane lysis the virus/vector can remain at the same position for several minutes before it subsequently starts moving to the cell interior. As such imaging Ad induced membrane lysis requires patience. Alternatively, we also performed synchronized infections, pre-binding viruses/vectors in the cold for 30 min (see alternative protocol). However, we disfavor this approach because the temperature shift provides focus aberrations making imaging more difficult and the results have proven to be more difficult to analyze (e.g. incubation at 4 °C results in microtubule de-polymerisation).
9. The nature of Gal-3 positive structures that carry virus/vector particles is currently not known. These Gal-3 positive structures could consist of damaged or resealed membrane structures. Similar to imaging the membrane lysis event, we have not identified any visual cue to predict when a virus/vector will escape the Gal-3 positive structure. However, post-escape the particles engage in fast movements resembling microtubule-based transport.

Acknowledgements

Part of this work was supported by Equipe FRM 2011 Projet DEQ 20110421299 (H.W.). C.W. acknowledges funding from the NIH (AI082430) and American Heart Association (2261306). A.M.B. acknowledges support from the NIH (AI007508). We acknowledge the Bordeaux imaging centre (BIC) for help in setting up the live cell imaging acquisition. H.W. is an INSERM fellow.

References

1. Tsai B (2007) Penetration of nonenveloped viruses into the cytoplasm. Annu Rev Cell Dev Biol 23:23–43
2. Ginsberg HS, Horswood RL, Chanock RM et al (1990) Role of early genes in pathogenesis of adenovirus pneumonia. Proc Natl Acad Sci USA 87(16):6191–6195
3. Aldhamen YA, Seregin SS, Amalfitano A (2011) Immune recognition of gene transfer vectors: focus on adenovirus as a paradigm. Front Immunol 2:40
4. Wiethoff CM, Wodrich H, Gerace L et al (2005) Adenovirus protein VI mediates membrane disruption following capsid disassembly. J Virol 79(4):1992–2000
5. Wodrich H, Henaff D, Jammart B et al (2010) A capsid-encoded PPxY-motif facilitates adenovirus entry. PLoS Pathog 6(3):e1000808
6. Maier O, Galan DL, Wodrich H et al (2010) An N-terminal domain of adenovirus protein VI fragments membranes by inducing positive membrane curvature. Virology 402(1): 11–19
7. Barlan AU, Danthi P, Wiethoff CM (2011) Lysosomal localization and mechanism of membrane penetration influence nonenveloped virus activation of the NLRP3 inflammasome. Virology 412(2):306–314
8. Barlan AU, Griffin TM, McGuire KA et al (2011) Adenovirus membrane penetration activates the NLRP3 inflammasome. J Virol 85(1):146–155
9. McGuire KA, Barlan AU, Griffin TM et al (2011) Adenovirus type 5 rupture of lysosomes leads to cathepsin B-dependent mitochondrial stress and production of reactive oxygen species. J Virol 85(20):10806–10813
10. Maier O, Marvin SA, Wodrich H et al (2012) Spatiotemporal dynamics of adenovirus membrane rupture and endosomal escape. J Virol 86(19):10821–10828
11. Paz I, Sachse M, Dupont N et al (2010) Galectin-3, a marker for vacuole lysis by invasive pathogens. Cell Microbiol 12(4):530–544
12. Dupont N, Lacas-Gervais S, Bertout J et al (2009) Shigella phagocytic vacuolar membrane remnants participate in the cellular response to pathogen invasion and are regulated by autophagy. Cell Host Microbe 6(2):137–149
13. Mittereder N, March KL, Trapnell BC (1996) Evaluation of the concentration and bioactivity of adenovirus vectors for gene therapy. J Virol 70(11):7498–7509
14. Rubin RW, Weiss GD (1975) Direct biochemical measurements of microtubule assembly and disassembly in Chinese hamster ovary cells. The effect of intercellular contact, cold, D2O, and N6, O2′-dibutyryl cyclic adenosine monophosphate. J Cell Biol 64(1):42–53
15. Greber UF, Webster P, Weber J et al (1996) The role of the adenovirus protease on virus entry into cells. EMBO J 15(8):1766–1777
16. Weber J (1976) Genetic analysis of adenovirus type 2 III. Temperature sensitivity of processing viral proteins. J Virol 17(2):462–471
17. Barry MA, Weaver EA, Hofherr SE (2010) Rescue, amplification, purification, and PEGylation of replication defective first-generation adenoviral vectors. Methods Mol Biol 651:227–239
18. Jager L, Hausl MA, Rauschhuber C et al (2009) A rapid protocol for construction and production of high-capacity adenoviral vectors. Nat Protoc 4(4):547–564
19. Tollefson AE, Kuppuswamy M, Shashkova EV et al (2007) Preparation and titration of CsCl-banded adenovirus stocks. Methods Mol Med 130:223–235

Chapter 16

Three-Dimensional Visualization of Virus-Infected Cells by Serial Sectioning: An Electron Microscopic Study Using Resin Embedded Cells

Martin Schauflinger, Clarissa Villinger, and Paul Walther

Abstract

In this paper we show how to obtain a three-dimensional model of virus-infected cells by serial sectioning of resin embedded samples and transmission electron microscopic imaging. The method bases on sample fixation by high pressure freezing and processing by freeze substitution with the goal to preserve the structures of interest close to the natural state, as previously described (Walther et al., High pressure freezing for scanning transmission electron tomography analysis of cellular organelles. In: Mossman BT, Taatjes DJ (eds) Cell imaging techniques, vol 931, Methods in molecular biology. Humana Press, Totowa, NJ, pp 525–535, 2013). Advantages of serial sectioning compared to that of other tomographic methods are as follows: No special and expensive additional equipment is required. Relatively large volumes, such as whole cells, can be three-dimensionally reconstructed in a reasonable amount of time. Serial sectioning is a non-destructive method; the sections can be stored, re-imaged, or processed for immunogold labeling when more specific data are requested or when new scientific questions are raised (e.g., higher magnifications, protein distributions). We have recently used this method to obtain a three-dimensional model of the complete assembly complex of an HCMV infected cell, which allowed a detailed insight into this virally induced compartment (Schauflinger et al., Cell Microbiol 15(2):305–314, 2013).

Key words Transmission electron microscopy, Serial sectioning, Three-dimensional visualization, Viral assembly complex, Virus morphogenesis

1 Introduction

Electron microscopy is a valuable tool for virus research, since it allows to directly observe viral particles and their interactions with cellular compartments. Three-dimensional visualization of virus-infected cells is helpful to address problems like the distribution of virus particles within the host cell. A number of methods exist for three-dimensional data acquisition using the electron microscopy, e.g., transmission or scanning transmission electron microscopy (TEM or STEM) tomography [1, 2]. However, these techniques require special and expensive equipment and only relatively small

Susanne M. Bailer and Diana Lieber (eds.), *Virus-Host Interactions: Methods and Protocols*, Methods in Molecular Biology, vol. 1064, DOI 10.1007/978-1-62703-601-6_16, © Springer Science+Business Media, LLC 2013

portions of cell volumes can be recorded with a maintainable effort. More recent approaches for three-dimensional data acquisition are focused ion beam/scanning electron microscopy (FIB/SEM) tomography [3] and serial block-face microscopy [4]. Both methods base on the scanning of a block face which is processed automatically inside a scanning electron microscope (SEM) by a focused ion beam or a diamond knife, respectively. The application of SEM may allow to obtain resolutions in *X* and *Y* direction which suffice to depict lipid membranes [5]. Yet, these methods also require special EM equipment. In contrast, serial sectioning is a classical method for three-dimensional visualization and has been developed since the late 1950s [6]. The basic idea of serial sectioning is to resin embed an object of interest, to produce consecutive ultrathin sections of this object, to select an area of interest on the sections (e.g., a particular cell), to image this area in all sections, and to align the micrographs to obtain a stack of images, which can be computed into a three-dimensional representation. Since the single images are standard transmission electron micrographs, the resolution in *X* and *Y* direction allows to display even single bilayers of lipid membranes [7]. The resolution in *Z* direction depends on the thickness of the sections. Hence, it is limited to ~50 nm [8]. Good quality three-dimensional images of relatively large volumes (e.g., whole eukaryotic cells) can be obtained with this approach in a reasonable amount of time. More importantly, it requires only basic equipment for transmission electron microscopy (TEM). Additionally, working with physical sections of a sample enables immunogold labeling. Thus, serial sectioning is a valuable method to address virological questions in three dimensions. For example, a detailed view on the distribution and the (envelopment) state of viral particles within a host cell can be obtained using this technique [9]. Here, we describe how to obtain three-dimensional representations of virus-infected cells.

2 Materials

2.1 Serial Sectioning

1. Virus-infected cells embedded in EPON, as described previously [10–12].
2. Inverted light microscope Wilovert (Helmut Hund GmbH, Wetzlar, Germany).
3. Fine and sharp needle suitable to scratch markings on a resin block.
4. Ultramicrotome Ultracut (Leica Microsystems, Wetzlar, Germany) equipped with a diamond knife (Diatome, Biel, Switzerland).
5. Fretsaw equipped with an extra fine blade (commercially available).

6. Single-edge razor blade (Teletex Medical, Somerset, USA).
7. Formvar-coated single slot copper grids, Ø 3.05 mm, slot dimensions 2 mm × 0.5 mm (Plano GmbH, Wetzlar, Germany).
8. Electric glow discharger (Edwards high vacuum, Leica Microsystems, Wetzlar, Germany).
9. Filter paper (Whatman, Maidstone, UK).
10. Custom made piece of polystyrene (dimensions ca. 70 mm × 5 mm × 5 mm) with an acute angle at one of the long ends.
11. Gridbox (e.g., Ted Pella Inc, Redding, CA, USA).

2.2 Image Acquisition

1. Transmission electron microscope JEM-1400 (Jeol, Tokyo, Japan) equipped with a CCD camera.

2.3 Image Processing and Three-Dimensional Reconstruction

1. Image processing software (we used Adobe® Photoshop®).
2. Three-dimensional visualization software. For image stacking, alignment, 3D reconstruction and data presentation we used Avizo® Fire 6.3 (Visualization Science Group Inc., Mercury Computer Systems, Burlington, USA). In case of distorted sections we used the IMOD [13] and Fiji (ImageJ) software package with suitable plugins. IMOD and ImageJ are supported by Windows, Mac, or Unix systems and can be downloaded as freeware from the IMOD Web site (http://bio3d.colorado.edu/imod/) or the ImageJ Web site (http://rsb.info.nih.gov/ij/download.html).

3 Methods

For this approach of serial sectioning, resin embedded samples are required like they are used routinely for standard transmission electron microscopy. For best structural conservation of the samples at a defined physiological state we recommend to use a protocol basing on high pressure freezing and freeze substitution as recently reviewed by McDonald [14]. We have optimized these protocols for the study of virus-infected cells [10–12]. Any adherent cell line can be used with this protocol, including primary (e.g., epithelial) cells. In short, we seed a monolayer of cells with 80 % confluence on carbon-coated 3 mm sapphire disks and infect the cells the next day with the virus of interest (e.g., we seed 2.4×10^4 HFFs per cm^2, and infect the cells with HCMV at an MOI of 0.5–1 the following day). The carbon coat on the sapphire disks supports cell attachment, and serves as predetermined breaking layer after sample embedment in Epon. Next, we cryofix the cells by high-pressure freezing at the time point of interest, dehydrate the sample in a freeze substitution process and embed the cells in EPON. When the sapphire disk is removed from the polymerized EPON, the virus-infected cells are located directly underneath the smooth surface of the resin block.

3.1 Serial Sectioning

1. Choose an area of interest on the surface of the resin block by examining the shapes of the embedded cells with an inverted light microscope. It is usually most effective to choose areas with a high density of cells. Mark this area with a fine needle by carefully scratching its rough outline into the surface. Keep in mind that the size of the sections must not be more than approximately 200 μm × 250 μm.
2. Claw the resin block into the trimming holder of the microtome so that the relevant surface is facing up and can be observed through the microtome's binocular. Dissect the resin block carefully with the help of a fretsaw so that it will fit into the microtome holder (*see* **Note 1**). Do not harm the relevant surface area.
3. Firmly claw the sawed resin block into the microtome holder with the relevant surface facing up. First, remove the surface around the marked area of interest with a razor blade (*see* **Note 2**). Trim the remaining surface of the resin block to a trapezoid shape (around 200 μm high and 250 μm wide) with a sharp razor blade by always cutting off only thin slices of resin at a time (*see* **Note 3**). The upper and lower edges of the trapezoid have to be exactly parallel (Fig. 1a). Trim the parallel edges of the trapezoid first. Then trim the other two edges. Remove the resin flakes with blasts of dry air.
4. Clamp the sample holder tightly into the cutting arm of the microtome, so that the longer base of the parallel trapezoid edges faces downwards and is parallel to the diamond knife edge (*see* **Note 4**). Carefully bring the diamond knife close to the trimmed edge without touching it.
5. Fill the boat of the diamond knife shank with distilled water so that the water is level with the cutting edge. The edge of the diamond knife needs to be completely moistened (*see* **Note 5**). First, fill water into the boat until the surface is convex. Observe the water surface next to the diamond knife through the binocular and adjust the water level with a syringe until the headlight of the microtome is reflected well.
6. Very accurately align the cutting edge of the diamond knife with the sample surface (*see* **Note 6**). A perfect section ribbon can only be obtained when the upper and the lower edges of the trapezoid are exactly parallel to the knife edge.
7. Cut ultrathin sections in ribbons of five sections (*see* **Note 7**). The ribbons will float on the water surface (Fig. 1b).
8. Mount the ribbons on formvar-coated single slot grids (Fig. 1c) (*see* **Note 8**). First, use forceps to slowly immerse the

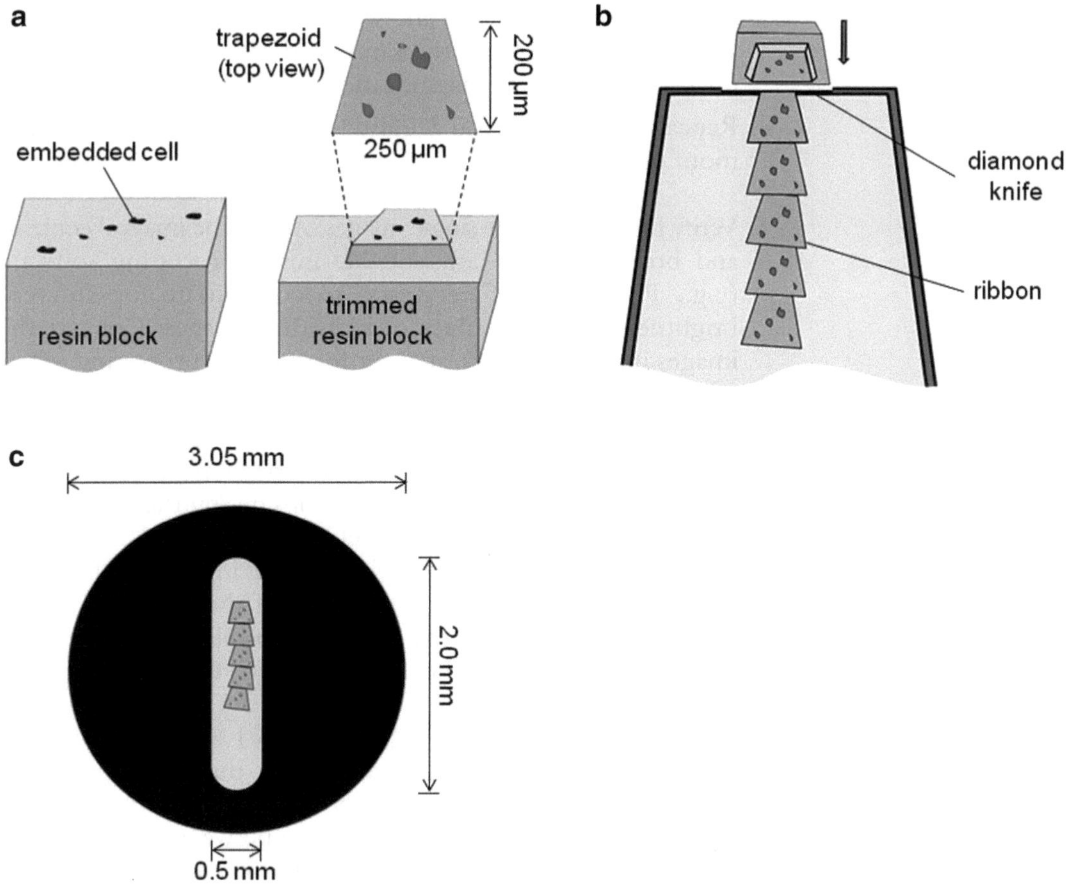

Fig. 1 (**a**) A resin block with embedded cells is prepared for serial sectioning. A trapezoid of approximately 250 µm × 200 µm is left over after the trimming of the resin block's surface. The cells which have been chosen for further investigation are located underneath the surface of this trapezoid. (**b**) Top view on the diamond knife shank after the sectioning process. The ribbon of ultrathin sections is collected on the surface of the water within the shank. (**c**) The ribbon is mounted on a formvar-coated single slot grid

formvar-coated grid into the water (*see* **Note 9**) in an angle of approximately 45°. Then place the ribbon close to the formvar film. During this process, control the orientation of the ribbon with a suitable tool in your other hand (we us an eye lash which was glued to the tip of a thin wooden stick). Slowly lift the grid out of the water to attach the sections onto the grid. Let the grids dry at room temperature and store them preferably in a grid box (*see* **Note 10**).

3.2 Image Acquisition

1. Insert the first grid into the transmission electron microscope and search for the very first section. Choose a motif of interest and adapt the magnification according to your motif (*see* **Note 11**). Acquire an image.

2. Go to the second section and search for the same motif on this section. Try to capture the same field of view and acquire an image with the same magnification as in **step 1**.
3. Repeat **step 2** until you acquired (at least) one image of your motif on every section.

3.3 Image Processing

1. Verify the order of the single images. Adjust the images' contrast and brightness with any suitable image processing software (e.g., Photoshop) if necessary, to compensate for different brightness/contrast values of individual images. Convert the images into grayscale format for further data processing.
2. The reconstruction software will ask you to adjust the distance between each image of the stack according to the thickness of the physical sections. For that you need to determine the amount of pixels in Z direction by considering the thickness of the physical section and the pixel length in X and Y direction (both in nm). The length of 1 pixel (in nm) in X and Y direction, respectively, is defined by the length of the depicted cellular area (in nm) divided by the amount of pixels in that direction. Next, calculate the number of pixels which equal the thickness of your serial sections.

   Example: Your image size is 1,024 × 1,024 pixels and displays an area of 10 μm × 10 μm. Hence, the length of 1 pixel is equivalent to ~9.77 nm. If the physical thickness of your serial sections is 100 nm, the distance between each image of the stack is equivalent to ~10 pixels. In this case, when loading the images into the software, specify the number of pixels as $X = 1{,}024$, $Y = 1{,}024$, and $Z = 10$ to adjust the height of the stack.
3. Use any suitable three-dimensional visualization software, e.g., Avizo, or ImageJ, to create a stack file containing the single images in their natural order. Use the information from the previous step to adjust the height of the stack accordingly.
4. Align the images within the stack with any suitable three-dimensional visualization software, e.g., Avizo. Automatic alignment is possible in Avizo. Usually the alignment has to be optimized manually by rotating and shifting the images one by one (Fig. 2a). Start alignment in the middle of the stack and continue towards the top and the bottom of the stack (*see* **Note 12**).

3.4 De-warping of Distorted Images

Sometimes distortions of the images may occur, e.g., due to deformation of the resin sections. Thus, if such optical distortions occur it is not possible to well align all corresponding structures on two subsequent images, and the images need to be de-warped. For this we recommend a combination of free software (i.e., IMOD, and ImageJ/FIJI). The IMOD user's guide (http://bio3d.colorado.

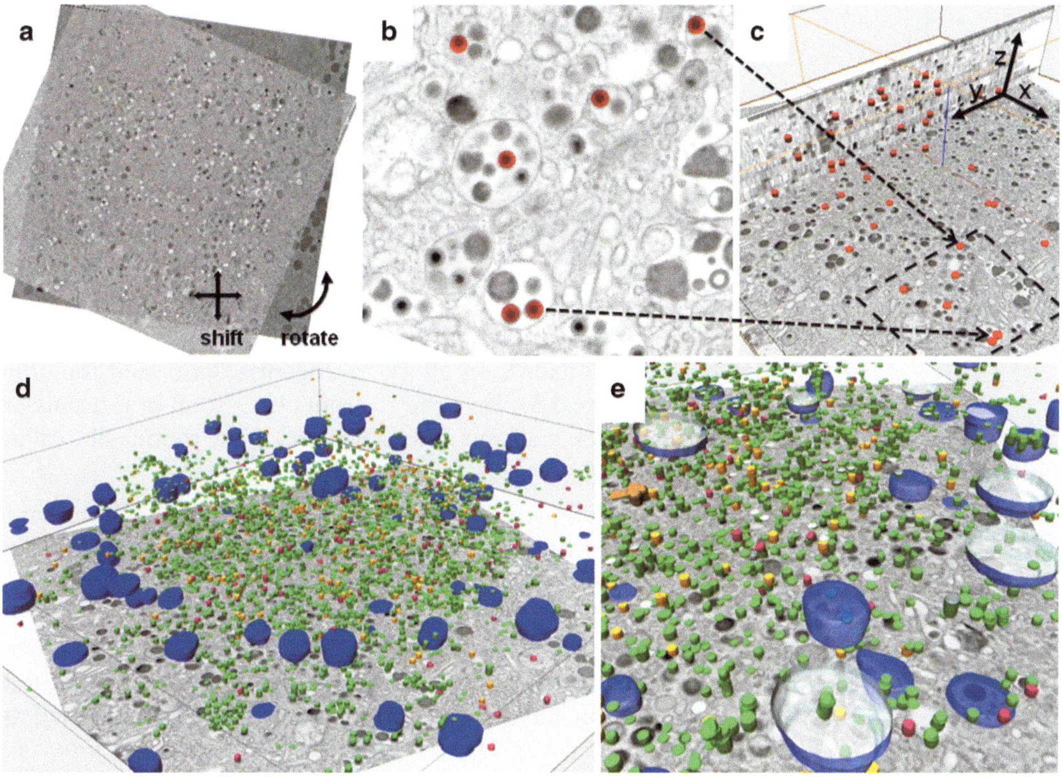

Fig. 2 A fibroblast was infected with human cytomegalovirus, cryo-fixed at day 5 post infection, embedded in EPON and serial sectioned. Images of the viral assembly complex in its cytoplasm were acquired from every section throughout the entire volume. The single images were superimposed to a virtual stack. (**a**) Two images of the stack are aligned by shifting and rotating the superimposed image. (**b**) Viral particles are segmented on a TEM micrograph. (**c**) This is repeated for the other micrographs of the stack and can then be displayed in three dimensions. (**d**, **e**) Three dimensional model of the assembly complex after surface generation in Avizo superimposed upon a TEM micrograph. *Green*, *red*, and *orange*: viral particles in different envelopment stages. *Blue*: multivesicular bodies. (**d**) Overview of the human cytomegalovirus assembly complex. (**e**) The model of the assembly complex after zooming in. The surface of the blue multivesicular bodies is displayed semi-transparently and cut open in order to show virus particles inside (application of *SurfaceCut*)

edu/imod/doc/guide.html) provides further information for the following steps.

1. Create a stack with ImageJ/FIJI by running the macro convert_tif_to_stack. The resulting file is called *stack.tif*. Convert it to *stack.st* with IMOD by using the shell commands "tif2mrc stack.tif stack.mrc" and "newstack stack.mrc stack.st". Store all files of this protocol in the same folder.
2. Load *stack.st* into MIDAS (plugin of the IMOD software package for manual alignment) with the shell command "midas stack.mrc". Manually perform rough alignment by only shifting and rotating the images (*see* **Note 13**). The model file is called *stack.xf*. Do save the adjustments frequently.

3. Convert *stack.xg* to *stack.xf* by the shell command "xftoxg stack.xf". Then align the images by applying the transformations to the stack using the shell command "newstack -xform stack.xg stack.mrc aligned.mrc". The resulting file is called *aligned.mrc.*
4. For further processing convert *aligned.mrc* to *aligned.st* with the command "newstack aligned.mrc aligned.st" and create a copy of *aligned.st* in the same folder and rename it *aligned.preali.*
5. Open the IMOD-plugin *3dmod* by entering "3dmod" in the command shell. Seed artificial landmarks (i.e., fiducials) in 3dmod by selecting *aligned.mrc* as image file (*see* **Note 14**). Check "bead fixer", un-check "autocentre" and seed the fiducials (*see* **Note 15**). Save the contours (i.e., total of fiducials on several planes marking one structure) in a model file called *fiducials.seed.*
6. Create the aligned and de-warped image stack. First, enter "xfmodel" in the shell, and enter the following in the dialogue window: "aligned.preali" as image file, "fiducials.seed" as input model file, and "aligned.xf" as name of the new file of f transforms. Enter "xftoxg" to open another dialogue to create the g transforms. Enter "aligned.xf" as input file and "aligned.xg" as output file. Finally, enter "newstack -xform aligned.xg aligned.st final.mrc" to create the finished stack file *final.mrc* (*see* **Note 16**).

3.5 Three-Dimensional Reconstruction/Segmentation

1. Visually scan through your aligned stack to search for interesting structures which are worth segmenting. Outline the structures of interest (e.g., membranes and virus particles) in every section manually using any suitable three-dimensional visualization software, e.g., Avizo (Fig. 2b, c) (*see* **Note 17**).
2. Generate the surfaces of the labeled structures and display the surfaces in three-dimensional visualization software (Fig. 2d, e) (*see* **Note 18**).

4 Notes

1. Start the sawing process carefully by moving the saw forward only until the slit is deep enough to saw without the danger of damaging the surrounding surface area. When sawing, do not apply too much pressure but let the saw blade do the work by using the whole length of the saw blade. When one resin block with embedded cells contains several interesting areas, it can be dissected into several pieces. Using a finer saw blade helps to retain more of the surface area.
2. Observe the trimming process through the binocular. Turn the resin block into the best suitable directions to make cutting

easier, i.e., the edge being trimmed should always face the experimenter and the trimming movement should be done towards him. Stabilize your hands by resting them on the holder. Use the thumbs and index fingers from both hands to hold and move the razor blade. A razor blade with only one edge allows safe working.

3. Always use a new razor blade to trim the edges of the trapezoid to obtain accurate and clean edges.
4. The resin block should be clamped close to the top (surface) side to avoid vibrations during the sectioning process. Vibrations will result in uneven sections.
5. If the water does not moisten the cutting edge completely, take a small acute-angled piece of polystyrene and move its tip carefully over the cutting edge.
6. Approach the diamond knife carefully towards the surface of the resin block until a chink of light appears (use the bottom light only). By its shape the alignment between the diamond knife and the resin surface can be optimized. It needs to keep the same width while moving the sample pass the diamond knife. Alignment has to be done very carefully. Remember to move the knife away from the resin surface before you do adjustments to avoid accidental damaging of the knife and the surface.
7. The section thickness determines the section stability, but also the image quality and the resolution in *Z* direction. We recommend to make 100 nm sections, which is a good trade-off between these factors (e.g., typical HCMV virions exhibit diameters of about 200 nm and thus each virus particle within the stack can be captured).
8. A cheaper alternative to commercially available formvar-coated grids is to coat the grids in the own lab, which is possible with minimal technical equipment. Various protocols for this method can be found on the internet.
9. If the formvar-coated grids repel the water, glow discharge the grids repeatedly with the help of a glow discharge device. Also, carefully remove excessive formvar film if necessary.
10. Carefully remove the water from the forceps and the grid with filter paper after you collected the ribbons, and only then release the grids. The grids need to be completely dry prior to imaging. We recommend to use a labeled grid box to store your grids. This helps to keep track of the sections' order and to store the grids safely without damaging the formvar film.
11. Obtain an overview about the quality, distribution and size of the cells throughout the volume prior to image acquisition. This is important for choosing a suitable cell and the right magnification to fit the entire area of interest (e.g., a whole cell) onto the images in every section.

12. When changing the direction in which alignment is performed in Avizo also switch the icon *transform upper slice* into *transform lower slice* or vice versa.
13. Make sure you do not stretch the sections since there is no undo-button.
14. Fiducials can also be seeded on the unaligned images. However, previous alignment eases this process. To be able to open the file *aligned.st* in 3dmod (IMOD-plugin), you first need to start etomo with the command "etomo". Use *aligned.st* as dataset. Pretend to load data from a tilt series but only enter small values (such as 0.00001) for starting angle, increment, etc.
15. Suitable spots for seeding fiducials are centers of circular structures which extend over more than one section (e.g., endosomes). Use the key "S" to save the contours and "N" to create a new contour for a new structure.
16. If you want to perform reconstruction in Avizo, divide the stack into single images using FIJI.
17. Outlining in Avizo is most efficiently done by using the *brush* tool in combination with a commercially available pen tablet. The *magic wand* tool may be used when the structure of interest exhibits a sufficient difference in grey values from the background. Consult the respective software manual for other suitable labeling options. Fill the outlined structures with the "F" key to gain nicer results from surface generation. Segment all structures in one common label field.
18. When using Avizo apply *SurfaceGen* and *SurfaceView*. Prior to surface reconstruction subdivide the common label field into one label field for each structure (keep the common label field for further changes). This allows individual label processing. For example for surface generation it can become useful to reduce the data size of some *.label*-files by applying the *Resample* module or *smooth labels* in the label field. Note that *smooth labels* slightly changes the boundaries of the labels. Take advantage of the various settings and possibilities to present your data well (e.g., transparent surfaces, *SurfaceCut*, *DemoMaker*, *CameraRotate*).

References

1. Baumeister W (2004) Mapping molecular landscapes inside cells. Biol Chem 385:865–872
2. Hohmann-Marriott MF, Sousa AA, Azari AA et al (2009) Nanoscale 3D cellular imaging by axial scanning transmission electron tomography. Nat Methods 6:729–731
3. Knott G, Marchman H, Wall D et al (2008) Serial section scanning electron microscopy of adult brain tissue using focused ion beam milling. J Neurosci 28:2959–2964
4. Denk W, Horstmann H (2004) Serial block-face scanning electron microscopy to reconstruct three-dimensional tissue nanostructure. PLoS Biol 2:1900–1909
5. Villinger C, Gregorius H, Kranz C et al (2012) FIB/SEM tomography with TEM-like

resolution for 3D imaging of high-pressure frozen cells. Histochem Cell Biol 138:549–556

6. Sjöstrand FS (1958) Ultrastructure of retinal rod synapses of the guinea pig eye as revealed by three-dimensional reconstructions from serial sections. J Ultrastruct Res 2:122
7. Walther P, Ziegler A (2002) Freeze substitution of high-pressure frozen samples: the visibility of biological membranes is improved when the substitution medium contains water. J Microsc 208:3–10
8. Merchán-Pérez A, Rodriguez JR, Alonso-Nanclares L et al (2009) Counting synapses using FIB/SEM microscopy: a true revolution for ultrastructural volume reconstruction. Front Neuroanat 3:1–14
9. Schauflinger M, Villinger C, Mertens T et al (2013) Analysis of human cytomegalovirus secondary envelopment by advanced electron microscopy. Cell Microbiol 15(2):305–314
10. Walther P, Schmid E, Höhn K (2013) High pressure freezing for scanning transmission electron tomography analysis of cellular organelles. In: Mossman BT, Taatjes DJ (eds) Cell imaging techniques, vol 931, Methods in molecular biology. Humana, Totowa, NJ, pp 525–535
11. Buser C, Walther P (2008) Freeze-substitution: the addition of water to polar solvents enhances the retention of structure and acts at temperatures around −60 °C. J Microsc 230:268–277
12. Schauflinger M, Fischer D, Schreiber A et al (2011) The tegument protein UL71 of human cytomegalovirus is involved in late envelopment processes and affects multivesicular bodies. J Virol 85:3821–3832
13. Kremer JR, Mastronarde DN, McIntosh JR (1996) Computer visualization of three-dimensional image data using IMOD. J Struct Biol 116:71–76
14. McDonald K (2007) Cryopreparation methods for electron microscopy of selected model systems. In: McIntosh JR (ed) Cellular electron microscopy, vol 79, Methods in cell biology. Academic, San Diego, CA, pp 23–56

Chapter 17

3D-Tissue Model for Herpes Simplex Virus-1 Infections

Ina Hogk, Steffen Rupp, and Anke Burger-Kentischer

Abstract

Infection with herpes simplex virus type 1 (HSV-1) causes the most common skin disease. Various test systems have been established to recapitulate this cyclical pathway of productive infection, latency, and reactivation. Most studies of latency and reactivation are conducted in animal models. However, the small number of neurons which harbor the viral genome, the complexity of the *in vivo* setting, and ethical constraints place limits on animal studies. So far, no *in vitro* model which resembles natural latency exists. Here, we describe the first *in vitro* HSV-1 infection model based on a human skin equivalent. The 3D infection model is generated using the human keratinocyte cell line HaCaT grown on a collagen substrate containing human primary fibroblasts and in addition a quiescently HSV-1 infected neuronal component.

Key words Herpes simplex virus type 1 (HSV-1), *In vitro* infection model, Skin equivalent, Latency, Herpes viral reactivation

1 Introduction

Following primary infection, the Herpes simplex virus type 1 (HSV-1) establishes life-long latency—a characteristic feature of all herpes viruses—within the trigeminal ganglion. The virus enters the nerves at the primary infection site, migrates into the cell body of the neuron where the circular viral genome can persist as an episomal molecule in a latent state [1–3]. At this stage, except for a series of Latency Associated Transcripts (LATs) which are produced, the viral lytic gene expression is silenced [2, 4–7]. The mechanism of latency and the subsequent reactivation are still not completely clarified.

Advances in understanding the mechanisms by which HSV-1 establishes and reactivates from latency are generally received by two-dimensional monolayer cell cultures or animal models [5, 8–20]. The insights gained thereby however are of limited value, since two-dimensional cell systems do not adequately mimic the three-dimensional environment of the native skin. Furthermore, they lack the crucial neuronal latency-forming component and thus are unable to provide information on mechanisms that lead to latency or reactivation [21–23].

Susanne M. Bailer and Diana Lieber (eds.), *Virus-Host Interactions: Methods and Protocols*, Methods in Molecular Biology, vol. 1064, DOI 10.1007/978-1-62703-601-6_17,

In contrast, the new HSV-1 infection model described here shows a significant improvement by integrating a neuronal cell line (PC12) within the dermal layer of the 3D skin equivalent. The basic structure of this organ-similar model is an epidermal-layer formed by keratinocytes applied to a dermal-layer built by a collagen gel substrate containing human primary fibroblasts. Cultivation at the air-liquid interface allows epidermal stratification and epidermal–dermal interactions to occur [24, 25]. As a central feature of this novel *in vitro* skin model we were able to integrate a quiescent HSV-1 infected neuronal component in order to mimic the *in vivo* environment. Thereby, the pheochromocytoma cell line PC12 was used to establish a quiescently infected neuronal cell line. Thus, the infected PC12 cells stay dormant within the 3D infection model while spontaneous reactivation is absent. Additionally, in this protocol we describe the successful reactivation of the virus within the 3D HSV-1 infection model using UVB-light (312 nm) at an intensity of 1,000 mJ/cm^2.

However, the reactivation process is still not fully reproducible. This could be explained by the fact that viral production occurred in a minority of latently infected cells. Studies derived from animal experiments suggest that only a low percentage (4–35 %) of HSV-1 infected neurons actually contain latent HSV-1-DNA [26, 27]. Another crucial point could be the freezing procedure which may have detrimental influence on the viral activity. Thus, the infection conditions have to be further refined in order to achieve a reproducible reactivation within the 3D HSV-1 infection model.

Nevertheless, here we describe a method to generate a complex *in vitro* 3D infection model including quiescently infected neuronal cells which prospectively will provide a basis to investigate the molecular mechanisms involved in establishing, maintaining, and controlling latency as well as to test pharmaceutical products.

2 Materials

Use commercially available cell culture plasticware by Corning, Falcon, Sarstedt, and Nunc (or equivalent) for maintenance of monolayer cell cultures unless otherwise specified. Use ultrapure water (purified deionized water with a conductivity of 18 MΩ cm at 25 °C) and store all solutions at 4 °C unless indicated otherwise. All prepared solutions, glassware and pipette tips used in cell culture should be sterilized prior to use.

2.1 Virus and Cell Lines

1. HSV-1, HF strain (American Type Culture Collection; ATCC VR-260) (*see* **Note 1**).
2. Human primary keratinocytes (*see* **Note 2**) and primary fibroblasts from human foreskin tissue (*see* **Note 3**).

3. Immortalized human keratinocytes (HaCaT) obtained from the German Cancer Research Center (Heidelberg, Germany).
4. Rat adrenal pheochromocytoma cell line PC12 (American Type Culture Collection; ATCC CRL-1721) (*see* **Note 4**).

2.2 Cell Culture Media and Components

1. Fibroblast growth media: Dulbecco's modified Eagle's medium (DMEM) with 4.5 g/L glucose supplemented with 2 mmol/L L-glutamine, penicillin (50 units/mL), streptomycin (0.05 mg/mL), and 10 % heat-inactivated fetal calf serum (FCS).
2. HaCaT growth media: Dulbecco's modified Eagle's medium (DMEM) with 4.5 g/L glucose supplemented with 2 mmol/L L-glutamine, penicillin (50 units/mL), streptomycin (0.05 mg/mL), and 10 % heat-inactivated fetal calf serum (FCS).
3. Keratinocyte growth media (KGM): Basal Medium supplemented with penicillin (50 units/mL), streptomycin (0.05 mg/mL) and with the provided supplement solutions (Keratinocyte Growth Medium 2 Kit). Final supplement concentrations: bovine pituitary extract (BPE) 0.004 mL/mL, epidermal growth factor (hEGF, recombinant human) 0.125 ng/mL, insulin (recombinant human) 5 μg/mL, hydrocortisone 0.33 μg/mL, epinephrine 0.39 μg/ mL, transferrin 10 μg/mL, and $CaCl_2$ 0.06 mM.
4. PC12 growth media: RPMI 1640 medium supplemented with 2 mmol/L L-glutamine, penicillin (50 units/mL), and streptomycin (0.05 mg/mL) and 10 % heat-inactivated horse serum (HS) and 5 % heat-inactivated fetal calf serum (FCS).
5. Murine nerve growth factor beta (mbeta-NGF).
6. Horse serum (HS), New Zealand origin (*see* **Note 5**).
7. Fetal bovine serum (FCS), qualified, EU approved, South American origin (*see* **Note 5**).
8. Phosphate-buffered saline (PBS) solution (10×), pH 7.4. Prepare a 1× PBS solution using ultrapure water. Solution should be autoclaved prior to use and stored at room temperature.
9. 0.5 % Trypsin–EDTA (10×). Prepare also a 0.05 % trypsin–EDTA (1×) solution in sterile phosphate-buffered saline (PBS).
10. Ca^{2+}/Mg^{2+} free phosphate-buffered saline, pH 7.4: 137 mM NaCl, 2.7 mM KCl, 1.5 mM KH_2PO_4 anhydrous, 8.1 mM $Na_2HPO_4{\cdot}2H_2O$, and 1 mM EDTA in ultrapure water. Solution should be autoclaved prior to use and stored at room temperature.
11. Collagen-I coating solution: prepare a 50 ng/mL collagen solution in 0.1 % acetic acid (*see* **Note 10**).
12. Sodium citrate buffer, pH 3: 10 mM KCl, 40 mM citrate, 135 mM NaCl in ultrapure water. Solution should be autoclaved prior to use and stored at room temperature.

Table 1
Recommended compositions for the freezing medium

Cell type	Growth medium (%)	DMSO (%)	Serum	
			FCS (%)	HS (%)
Primary fibroblasts	65	5	30	–
Primary keratinocytes	60	10	30	–
PC12 cells	65	5	10	20

13. Acycloguanosine: 100 μM (*see* **Note 6**).
14. Freezing medium: prepare freezing medium dependent on each cell type according to Table 1.

2.3 Organotypic Cell Culture Media and Components

1. Submerse media: Keratinocyte Growth Medium (KGM) (*see* Subheading 2.2, **item 2**) supplemented with descending FCS concentration: 5 %, 2 % and 0 %, respectively.
2. Airlift media: Keratinocyte Growth Medium (KGM) supplemented with all provided stock solutions except for BPE and hEGF (*see* Subheading 2.2, **item 2**) and additional 1.88 mM $CaCl_2$ (*see* **Note 7**).
3. Cell culture inserts, multidish 24-well, pore size 8 μm.
4. Serological pipette, 5 mL, sterile.
5. Neutralization solution: add 7.5 mL 3 M HEPES (4-(2-hydroxyethyl)-1-piperazineethanesulfonic acid) solution, pH 7.8 (*see* **Note 8**), 1.25 mL of a 5 mg/mL chondroitin-4-sulfate solution, 1.25 mL of a 5 mg/mL chondroitin-6-sulfate solution and 7.5 mL heat-inactivated fetal calf serum (FCS) to 232.5 mL DMEM 2× medium. Solution should be sterile-filtered prior to use.
6. Type I collagen: use either commercially available collagen type I solution in a concentration of 6.0 mg/mL or extract collagen from rat tail tendons and dissolve the collagen fibers in acetic acid and adjust the final concentration of 6.0 mg/mL (*see* **Note 9**).
7. Crosslinker Bio-Link BLX 312.
8. Puradisc FP 30 mm Syringe Filter, Cellulose Acetate, 0.2 μm, sterile.

3 Methods

Carry out all procedures under sterile conditions. All incubations should be performed in a humidified, temperature-controlled CO_2 incubator at 37 °C and 5 % CO_2. For maintenance of the cells it is recommended to subculture them as soon as they reach confluence (80 %), generally every 3–4 days. The cells should be subcultured at least once before using them for experiments. Diligently follow the Biosafety Level disposal regulations when disposing viral and human materials.

3.1 Generation of a Quiescently HSV-1 Infected Neuronal Cell Line

1. Seed PC12 cells at a density of 3.9×10^4 cells/cm^2 on a collagen 1 pre-coated (*see* **Note 10**) cell culture flask containing pre-warmed PC12 growth medium and allow them to settle overnight.
2. Replace the growth medium with fresh medium containing 100 μM acycloguanosine (*see* **Note 6**) and 100 ng/mL murine nerve growth factor beta (mbeta-NGF), and cultivate for at least 1 week (*see* **Note 11**).
3. After the acycloguanosine pretreatment for 1 week remove medium and floating cells.
4. Add an appropriate volume of Ca^{2+}/Mg^{2+} free phosphate-buffered saline (e.g., 2 mL to a 75 cm^2 flask) and incubate for 3–5 min at 37 °C to detach cells from the surface of the cell culture flask (*see* **Note 12**).
5. Resuspend the cells in fresh growth medium (e.g., 8 mL in case of a 75 cm^2 flask) and transfer them in a new cell culture flask (not coated).
6. Determine cell density and total cell counts by a Coulter Counter or manually using a hemocytometer chamber.
7. Infect the cells with HSV-1 at an MOI (multiplicity of infection) of 20, and allow the virus to absorb for 2 h at 37 °C.
8. Remove unbound virus by washing the cells with sodium citrate buffer for 30 s, immediately thereafter centrifuge the cell suspension at $100 \times g$ for 1 min and resuspend the cell pellet in 20 mL of fresh growth medium to neutralize the low-pH treatment.
9. Centrifuge cells at $100 \times g$ for 3 min and resuspend the cell pellet in predetermined volume of chilled freezing medium (*see* Subheading 2.2, **item 12**) at a concentration of 3×10^6 cells/mL. Dispense aliquots of this suspension into cryovials (e.g., 1 mL in a 1.8 mL cryovial) and freeze cells in a controlled-rate freezing apparatus (Nalgene®Mr. Frosty) at −80 °C (*see* **Note 13**).
10. Transfer frozen vials to liquid nitrogen for long-term storage (*see* **Note 14**).
11. Verify the quiescent state of the HSV-1 infected PC12 cells either by PCR, transmission electron microscopy (TEM) or tissue culture infective dose 50 assay ($TCID_{50}$) before using them for further experiments. Thereby, in a quiescent state

Table 2
All components of the collagen gel solution. Example represents volumes for six skin models, 600 μL each

Model counts	Model surplus	Fibroblast cell count (0.25 × 10⁶/mL)	PC12 cell count (0.14 × 10⁶/mL)	Neutralization solution	Collagen-I (6 mg/mL)	Keratinocyte cell count
6	10	1.5×10^6/mL	0.84×10^6/mL	2 mL	4 mL	2×10^6/mL

viral DNA should be detectable whereas extracellular and intracellular virus activity and viral particles should not be found within the HSV-1 infected cells.

3.2 Generation of a 3D HSV-1 Infection Model

The present proportions are indicated for six skin models (600 μL each). Adjust the number of models according to your individual requirements. To generate the epidermal layer either primary isolated keratinocytes or the spontaneous immortalized human skin keratinocyte cell line HaCaT can be used [28].

1. Add the appropriate volume of collagen-I and neutralization solution in sterile tubes according to Table 2 and keep them on ice.
2. Transfer the required amount of cell culture inserts into a 24-well plate.
3. Harvest the cells (PC12 cells and primary fibroblasts) at desired cell counts according to Table 2.
4. Resuspend each cell pellet in the chilled neutralization solution and mix well.
5. Immediately thereafter, transfer the cell suspension in the appropriate volume of chilled collagen-I solution, mix well rapidly using a serological pipette without introducing air bubbles (*see* **Note 15**).
6. Pipette 600 μL of the chilled cell-collagen suspension in each tissue culture insert and transfer to 37 °C for 10–20 min to initiate polymerization of the collagen (*see* **Note 16**).
7. Overlay the polymerized fibroblast/PC12-collagen gel (*see* Fig. 1a) with DMEM medium supplemented with 10 % FCS and incubate overnight at 37 °C and 5 % CO_2.
8. After formation of the dermal layer [24 h (±2 h)] aspirate the medium completely, thereby avoid injury of the matrix.
9. To generate the epidermal layer, harvest the keratinocytes (*see* **Note 17**) and resuspend desired cell counts in 50 μL (per model) of submerse medium supplemented with 5 % FCS (*see* Subheading 2.3, **item 1**). Seed cell suspension onto each collagen gel matrix. *See* Table 2 for recommended cell counts.

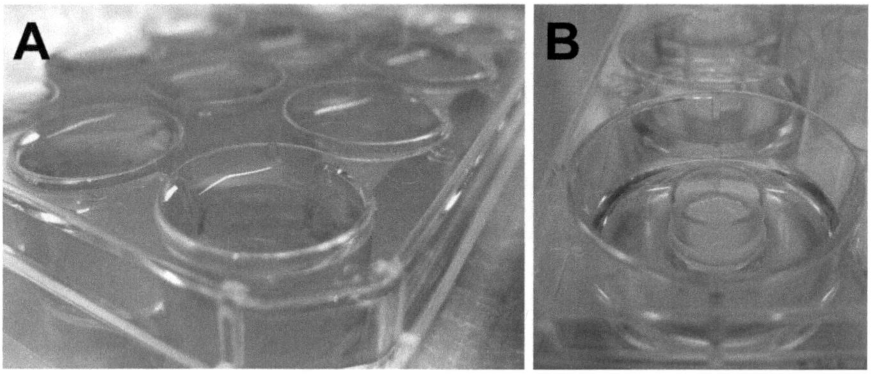

Fig. 1 Cultivation phase of a 3D skin model. (**a**) Submerge-phase cultivation and (**b**) airlift-phase cultivation

10. Transfer to a 37 °C incubator for 60 min and allow the keratinocytes to adhere.
11. Overlay the two-layered matrix with submerse medium supplemented with 5 % FCS.
12. After 2–3 days in culture with submerse medium supplemented with 5 % FCS decrease FCS concentration: aspirate the medium completely and overlay the collagen matrix with submerse medium supplemented with 2 % FCS.
13. After another 2–3 days in culture repeat **step 12** using serum free submerse medium (0 % FCS).
14. After the submerge-phase cultivation (5–7 days) (*see* Fig. 1a) lift the organotypic cell cultures to the air–liquid interface using airlift medium (*see* Subheading 2.3, **item 2**) for 12–14 days (*see* **Note 18**). For cultivation in the airlift-phase transfer the skin equivalents within the cell culture inserts into a 6-well cell culture plate using sterile forceps (*see* Fig. 1b). Replace medium every 2–3 days. Add the airlift medium carefully, laterally on the outside of the cell culture inserts without wetting the epidermal layer. Make sure that the medium level does not exceed the meniscus of the collagen matrix (*see* **Note 19**).
15. After the end of the airlift-phase the infection models can be used for irradiation via UVB light in order to reactivate the virus (*see* Subheading 3.3).
16. For histological and immunohistological analyses the 3D skin models (*see* Fig. 2) should be formalin-fixed and paraffin-embedded according to standard protocols.

3.3 In Vitro Reactivation of Quiescently Infected HSV-1

For reactivation of HSV-1 within the 3D infection model use the skin equivalents at the end of the airlift-phase (*see* Subheading 3.2, **step 15**). Perform the irradiation at room temperature in air atmosphere using a UVB lamp (Crosslinker Bio-Link BLX 312) which emits mainly radiation of 312 nm wavelength (UVB).

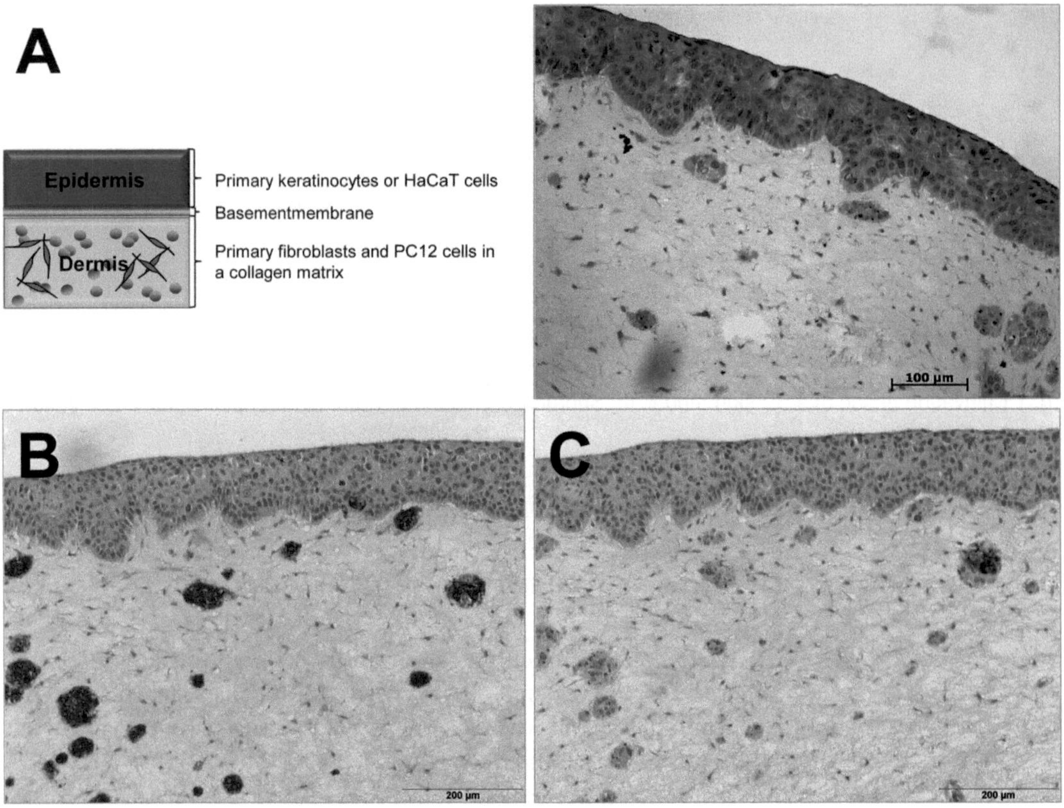

Fig. 2 Schematic figure and cross sections of the HSV-1 infection model with quiescently infected PC12 cells integrated into the dermal layer of the skin equivalent. (**a**) H&E staining of cross sections shows uniformly distributed PC12 cells within the dermal layer of the *in vitro* epidermal model. (**b**) Immunohistochemical staining using a monoclonal antibody raised against a specific neuronal marker (Tyrosine Hydroxylase; dilution 1:400) shows PC12 cells in *brown*. (**c**) Isotype Control (IgG2α) (Color figure online)

1. Add 1.2 mL airlift medium into a new 6-well cell culture plate.
2. Transfer the organotypic cell cultures within the cell culture inserts into the 6-well cell culture plate using sterile forceps.
3. Remove the lid of the cell culture plate and place it centrally under the UV source to ensure homogenous exposure (*see* **Note 20**).
4. Expose the *in vitro* skin models to UV light at an intensity of 1,000 mJ/cm^2 (*see* **Note 21**).
5. Aspirate the medium completely (avoid injuring the matrix), transfer the skin models back into the initial 6-well plate with replaced airlift medium and incubate the irradiated skin equivalents at 37 °C.
6. For additional exposure repeat **steps 1–5** after 24 h (±2 h).
7. Let the organotypic cell cultures incubate for another 6 days as described before (*see* Subheading 3.2, **step 14**).

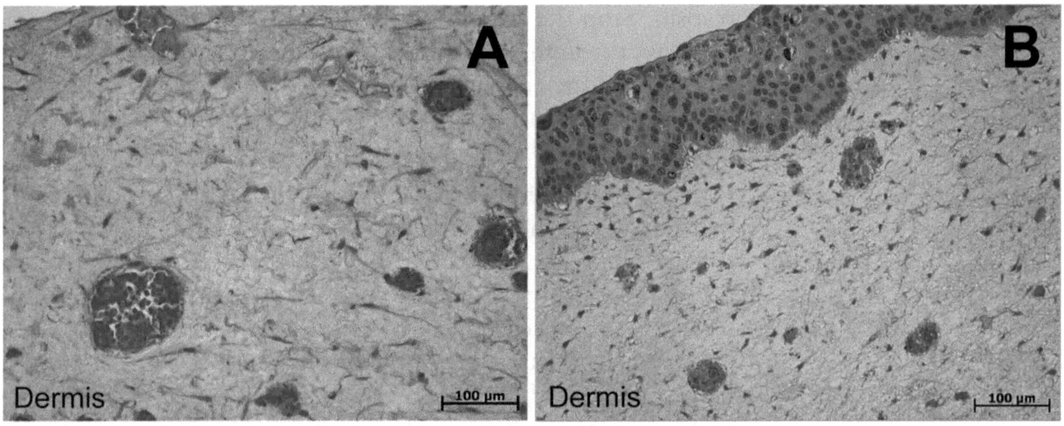

Fig. 3 Immunohistochemical detection of HSV-1 using a polyclonal rabbit anti-HSV-1 antibody (dilution 1:100) in cross section of the HSV-1 infection model. (**a**) Successful reactivation of HSV-1 (*red*) could be achieved after UVB-irradiation at 1,000 mJ/cm^2. Thereby, reactivation is shown within the PC12 cell clusters (*red*). (**b**) The non-irradiated infection model shows no virus reactivation within the integrated PC12 clusters (Color figure online)

8. For histological and immunohistological analyses of the UVB irradiated *in vitro* HSV-1 infection models (*see* Fig. 3), the skin equivalents should be formalin-fixed and paraffin-embedded according to standard protocols.

4 Notes

1. Human herpes virus is a Biosafety Level 2 (BSL-2) pathogen. Prepare the virus stock in African green monkey kidney (Vero-B4) cells (American Type Culture Collection; ATCC ACC 33) [29]. Aliquot and store the virus stock at −70 °C or colder. The virus titer can be determined by a standard plaque assay on a Vero-B4 cell monolayer [29]. The virus infectivity in supernatants can be determined by a $TCID_{50}$ (tissue culture infective dose 50) assay on Vero-B4 cells [30].
2. Foreskin full-thickness tissues should only be used from healthy males aged newborn to infant (up to 7 years) from elective circumcisions. The use of adult tissue is not recommended due to the reduced cell viability. To isolate the primary keratinocytes dissect the dermis, cut into small pieces using a scalpel under sterile conditions and incubate overnight with 2 units Dispase in PBS at 4 °C. Dispase is a protease which cleaves fibronectin, collagen IV, and to a lesser extent collagen-I and is therefore used to separate the epidermal from the dermal layer of the full-thickness tissue. To isolate the keratinocytes remove the epidermal layer from the dermis, cut the epidermis into very small pieces and incubate for 5 min at 37 °C with 0.05 %

trypsin. Stop the reaction by adding FCS. Resuspend the cells and pass through a 100 μm sieve and centrifuge at 400 × *g* for 3 min. Seed the cells at a density of 2.6×10^4 cells/cm^2 in KGM supplemented with 10 % FCS. Change media after 24 h to remove tissue debris.

3. Primary human fibroblasts can be isolated by the outgrowth method using the dermal layer of the foreskin full-thickness tissues (*see* **Note 2**). Therefore, cut dermis into small pieces of 0.5–1 mm^3 using a scalpel under sterile conditions. Place the small tissue pieces in a sterile culture dish (3-cm Ø), 5–6 pieces/dish and arrange them in a circle. Allow them to adhere before adding culture media (DMEM supplemented with 20 % FCS). During the next 3–10 days, the cells will grow out from the tissue pieces and will gradually form a confluent layer. Subculture the cells after they reach confluence using a 0.05 % trypsin solution. Seed the cells at a density of 1.3×10^4 cells/cm^2 in fibroblast growth media (DMEM supplemented with 10 % FCS).
4. PC12 cells attach to the surface when cultured on collagen I-pre-coated cell culture flasks. Prepare collagen I-coating solution at 50 ng/mL in 0.1 % acetic acid (*see* **Note 10**). Subculture the cells using Ca^{2+}/Mg^{2+}-free phosphate-buffered saline. Seed the cells at a density of 3.9×10^4 cells/cm^2 in PC12 growth media.
5. Perform heat-inactivation of the serum at 56 °C for 30 min. Ensure there is sufficient water within the water bath to immerse the bottle above the level of serum. Do not allow the water level to reach the neck of the bottle. Mix gently every 5 min to assure uniform heating. Aliquot (50 mL) and store at −20 °C.
6. To prepare a 10 mM acycloguanosine stock solution dissolve 20 g acycloguanosine in 45.16 mL dimethyl sulfoxide (DMSO). Due to the light sensitivity of DMSO it is recommended to store the stock solution in the dark by wrapping the tube/bottle in aluminum foil. Use acycloguanosine at a final concentration of 100 μM: add 100 μL stock solution to 10 mL PC12 cell culture media.
7. To prepare a 1.88 mM $CaCl_2$ solution (final concentration after addition to 500 mL airlift medium) dissolve 0.138 g $CaCl_2 \cdot 2H_2O$ in 1 mL of ultrapure water. The solution should be sterile-filtered prior to adding to the airlift medium.
8. Add about 18 mL to a 100 mL cylinder. Weigh 35.7 g HEPES and transfer it gently to the cylinder without performing lumps. Add ten sodium hydroxide (NaOH) pellets/flakes and mix until the solution is clear. Adjust volume to 50 mL with ultrapure water and adjust pH with 1 N NaOH. HEPES can be dissolved faster when the solution is heated (37 °C) while mixed but

ensure to bring the solution to room temperature before adjusting pH. Solution should be sterile-filtered prior to use.

9. To extract collagen from rat tail tendons let frozen tails thaw to room temperature and soak in 70 % ethanol for 30 min. Then split the skin at the tail root and peel away from the tail. Cut off the distal and proximal quarters of the tail (~2–3 cm) and divide the remainder into three pieces. Dissect each tendon separately using a scalpel and tease the tendons with the blade to separate the fibers. Weigh the tendons (1 g tendon will give 100 mL collagen solution) and transfer them to a conical 50 mL tube and wash the tendons first with 70 % ethanol and afterwards repeatedly with PBS. When the tendons appeared clean transfer to a 0.1 % acetic acid solution (100 μL glacial acid per 1 g tendon in 100 mL water). Stir the mixture gently at 4 °C and 500 rpm on a magnetic stirrer for at least a week. Working at 4 °C, decant the mixture into 50 mL Falcon tubes and centrifuge at 3,000 × *g* for 15 min. In a further step, centrifuge the supernatant again at 17,700 × *g* for 1 h. The purity of the collagen can be determined by SDS-PAGE. Store the supernatant aliquots at −20 °C until use. The collagen solution prepared by this method has been shown to exhibit a high degree of purity.
10. Use collagen-I solution as described in **Note 9** to prepare a collagen-I coating solution dissolved in 0.1 % acetic acid at a concentration of 50 ng/mL. Dispense the appropriate amounts into the desired tissue culture Flask (e.g., 10 mL/75 cm^2 flask) and incubate for at least 30 min. Aspirate the coating solution and allow the flask to air-dry.
11. We observed that pretreating the cells with antiviral compound and murine nerve growth factor beta (mbeta-NGF) prior to infection is beneficial for generating quiescently infected cells.
12. The cells can be displaced easier from the flask by slapping the flask sharply against your hand one or two times.
13. For ideal cryopreservation, the freezing temperature should be decreased by 1 °C/min.
14. For best results we recommend transferring the vials to the vapor phase of a liquid nitrogen storage facility 2–4 days after the cells have reached −80 °C.
15. The pH of the collagen gel solution should be neutral, which is indicated by the pink/red color of phenol red in the medium.
16. In case of optimal polymerization the collagen gel will look cloudy.
17. Cell diameter should be at a range of Ø 16–22 μm to assure undifferentiated state. A high amount of differentiated cells

would lead to an inappropriate development of the epidermal layer.

18. A high concentration of extracellular calcium promotes keratinocyte differentiation. The elevation of extracellular calcium concentration induces an increase in intracellular free calcium in keratinocytes [31]. Thereby the epidermal keratinocytes differentiate into a multilayered epidermis including a stratum corneum [24].
19. Wetting the dermal layer would inhibit the formation of a multilayered epidermis.
20. For multiple dishes irradiate one dish at a time.
21. The length of the time will vary according to the strength of the UV bulbs. Therefore, it is not possible to estimate the UVB intensity from the duration unless the strength of the UV bulbs is known.

References

1. Decman V, Freeman ML, Kinchington PR, Hendricks RL (2005) Immune control of HSV-1 latency. Viral Immunol 18:466–473
2. Frampton AR Jr, Goins WF, Nakano K, Burton EA, Glorioso JC (2005) HSV trafficking and development of gene therapy vectors with applications in the nervous system. Gene Ther 12:891–901
3. Su YH, Moxley MJ, Ng AK, Lin J, Jordan R, Fraser NW et al (2002) Stability and circularization of herpes simplex virus type 1 genomes in quiescently infected PC12 cultures. J Gen Virol 83:2943–2950
4. Feldman LT (1994) Transcription of the Hsv-1 genome in neurons in-vivo. Semin Virol 5: 207–212
5. Perng GC, Dunkel EC, Geary PA, Slanina SM, Ghiasi H, Kaiwar R et al (1994) The latency-associated transcript gene of herpes simplex virus type 1 (HSV-1) is required for efficient in vivo spontaneous reactivation of HSV-1 from latency. J Virol 68:8045–8055
6. Perng GC, Slanina SM, Yukht A, Ghiasi H, Nesburn AB, Wechsler SL (2000) The latency-associated transcript gene enhances establishment of herpes simplex virus type 1 latency in rabbits. J Virol 74:1885–1891
7. Umbach JL, Kramer MF, Jurak I, Karnowski HW, Coen DM, Cullen BR (2008) MicroRNAs expressed by herpes simplex virus 1 during latent infection regulate viral mRNAs. Nature 454:780–783
8. Clements GB, Kennedy PGE (1989) Modulation of Herpes-Simplex Virus (Hsv) infection of cultured neuronal cells by nerve growth-factor and antibody to Hsv. Brain 112:1277–1294
9. Danaher RJ, Jacob RJ, Chorak MD, Freeman CS, Miller CS (1999) Heat stress activates production of herpes simplex virus type 1 from quiescently infected neurally differentiated PC12 cells. J Neurovirol 5:374–383
10. Danaher RJ, Jacob RJ, Miller CS (1999) Establishment of a quiescent herpes simplex virus type 1 infection in neurally-differentiated PC12 cells. J Neurovirol 5:258–267
11. Danaher RJ, Jacob RJ, Miller CS (2006) Reactivation from quiescence does not coincide with a global induction of herpes simplex virus type 1 transactivators. Virus Genes 33: 163–167
12. Danaher RJ, McGarrell BS, Stromberg AJ, Miller CS (2008) Herpes simplex virus type 1 modulates cellular gene expression during quiescent infection of neuronal cells. Arch Virol 153:1335–1345
13. Kumar M, Kaufman HE, Clement C, Bhattacharjee PS, Huq TS, Varnell ED et al (2010) Effect of high versus low oral doses of valacyclovir on herpes simplex virus-1 DNA shedding into tears of latently infected rabbits. Invest Ophthalmol Vis Sci 51:4703–4706
14. Moxley MJ, Block TM, Liu HC, Fraser NW, Perng GC, Wechsler SL et al (2002) Herpes simplex virus type 1 infection prevents detachment of nerve growth factor-differentiated PC12 cells in culture. J Gen Virol 83: 1591–1600

15. Penfold ME, Armati P, Cunningham AL (1994) Axonal transport of herpes simplex virions to epidermal cells: evidence for a specialized mode of virus transport and assembly. Proc Natl Acad Sci USA 91:6529–6533
16. Sears AE, Meignier B, Roizman B (1985) Establishment of latency in mice by herpes simplex virus 1 recombinants that carry insertions affecting regulation of the thymidine kinase gene. J Virol 55:410–416
17. Stanberry LR (1994) Animal-models and Hsv latency. Semin Virol 5:213–219
18. Su YH, Meegalla RL, Chowhan R, Cubitt C, Oakes JE, Lausch RN et al (1999) Human corneal cells and other fibroblasts can stimulate the appearance of herpes simplex virus from quiescently infected PC12 cells. J Virol 73:4171–4180
19. Su YH, Moxley M, Kejariwal R, Mehta A, Fraser NW, Block TM (2000) The HSV 1 genome in quiescently infected NGF differentiated PC12 cells can not be stimulated by HSV superinfection. J Neurovirol 6: 341–349
20. Tanaka S, Minagawa H, Toh Y, Liu Y, Mori R (1994) Analysis by RNA-PCR of latency and reactivation of herpes simplex virus in multiple neuronal tissues. J Gen Virol 75(Pt 10): 2691–2698
21. Hukkanen V, Mikola H, Nykanen M, Syrjanen S (1999) Herpes simplex virus type 1 infection has two separate modes of spread in three-dimensional keratinocyte culture. J Gen Virol 80:2149–2155
22. Syrjanen S, Mikola H, Nykanen M, Hukkanen V (1996) In vitro establishment of lytic and nonproductive infection by herpes simplex virus type 1 in three-dimensional keratinocyte culture. J Virol 70:6524–6528
23. Visalli RJ, Courtney RJ, Meyers C (1997) Infection and replication of herpes simplex virus type 1 in an organotypic epithelial culture system. Virology 230:236–243
24. Dieterich C, Schandar M, Noll M, Johannes FJ, Brunner H, Graeve T et al (2002) In vitro reconstructed human epithelia reveal contributions of Candida albicans EFG1 and CPH1 to adhesion and invasion. Microbiology 148:497–506
25. Mertsching H, Weimer M, Kersen S, Brunner H (2008) Human skin equivalent as an alternative to animal testing. GMS Krankenhhyg Interdiszip 3:Doc11
26. Maggioncalda J, Mehta A, Su YH, Fraser NW, Block TM (1996) Correlation between herpes simplex virus type 1 rate of reactivation from latent infection and the number of infected neurons in trigeminal ganglia. Virology 225:72–81
27. Sawtell NM (1997) Comprehensive quantification of herpes simplex virus latency at the single-cell level. J Virol 71:5423–5431
28. Schoop VM, Mirancea N, Fusenig NE (1999) Epidermal organization and differentiation of HaCaT keratinocytes in organotypic coculture with human dermal fibroblasts. J Invest Dermatol 112:343–353
29. Blaho JA, Morton ER, Yedowitz JC (2005) Herpes simplex virus: propagation, quantification, and storage. Curr Protoc Microbiol Chapter 14:Unit 14E 1
30. LaBarre DD, Lowy RJ (2001) Improvements in methods for calculating virus titer estimates from TCID50 and plaque assays. J Virol Methods 96:107–126
31. Hennings H, Kruszewski FH, Yuspa SH, Tucker RW (1989) Intracellular calcium alterations in response to increased external calcium in normal and neoplastic keratinocytes. Carcinogenesis 10:777–780

Chapter 18

In Vivo Visualization of Encephalitic Lesions in Herpes Simplex Virus Type 1 (HSV-1) Infected Mice by Magnetic Resonance Imaging (MRI)

Wali Hafezi and Verena Hoerr

Abstract

Herpes simplex encephalitis (HSE) is one of the most severe viral infections affecting the temporal lobes of the brain. Despite the improvements in diagnosis and antiviral drug treatment, one third of all patients fail to respond to therapy or subsequently suffer neurological relapse and develop long term neurological damage [1, 2]. Magnetic resonance imaging (MRI) is among the appropriate noninvasive tools for early diagnosis of viral central nervous system (CNS) infections. In this chapter we introduce a mouse model for HSE and describe a MRI protocol to characterize the pathogenesis of HSE over time.

Key words Herpes simplex virus type 1, Herpes simplex encephalitis, Mouse model, HSE pathogenesis, Viral infection, Magnetic resonance imaging

1 Introduction

Herpes simplex encephalitis (HSE) induced by Herpes simplex virus type 1 (HSV-1) is the most common cause of fatal sporadic encephalitis in immunocompetent individuals [3, 4]. Primary infections with HSV-1 are usually mild and often occur during childhood. Once the orolabial mucosal cells are infected, the virus replicates, is transmitted to the sensory nerve endings, and establishes life-long latency in the trigeminal ganglia [5].

HSV-1 can be reactivated by both internal (e.g., feverish infection) and external (UV-radiation; stress factors) stimuli and can result in recurrent infections such as herpes labialis or herpes corneae [6]. The virus can also be transported anterogradely to the brain and can infect frontal as well as temporal regions which results in necrotizing hemorrhagic encephalitis with subsequent cerebral edema [7].

To study the pathogenesis of HSV infections different mouse models have been established. However, there is an enormous

Susanne M. Bailer and Diana Lieber (eds.), *Virus-Host Interactions: Methods and Protocols*, Methods in Molecular Biology, vol. 1064, DOI 10.1007/978-1-62703-601-6_18, © Springer Science+Business Media, LLC 2013

variation in the outcome of the HSV-1 infection of the nervous system in laboratory animals which makes the choice of an appropriate animal model difficult. In mice the course of infection depends on various factors including the viral strain, infectious dose, mouse strain and age as well as prior immunization history of the animal [8].

In order to visualize the course of an infection in real time *in vivo* imaging techniques such as MRI, computer tomography (CT), fluorescence imaging, and positron emission tomography (PET) have been developed over the past decades [9–11]. Due to high spatial resolution and high tissue contrast, especially MRI represents a versatile method to investigate the stage of infection noninvasively. In addition to the detection of high resolution morphological structures, MRI provides detailed anatomical information. To detect infections and inflammatory response of the host, especially T2-weighted imaging sequences are frequently used to create high tissue contrast in foci of inflammation [12]. Besides morphological changes, different immune cells and enzyme activity can be monitored during an infection and tracked by MRI using contrast agents such as iron oxide particles or lanthanide complexes [13, 14]. Here we describe a mouse model of Herpes simplex encephalitis following up the inflammatory process by MRI.

2 Materials

All tissue culture work and virus work should be carried out in a class II lamina air flow hood. Always wear a lab coat, protective glasses and gloves when handling virus. To avoid contamination discard gloves if contamination is suspected. Use disposable cell culture flasks, dishes, plates, and pipettes exclusively.

2.1 Production of High Titer Virus Stock

1. Cell culture: BHK (baby hamster kidney) cells and Vero cells (kidney cells from African green monkey).
2. Virus strain: HSV-1 17syn+.
3. Cell culture roller bottle.
4. Minimal essential medium (MEM) containing 2 % FCS (fetal calf serum).
5. CMC (carboxymethylcellulose) medium: suspend 2 g of CMC in 100 mL bidest H_2O (2 %).
6. Prepare CMC medium for plaque titration in a final concentration of 1 %, by adding 1 volume of CMC medium 2 % to 1 volume of 2× medium.
7. Phosphate-buffered saline (PBS) without Ca^{2+} and Mg^{2+}.
8. 37 °C incubator for roller bottle.
9. 50 mL conical centrifuge tube.

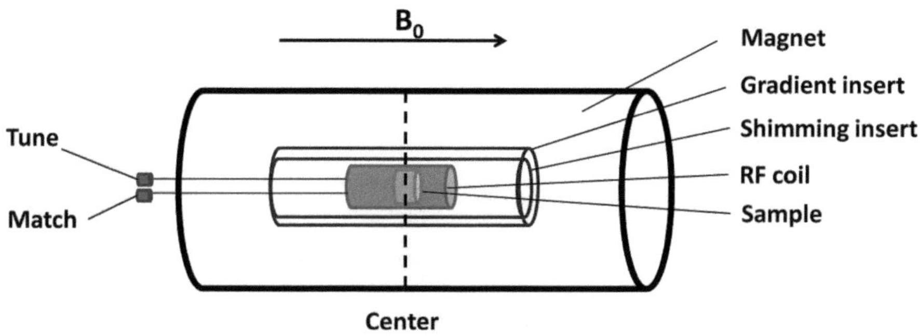

Fig. 1 Schematic setup of an *in vivo* small animal imaging system

10. Disposable pipettes.
11. 6-well tissue culture plates.
12. Benchtop centrifuge.
13. Ultracentrifuge.
14. Tissue culture (class II) hood.
15. 1.5 mL microcentrifuge tubes.
16. Gloves, lab coat, protective glasses.

2.2 Herpes Simplex Virus Type 1 (HSV-1) Infection in Mice

1. C57BL/6N female mice, 4 weeks old.
2. Ketamine 10 %: ketamine hydrochloride 115.34 mg/mL is equivalent to ketamine 100 mg/mL.
3. Xylazine 2 %: xylazine hydrochloride 23.3 mg/mL is equivalent to xylazine 20 mg/mL.
4. 1 mL tuberculin syringe 0.01–1 mL, needle (27 G × ¾″; 04 × 20).
5. Binocular loupe.
6. Cold-light source.
7. Heating pad.

2.3 MRI Hardware

1. *In vivo* preclinical and molecular MRI systems usually operate at static magnetic field strengths of 4.7, 7, 9.4 and 11.7 T (*see* **Notes 1** and **2**).
2. Magnets are equipped with shim coils, magnetic field gradient coils, tuning and matching capacitors and transmitter and receiver coils (Fig. 1).
3. Shim coils homogenize the magnetic field once the sample or animal is placed in the magnet.
4. Magnetic field gradients encode the spatial information in imaging systems. Therefore three sets of gradient coils create additional fields along the three main field directions.
5. The tuning capacitor changes the resonance frequency of the radiofrequency (rf) coil which has to be adjusted to that of the subject.

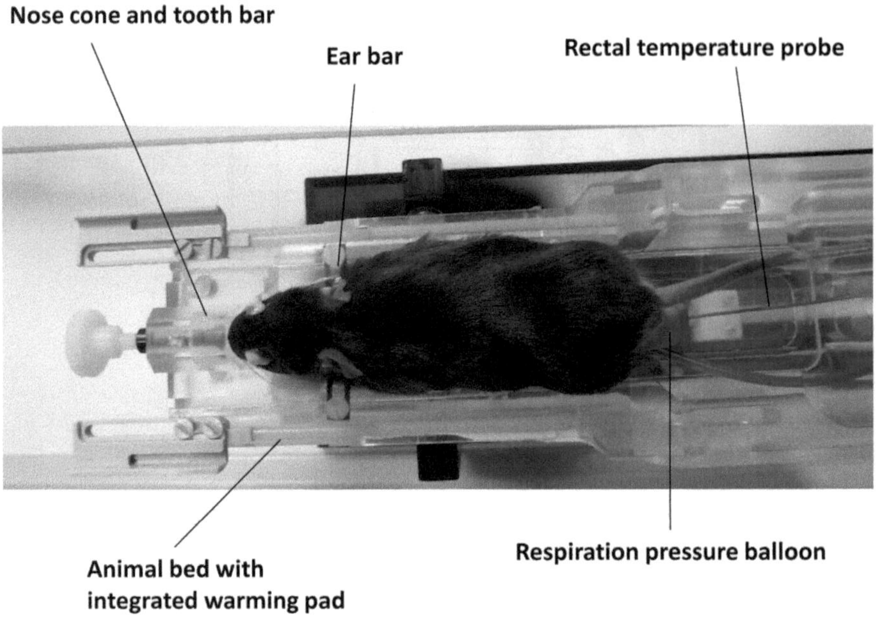

Fig. 2 Mouse setup for *in vivo* MRI

6. Matching a probe adjusts its impedance to that of the subject.
7. ^{1}H transmitter coils provide a radiofrequency magnetic field that excites the ^{1}H nuclei (*see* **Note 3**).
8. Receiver coils detect the decaying cosinusoidal signal from the excited spins (*see* **Note 3**).
9. There are two major categories of small animal MRI coils: volume coils and surface coils (*see* **Note 4**).
10. Even though volume coils provide the advantage of a homogeneous data acquisition throughout the transversal imaging plane, we preferred using a surface coil in our study and used a mouse head cryogenic probe taking advantage of increased SNR by a factor of 2.5 compared to commonly used surface coils.

2.4 Animal Handling

Animal handling and preparation represents a critical part in *in vivo* MRI and can strongly affect the image quality and reproducibility. The smaller the region of interest the more important animal handling becomes. Special attention should be paid to positioning (*see* **Note 5**), warming, anesthesia and physiological control. Important animal positioning, handling and monitoring equipment are illustrated in Fig. 2 (*see* **Note 6**).

1. MRI of animals is most commonly performed under inhalation anesthesia with isoflurane and a mixture of air, oxygen, or N_2O.
2. Animal bed: Position the animal and insert it to the center of the magnet and the coil.

3. Nose cone and tooth/bite bar: Fix and center the animal head.
4. Ear bars: Stabilize the animal head to reduce motion artifacts.
5. Warming pad: Keep the animal's core temperature stable.
6. Anesthesia integration:
   - Air and oxygen supply: Carrier gas for anesthetics.
   - Vaporizer: Mixes anesthetic with air or oxygen.
   - Gas flow regulation.
   - Tubing to connect vaporizer with nose cone on the animal bed.
   - Induction chamber: Box which is filled with anesthetic gas to induce anesthesia.
   - Anesthetic gas removal: Pump that removes non-breathed anesthetic gas.
7. Physiological sensor integration:
   - Respiration control: Sensitive pressure transducer connected to air-filled pillow or balloon.
   - Temperature control: Rectal temperature probe connected to an electrically isolated temperature amplifier.
8. Physiological monitoring system which measures and displays various physiological parameters.

3 Methods

3.1 Infection of BHK Cells

1. Grow 5×10^8 BHK cells to confluence in roller bottle (*see* **Note** 7).
2. Infect the cells at MOI (multiplicity of infection) of 0.01–0.05 by adding the virus (HSV-1 17syn+) in 20 mL medium containing 2 % FCS. Allow the virus to adsorb for 1 h at 37 °C.
3. Discard the virus inoculum and wash the cells with 30 mL of PBS (*see* **Note** 7). Add 60 mL of fresh MEM containing 2 % FCS to the cells.
4. Incubate the infected cells for 3–5 days in 37 °C incubator, until all cells show advanced cytopathic effect (CPE). CPE is characterized as any morphological change in the host cell due to infection (cell rounding, swelling or shrinking and detachment from the surface of plastic).
5. Transfer the supernatant to a 50 mL conical centrifuge tube and pellet the cells by centrifugation at $1{,}500 \times g$ for 30 min.
6. Decant the supernatant in two 30 mL Beckman centrifuge tubes and centrifuge in SW28 rotor for 90 min at $100{,}000 \times g$.
7. Discard the supernatant carefully; resuspend the virus pellet in 300 μL PBS (*see* **Note 8**) and store aliquots of 10 μL at −80 °C.

3.2 Titration of Virus Stock

1. Seed 3×10^5 Vero cells in 6-well dishes to form confluent monolayers by incubating at 37 °C, 5 % CO_2 and 95 % humidity.
2. Pipette 990 μL aliquots of medium containing 2 % FCS into first 1.5 mL microcentrifuge tube, and 900 μL aliquots of medium into the next seven tubes for each titration.
3. Carry out tenfold dilution series of virus stock up to 10^{-9}, starting with 10 μL of the undiluted virus into the first tube (1:100). Change the tip between each dilution step.
4. Remove the medium from cells and plate out 800 μL of each of the 10^{-4}–10^{-9} dilutions in one well, respectively. Start with highest dilution 10^{-9}, use the same tip for the entire dilution series.
5. Allow the virus to adsorb for 1 h, remove the virus inoculum, wash with 3 mL PBS and add 1× CMC medium to the cells (*see* **Note 9**).
6. Incubate the infected cells in 37 °C incubator for 3–4 days. Remove the 1× CMC medium and fix the cell monolayers with 3.7 % formalin for 10 min.
7. Stain the plaques for 10 min in 0.1 % crystal violet solution (*see* **Note 10**).
8. Wash the monolayers with tap water and dry at room temperature.
9. Count the number of plaques and calculate the titer as plaque forming units/mL (pfu) virus suspension.
10. Titration should be made in duplicate (*see* **Note 11**).
11. Determination of virus titer:
    Example:
    50 plaques in 10^{-8} dilution (800 μL virus inoculum):

$$\frac{50}{800} \times 1{,}000 = 62.5 \text{pfu} / \text{mL}.$$

    The final titer counts 6.25×10^9 pfu/mL.

3.3 General Anesthesia and Infection of the C57BL/6N Mouse

1. Fill up the syringe in the following order: 0.2 mL xylazine 2 % and 0.6 mL ketamine 10 % (=4 mg xylazine and 60 mg ketamine in 0.8 mL). Mix the xylazine/ketamine solution by inverting the syringe several times. Change the needle before injection (*see* **Note 12**).
2. Weigh the mouse prior to narcosis and infection with HSV-1 (*see* **Note 13**).
3. Inject 0.03 mL of the xylazine/ketamine solution i.p. (intraperitoneally) (6 mg/kg xylazine and 90 mg ketamine/kg; kg refers to body weight) and wait for 10 min until the mouse is narcotized (*see* **Note 14**).
4. Dilute the virus stock in PBS to 1×10^6 pfu of HSV-1 in 10 μL (1×10^8 pfu/mL).

5. Place the mouse in the right lateral recumbent position and carefully pipette 10 μL of virus inoculum into the left nostril by using a 10 μL tip and binocular loupe (*see* **Note 15**).
6. Approximately 2 h after injection the mouse recovers from anesthesia. Keep the mouse warm during recovery phase by placing it on the heating pad (*see* **Note 16**).
7. Place the mouse back in the cage after waking up from anesthesia.
8. Weigh and observe the infected mouse daily and record the behavior (anxiety, floppiness, huddle posture, shakiness, and shivering) and local aspect, e.g., changes at the infected nostril (local infection).
9. On day 3–4 post infection the mouse shows the first signs of HSV-encephalitis (loss of body weight approximately 10 %, shivering) (*see* **Note 17**).
10. Prepare the mouse for MRI on day 7–9 post infection.

3.4 Animal Preparation and Handling for MRI Scanning

During MRI measurements, animals are anesthetized with 1–2 % isoflurane and are monitored for core body temperature and respiration rate using a MRI compatible monitoring system (SA Instruments).

1. The mouse should be kept in anesthesia no longer than 1–2 h (*see* **Note 18**).
2. Before one starts the experiment the following points should be checked and the required equipment should be in place:
   - MR scanner ready and appropriate hardware built in.
   - Isoflurane vaporizer filled.
   - Supply lines for anesthesia connected to the mouse cone.
   - Warming bed switched on.
3. Place the mouse into the induction chamber and flood the container with air (flow 1 L/min) and isoflurane (3–5 %).
4. Wait until the mouse is sleeping (approximately 2 min) and the respiration frequency is going down.
5. Put the mouse on the mouse bed with integrated warming system (*see* **Note 19**).
6. Cover the mouse eyes with eye ointment (*see* **Note 20**).
7. Fix the mouse in the nose cone using a bite/tooth bar.
8. Check the connecting tubing from the vaporizer to the nose cone.
9. Adjust the isoflurane flow to 2–2.5 %.
10. Fix the mouse head with ear bars.
11. Place the air-filled respiration pillow/balloon under the abdomen.
12. Connect the respiration pillow to the pressure transducer.

13. Fix the mouse on the mouse bed using adhesive strips.
14. Check respiration and trigger signal on the physiological monitoring system (*see* **Note 21**).
15. Adjust isoflurane flow to a breathing rate of 60–70/min (*see* **Note 22**).
16. Use a rectal temperature probe to record body temperature (*see* **Note 23**).
17. Place the mouse in the magnet.

3.5 MR Imaging

For the following imaging protocol a Bruker Biospec 9.4 T/20 MR system was used equipped with a 0.7 T/m gradient system and a cryogenic probe which provides highest resolution. The system was interfaced to a console running ParaVision software 5.1.

1. Generate a "New Patient" defining a patient name, study name and user name.
2. Generate a "New Scan" and select the scan protocol.
3. Select the scan protocol "A_TRIPILOT_GE_bas" (*see* **Note 24**).
4. Press the "Edit Method" button in the spectrometer control tool and adjust the field of view to 30 mm × 30 mm.
5. Press "GOP" and start the measurement.
6. If the position of the rf coil and the animal head is adequate tune and match the rf coil.
7. Clone the "A_TRIPILOT_GE_bas" protocol.
8. Start adjustment routines and press "Traffic Light" (*see* **Note 25**).
9. The image is acquired automatically.
10. Load a new scan and select a "RARE" protocol (*see* **Note 26**).
11. Create a customized T2-weighted RARE protocol and adjust NMR parameters in "Edit Method" as in Table 1 (*see* **Note 27**).
12. Save the protocol.

Table 1
MR parameters for T2-weighted RARE measurements

Repetition time	4,500 ms
Echo time	40 ms
Field of view	2 cm × 2 cm
Matrix size	256 × 256
Slice thickness	0.6 mm
Slices	12
RARE factor	12

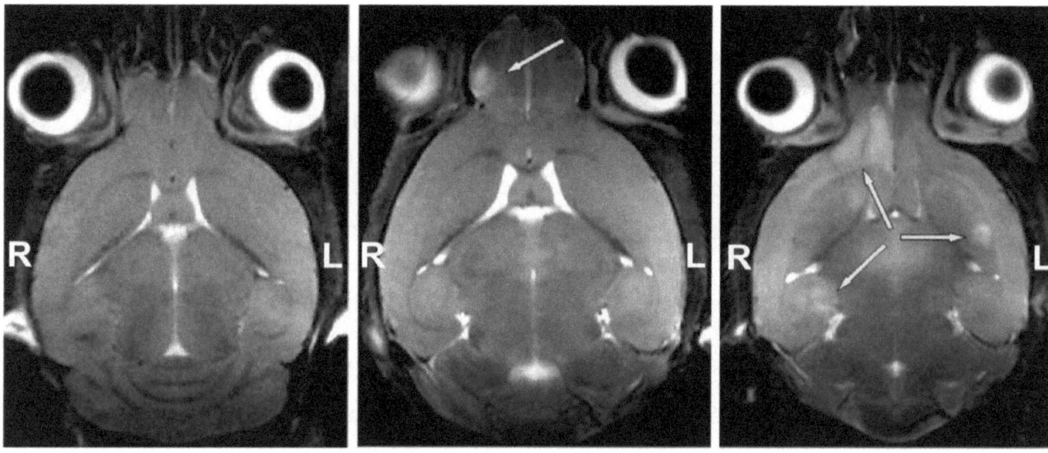

Fig. 3 Herpes simplex lesions by MRI. Coronal T2-weighted RARE images through the mouse brain. *Left*: Brain of a healthy mouse. *Middle*: Day 7 of an HSV-1 infected mouse showing a small inflammatory lesion at the *bulbus olfactorius* (*yellow arrow*). The animal had normal body weight, however, showed reduced activity. *Right*: Day 7 of an HSV-1 infected mouse showing severe disease symptoms: high loss in body weight, tremor, and seizures. The MR image shows expanded lesions from the *bulbus olfactorius* to the *ipsi* and the contralateral *cortex cerebri* (*yellow arrows* at the *right* (*R*) and *left* (*L*) side of the brain)

13. Open the geometry editor and load the previously acquired "A_TRIPILOT_GE_bas" scan as reference scan.
14. Adjust the field of view and the number of slices to the brain volume.
15. Press "Traffic Light" and acquire the customized RARE image (Fig. 3).

3.6 Diagnosis

Herpes simplex encephalitis leads to cerebral inflammatory lesions with edema formation which is detected as hyperintensities in T2-weighted RARE images (Fig. 3).

4 Notes

1. High magnetic field strengths benefit the signal to noise ratio (SNR) which is defined as the mean signal in a region of interest divided by the standard deviation of the background noise. High SNR can be exploited to improve spatial resolution and to reduce the scan time.
2. Most *in vivo* imaging systems consist of horizontal magnets. When vertical magnets are used a decreased cerebral microvascular blood flow (CBF) has to be considered [15].
3. Rf coils represent the core unit in the MR hardware setup. In order to obtain good SNR, size, transmission, and reception should be optimized for each experiment.

4. Conventionally, small animal transmit-receive volume coils, transmit-receive surface coils, and separate transmit and receive coils are used. Volume coils provide high field homogeneity over the sample for excitation and reception while surface coils exhibit only low magnetic field penetration depth and homogeneity, but provide high SNR close to the coil loop [16].
5. For good image quality and to avoid motion artifacts, the region of interest has to be fixed and the animal has to be placed into the center of the magnet and to the sensitivity field of the rf coil.
6. Examples of manufacturers providing animal handling products:

   ASI Instruments (http://www.asi-instruments.com/).

   m2m Imaging (http://www.m2mimaging.com/).

   Agilent (http://www.chem.agilent.com/Library/default.aspx).

   Bruker (http://www.bruker-biospin.com/mri_accessories.html).

   Rapid Biomedical (http://www.rapidbiomed.com/pages/english/animal-coils/equipment/air-heated-holder.php).

   Minerve (http://www.bioscan.com/molecular-imaging/minerve).

   Numira Biosciences (http://www.numirabio.com/site/).
7. Avoid dehydration of the cells by rotating the roller bottle gently.
8. Resuspend the virus particles (pellet) by pipetting up and down carefully (avoid frothing).
9. CMC medium is viscous and prevents the spread and secondary infection by progeny virus.
10. Crystal violet is a harmful substance. For preparation of crystal violet solution always wear safety glasses, gloves, lab coat and a fine-particle mask.
11. Titration should always be made in duplicates from different −80 °C aliquots (same virus stock). The mean of counted plaques of the same dilution steps should be used for calculation of the virus titer.
12. Change the needle before injection of the xylazine/ketamine solution to avoid the injection of more xylazine into the mouse.
13. The control of the body weight of the mouse prior to infection and daily post infection are an useful and important indicator for illness or health of the mouse during the experiment. Loss of body weight indicates lack of food and water ingestion.

14. 10 min post injection of anesthesia the mouse is analgized, the muscles are relaxed and the mouse shows a weak respiratory depression.
15. 10 μL is the maximum of fluid which can be pipetted into the nostril, more volume could be inhaled into the lungs or pour out of the nostril. Using the loupe eases the infection into the tiny nostril.
16. Xylazine causes hypothermia (decrease of body temperature), the mouse should be kept warm on a heating pad during anesthesia. It is recommended to anesthetize the mouse for the infection of mucosa for 1 h to prevent sneezing.
17. Developing encephalitis, the mouse is not anymore able to eat or drink water. It is essential to place water soaked pellets or 0.5 % agarose at the bottom of the cage as an alternative.
18. After measuring animals under isoflurane anesthesia, animals should be kept without anesthesia for 1–2 days.
19. The warming pad can consist of either a thermostatic heating blanket or heating pipelines integrated in the animal bed. It can operate either electrically or by circulating hot water.
20. Rodents do not close their eyes during anesthesia. To protect the eyes from drying-out, eye ointment should to be used.
21. In order to minimize motion artifacts in the MR images, the acquisitions are synchronized with breathing motion, monitored by an air-filled respiration pillow/balloon under the abdomen.
22. Isoflurane should be applied in a concentration as low as possible and should be adjusted by monitoring the breathing rate and breathing pattern. Breathing rates below 70/min indicate an unnecessary high isoflurane concentration, while breathing rates above 120/min indicate an insufficient concentration.
23. Under anesthetics the body temperature of rodents decreases and becomes sensitive to the room temperature. The animal's body temperature has to be maintained carefully and should be kept at 37 °C using warming blankets or warming pads integrated in the animal bed.
24. Scout images acquired by "A_TRIPILOT_GE_bas" scan protocols are required to check the animal position inside the coil and to provide a localization reference to plan the customized experiments. Scout images are acquired in three (or more) orthogonal slices in axial, coronal, and sagittal orientation.
25. The "Traffic Light" button initializes standard adjustments such as: shimming, basic frequency, reference pulse gain, receiver gain.

Table 2
Correlation between repetition time (TR), echo time (TE), and image contrast for a spin echo sequence

TR	TE	Contrast
Short	Short	T1
Short	Long	T1 and T2
Long	Short	Proton density
Long	Long	T2

26. RARE stands for "rapid acquisition with relaxation enhancement" [17] and is the name of a spin echo pulse sequence. It can be used to acquire T1- or T2-weighted images.
27. With appropriate scan parameters, spin echo pulse sequences can be used to obtain different image contrasts: proton density-, T1-, or T2-weighted images (Table 2).

Acknowledgment

This project was supported by the Innovative Medical Research (IMF) Münster, Germany (HÖ 211112).

References

1. Lellouch-Tubiana A, Fohlen M, Robain O, Rozenberg F (2000) Immunocytochemical characterization of long-term persistent immune activation in human brain after herpes simplex encephalitis. Neuropathol Appl Neurobiol 26(3):285–294, nan243 [pii]
2. Whitley RJ, Gnann JW (2002) Viral encephalitis: familiar infections and emerging pathogens. Lancet 359(9305):507–513. doi:10.1016/S0140-6736(02)07681-X, S0140-6736(02)07681-X [pii]
3. Roos KL (1999) Encephalitis. Neurol Clin 17(4):813–833
4. Whitley RJ, Kimberlin DW, Roizman B (1998) Herpes simplex viruses. Clin Infect Dis 26(3):541–553, quiz 554–545
5. Chen SH, Yao HW, Huang WY, Hsu KS, Lei HY, Shiau AL (2006) Efficient reactivation of latent herpes simplex virus from mouse central nervous system tissues. J Virol 80(24):12387–12392. doi:10.1128/JVI.01232-06, JVI.01232-06 [pii]
6. Whitley RJ (1996) Herpes viruses, chapter 68. In: Baron S (ed) Medical microbiology, 4th edn. University of Texas Medical Branch at Galveston, Galveston, TX
7. Boos J, Esiri MM (1986) Sporadic encephalitis I. Viral encephalitis: pathology, diagnosis, and management. Blackwell, Boston, MA
8. Sancho-Shimizu V, Zhang SY, Abel L, Tardieu M, Rozenberg F, Jouanguy E, Casanova JL (2007) Genetic susceptibility to herpes simplex virus 1 encephalitis in mice and humans. Curr Opin Allergy Clin Immunol 7(6):495–505. doi:10.1097/ACI.0b013e3282f151d2, 00130832-200712000-00006 [pii]
9. Burgos JS, Guzman-Sanchez F, Sastre I, Fillat C, Valdivieso F (2006) Non-invasive bioluminescence imaging for monitoring herpes simplex virus type 1 hematogenous infection. Microbes Infect 8(5):1330–1338. doi:10.1016/j.micinf.2005.12.021, S1286-4579(06)00045-1 [pii]
10. Lundberg P, Ramakrishna C, Brown J, Tyszka JM, Hamamura M, Hinton DR, Kovats S, Nalcioglu O, Weinberg K, Openshaw H, Cantin EM (2008) The immune response to herpes simplex virus type 1 infection in suscep-

tible mice is a major cause of central nervous system pathology resulting in fatal encephalitis. J Virol 82(14):7078–7088. doi:10.1128/JVI.00619-08, JVI.00619-08 [pii]

11. Liu H, Ren G, Liu S, Zhang X, Chen L, Han P, Cheng Z (2010) Optical imaging of reporter gene expression using a positron-emission-tomography probe. J Biomed Opt 15(6):060505. doi:10.1117/1.3514659
12. Misra UK, Kalita J, Phadke RV, Wadwekar V, Boruah DK, Srivastava A, Maurya PK, Bhattacharyya A (2010) Usefulness of various MRI sequences in the diagnosis of viral encephalitis. Acta Trop 116(3):206–211. doi:10.1016/j.actatropica.2010.08.007, S0001-706X(10)00209-3 [pii]
13. Lee JS, Kang HJ, Gong G, Jung HD, Lim KH, Kim ST, Lim TH (2006) MR imaging of in vivo recruitment of iron oxide-labeled macrophages in experimentally induced soft-tissue infection in mice. Radiology 241(1):142–148. doi:10.1148/radiol.2403051156, 241/1/142 [pii]
14. Hanaoka K, Kikuchi K, Terai T, Komatsu T, Nagano T (2008) A Gd3+-based magnetic resonance imaging contrast agent sensitive to beta-galactosidase activity utilizing a receptor-induced magnetization enhancement (RIME) phenomenon. Chemistry 14(3):987–995. doi:10.1002/chem.200700785
15. Foley LM, Hitchens TK, Kochanek PM, Melick JA, Jackson EK, Ho C (2005) Murine orthostatic response during prolonged vertical studies: effect on cerebral blood flow measured by arterial spin-labeled MRI. Magn Reson Med 54(4):798–806. doi:10.1002/mrm.20621
16. Haase A, Odoj F, von Kienlin M, Warnking J, Fidler F, Weisser A, Nittka M, Rommel E, Lanz T, Kalusche B, Griswold M (2000) NMR probeheads for in vivo applications. Concepts Magn Reson 12(6):361–388
17. Hennig J, Nauerth A, Friedburg H (1986) RARE imaging: a fast imaging method for clinical MR. Magn Reson Med 3(6):823–833

Chapter 19

Detection of Antigen-Specific T Cells Based on Intracellular Cytokine Staining Using Flow-Cytometry

Tina Schmidt and Martina Sester

Abstract

CMV-specific T cells may be detected and quantified after antigen-specific stimulation based on the induction of cytokines as a readout system. Secreted cytokines may be detected from the supernatant of stimulated cells using ELISA. Alternatively, antigen-specific cytokine-secreting cells may be enumerated using an ELISPOT assay. These assays generally rely on the detection of IFNγ and do not allow for a simultaneous assessment of several cytokines on a single cell basis. Here we describe a flow-cytometry based method to analyze CMV-specific CD4 T cells after specific stimulation with a whole antigen lysate. In this assay, cytokine secretion from stimulated cells is blocked by the addition of brefeldin A. Using a panel of fluorescently labeled antibodies, not only intracellularly accumulated cytokines but also phenotypical characteristics of specifically activated T cells may be quantified in a multiparameter staining approach.

Key words Interferon γ, Cytokines, T cell, Immune response, Flow-cytometry, Antigen

1 Introduction

When a pathogen enters the human body, pathogen-derived antigens are processed by antigen presenting cells (APCs) and presented on their surface. T lymphocytes that are specific for the respective antigen bind to the APC via their T-cell receptor which is immediately followed by T-cell activation and the induction of several cytokines within stimulation times of 6–24 h.

There are three widely applied cytokine-based methods to assess antigen-specific T-cell reactivity. As interferon γ (IFNγ) is commonly analyzed as a readout system, these assays have been termed interferon γ release assays (IGRAs). ELISA-based or ELISPOT-based IGRA formats are commercially available for the detection of immunity towards pathogens, such as Cytomegalovirus (CMV) or *Mycobacterium tuberculosis*. In the ELISA approach, cytokines are quantified from the supernatant of stimulated samples and used as a surrogate to determine the number of antigen-specific T cells. In the ELISPOT assay, peripheral blood mononuclear

Susanne M. Bailer and Diana Lieber (eds.), *Virus-Host Interactions: Methods and Protocols*, Methods in Molecular Biology, vol. 1064, DOI 10.1007/978-1-62703-601-6_19, © Springer Science+Business Media, LLC 2013

cells (PBMCs) are stimulated in a microtiter plate, where the cytokine induced by each sensitized T cell is locally captured by plate-bound anti-cytokine antibodies. Spots of locally bound cytokines are subsequently colorimetrically detected and represent individual footprints of specifically activated cells. Finally, pathogen-specific T cells may be quantified by multiparameter flow-cytometry, using cytokine staining after intracellular accumulation. Unlike ELISA or ELISPOT assays, it allows for the simultaneous analysis of several cytokines on the single cell level and may readily be combined with additional cell surface markers to directly characterize antigen-specific T cells on a phenotypical basis.

Concerning sample material, the ELISPOT assay requires prior isolation of PBMCs, whereas ELISA and flow-cytometry may be performed directly from whole blood. The use of whole blood directly *ex vivo* closely resembles the situation *in vivo* and requires small sample volumes of <1 mL only. In addition, the flow-cytometric approach may be performed after stimulation times as short as 6 h, whereas ELISA or ELISPOT assay require incubation times of 18–24 h.

The flow-cytometric assay is described in detail in this chapter using CMV-specific CD4 T cell analysis as an example. The flow-cytometry based method is a rapid, very sensitive and effective method for the assessment of the individual immunocompetence towards CMV and other pathogens [1–3]. In detail, lymphocytes from whole blood are stimulated by the addition of CMV antigens in the presence of co-stimulatory antibodies. Addition of brefeldin A after 2 h leads to impairment of the vesicular export which causes intracellular accumulation of the induced cytokines. After an additional 4 h of stimulation, the cells are fixed and processed for flow-cytometric analysis by staining of intracellular cytokines and surface molecules with fluorescently labeled antibodies. The percentage of cytokine-positive T cells may subsequently be quantified and phenotypically characterized on a single cell basis. In this protocol, fluorescently labeled antibodies towards CD4, CD69, and IFNγ are used, but additional antibodies may be used to characterize specific T cells in more detail.

2 Materials

2.1 Buffers and Solutions

Prepare all solutions using ultrapure water and analytical grade reagents. Prepare all reagents at room temperature and store them at 4–8 °C. To increase maximum storage time, keep all buffers and solutions sterile.

1. Brefeldin A (BFA) solution: 5 mg/mL BFA (Sigma) in ethanol, stored in 10 μL aliquots at −20 °C.
2. Lysing solution: 10× Lysing Solution (Becton Dickinson). Prepare 1× solution with water.

3. EDTA stock solution: 500 mM EDTA in water, pH 8. Add 70 mL water to a graduated glass bottle. Weigh 18.61 g EDTA and transfer it to the bottle. Dissolve the EDTA by incubating at 56 °C in a shaking water bath. Adjust pH with HCl (*see* **Note 1**). Make up to 100 mL with water.
4. EDTA working solution: 20 mM EDTA in water, pH 8. Mix 2 mL of the EDTA stock solution with 48 mL water.
5. FCS for work: Filter the FCS through a high efficiency leukocyte removal filter and store 50 mL aliquots at −20 °C.
6. NaN_3 stock solution: 10 % NaN_3 in water.
7. FACS buffer: PBS, 5 % FCS, 0.5 % BSA, 0.07 % NaN_3. Add about 200 mL PBS to a 1 L graduated cylinder. Weigh 5 g of BSA and transfer it to the cylinder. Add 50 mL of the filtered FCS (one aliquot) and 7 mL of the NaN_3 stock solution. Make up to 1 L with PBS. Mix and filter the buffer through a folded filter (grade 3 hw) into a glass bottle.
8. Saponin stock solution: PBS, 5 % saponin, 0.07 % NaN_3. Weigh 2 g saponin into a 50 mL Falcon tube. Add 39.7 mL PBS and 280 μL NaN_3 and mix (*see* **Note 2**).
9. Saponin buffer: FACS buffer, 0.1 % saponin. Mix 490 mL of the FACS buffer with 10 mL of the saponin stock solution.
10. PFA stock solution: PBS, 4 % PFA, pH 7.4–7.6. Add 200 mL PBS to a graduated brown glass bottle (*see* **Note 3**). Weigh 8 g PFA and transfer it to the bottle. Dissolve the PFA by incubating at 56 °C in a shaking water bath. Adjust pH with NaOH.
11. PFA working solution: PBS, 1 % PFA. Mix 45 mL PBS with 15 mL of the PFA stock solution in a brown glass bottle (*see* **Note 4**).

2.2 Antigens and Antibodies

1. CMV control antigen (Virion Serion, Würzburg, Germany), store at −80 °C (*see* **Note 5**).
2. CMV antigen (Virion Serion), store at −80 °C (*see* **Note 5**).
3. *Staphylococcus aureus* enterotoxin B (SEB) stock solution: 1 mg/mL in water, store at −20 °C (*see* **Note 5**).
4. Co-stimulatory antibody mix: 0.33 mg/mL mouse anti-human CD28 antibody (0.5 mg/mL, clone L293, BD Biosciences, Heidelberg, Germany), 0.33 mg/mL mouse anti-human CD49d antibody (1 mg/mL, clone 9F10, BD Biosciences). Mix 20 μL anti-CD28 antibody with 10 μL anti-CD49d antibody and store it at 4 °C.
5. Staining antibodies: Mouse anti-human CD4-APC (0.006 mg/mL, clone SK3), mouse anti-human CD69-PerCP (0.05 mg/mL, clone L78), and mouse anti-human IFNγ-FITC (0.5 mg/mL, clone 4S.B3, all BD Biosciences).

2.3 Specific Technical Equipment

1. Flow-cytometer: FACS Canto II (BD Biosciences).
2. Flow-cytometry software: BD FACSDiva Software Version 6.1.3 (BD Biosciences).
3. Tubes for flow-cytometry (FACS tubes): 5 mL, 75 mm × 12 mm, PS (Sarstedt, Nümbrecht, Germany).
4. Serological pipettes (2 mL).

3 Methods

Each blood sample is analysed using CMV antigen as a stimulus along with positive and negative controls. Carry out all procedures at room temperature unless otherwise specified. You need blood cells of about 150–225 µL whole blood for every FACS staining. In the following, the protocol is described for a total of two concurrent staining preparations per each of the three stimulation samples (negative control, positive control and antigen-specific stimulation).

3.1 Stimulation Reaction

1. Prepare three 15 mL polypropylene tubes. Provide one of the tubes with 4.05 µL of the co-stimulatory antibody mix (αCD28 and αCD49d antibodies), add 1,350 µL of heparinized whole blood (*see* **Note 6**) with a serological pipette and mix thoroughly by pipetting up and down (*see* **Note 7**). Then distribute equal amounts of the mixture (450 µL each) to the three tubes.
2. Add the different antigens to the tubes: tube 1 = negative control, add 14.4 µL CMV control antigen from commercially available stock; tube 2 = antigen of interest, add 14.4 µL of CMV antigen from commercially available stock; tube 3 = positive control, add 1.125 µL SEB stock. Shortly mix on the Vortex mixer and incubate the tubes in an upright position at 37 °C and 5 % CO_2, do not fasten the lids tightly to allow for aeration.
3. After 2 h, add 0.9 µL brefeldin A solution directly into the blood of each sample, mix again shortly on the Vortex mixer and incubate as before for additional 4 h at 37 °C and 5 % CO_2.

3.2 Fixation Reaction

1. Add 45 µL EDTA working solution to each sample, mix on a Vortex mixer for 10 s and incubate for 15 min.
2. Add 4 mL of 1× Lysing solution to each sample and incubate for additional 10 min (*see* **Note 8**).
3. Centrifuge for 10 min at 320 × *g*, aspirate and discard the supernatant (all but about 50 µL) and resuspend the cell pellet thoroughly in the remaining supernatant (*see* **Note 9**).

Table 1
Antibody mix for staining of the cells

Component	Per reaction (μL)	For three reactions (multiplied by 3.24)[a], (μL)
Anti-CD4-APC antibody	0.5	1.62
Anti-CD69-PerCP antibody	2.0	6.48
Anti-IFNγ-FITC antibody	0.5	1.62
Saponin solution (5 %)	1.0	3.24
FACS buffer	46.0	149.04

[a] *See* **Note 14**

4. Add 2 mL of FACS buffer and repeat **step 3**, but this time, remove the supernatant completely (*see* **Note 10**).
5. Add 400 μL of FACS buffer and store the cells at 4 °C until further processing (*see* **Note 11**).

3.3 Staining of the Cells

1. Thoroughly mix the fixed cells on a Vortex mixer and transfer 200 μL into a FACS tube (*see* **Note 12**).
2. Add 2 mL saponin buffer and incubate the samples for 10 min.
3. In the meantime prepare the antibody mix for staining (*see* **Note 13**). Mix the components as indicated in Table 1 (for three stimulatory reactions) and store the mix at 4 °C in the dark until it is needed (**step 5**).
4. Centrifuge for 10 min at 320 × *g* and take off the supernatant completely by aspiration (*see* **Note 15**).
5. Add 50 μL of the prepared antibody mix (*see* **step 3**), mix shortly on a Vortex mixer and incubate for 30–45 min in the dark (*see* **Note 16**).
6. Add 3 mL FACS buffer, centrifuge the samples for 10 min at 320 × *g*, and discard the supernatant completely.
7. Resuspend the stained cells in 300 μL PFA and store it at 4 °C in the dark until flow-cytometric measurement (*see* **Note 17**).

3.4 Analysis of the Cells in the Flow-Cytometer

1. Count at least 10,000 cells of the population of interest (here CD4 T cells).
2. Quantify the frequency of antigen-specific CD4 T cells using BD FACSDiva Software Version 6.1.3 as follows (*see* also Fig. 1):
   (a) First gate the lymphocyte population according to their size and granularity using forward scatter (FSC) and side scatter (SSC), respectively.

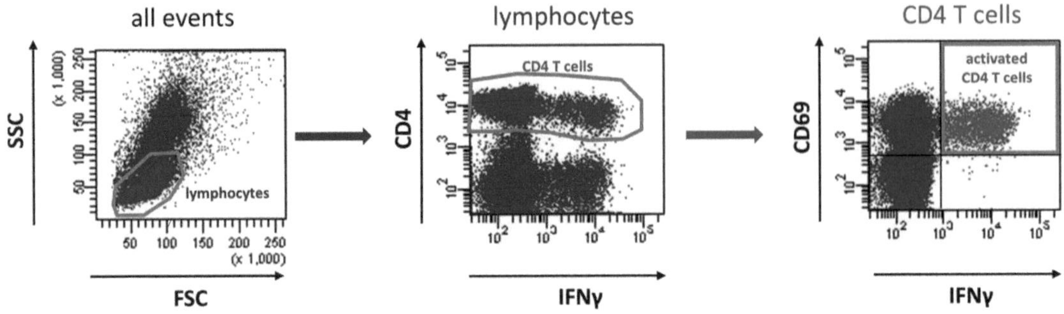

Fig. 1 Gating strategy for determination of antigen-specific CD4 T-cell frequencies. First depict all events (*left dot plot*) and identify the lymphocyte population (*marked area*) using size (defined by the forward scatter, FSC) and granularity (defined by the side scatter, SSC). Then depict only the lymphocytes (*dot plot* in the *middle*) and gate the cells with a clear signal of the APC-labeled anti-CD4 antibody (*marked area*). Finally display these CD4 T cells according to their CD69-PerCP and their IFNγ-FITC signal (*right dot plot*). Cells that are co-expressing the activation marker CD69 and the cytokine IFNγ are defined as activated cells (*upper right quadrant*). The frequency of antigen-specific CD4 T cells is calculated by the number of activated cells divided by all CD4 T cells

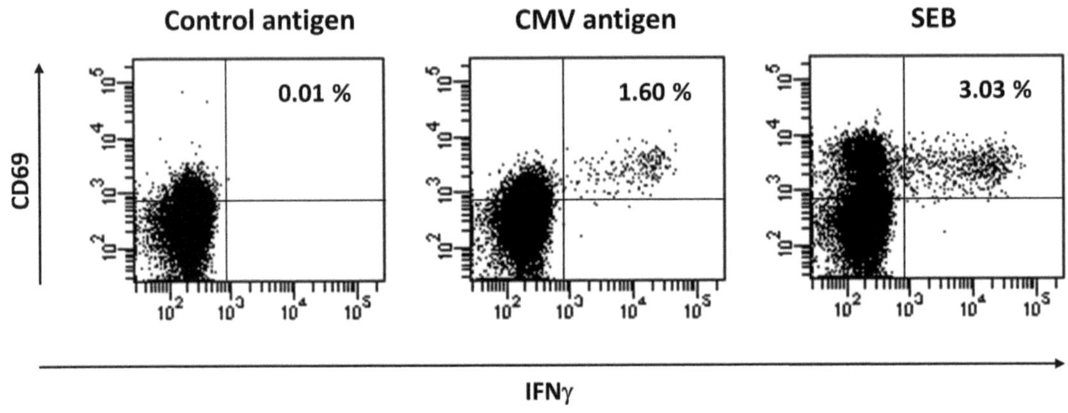

Fig. 2 Representative example of a stimulation result of a CMV-positive whole blood sample. Whole blood samples were stimulated with the CMV control antigen (negative control, *left dot plot*), the CMV antigen (*dot plot* in the *middle*) and the *Staphylococcus aureus* enterotoxin B (SEB, positive control, *right dot plot*), respectively, and reactive CD4 T-cells (CD69+/IFNγ+) were analyzed using flow-cytometry. The numbers indicate the percentage of reactive CD4 T-cells. To calculate the frequency of CMV-specific CD4 T cells, subtract the frequencies of reactive cells in the negative control from those stimulated with the CMV antigen (i.e., 1.60–0.01 % = 1.59 %)

(b) Then identify CD4 T cells by the fluorescence signal of the respective antibody.

(c) Identify the specifically stimulated CD4 T cells by displaying the activation marker CD69 and the IFNγ signal.

(d) Determine the frequency of specifically stimulated CD4 T cells by subtracting the frequency of the negative control from that of the antigen-specifically stimulated sample (*see also* Fig. 2).

4 Notes

1. As EDTA is difficult to dissolve, it may be necessary to adjust the pH first and to shake overnight, to obtain complete dissolution of the EDTA.
2. As saponin is a detergent, mix gently to avoid extensive foaming.
3. Keep the PFA solutions in the dark.
4. Do not store the PFA working solution for longer than 2 weeks.
5. For easier handling and a greater day-to-day reproducibility, we recommend storing the antigens in aliquots (for example 100 μL for control- and CMV-antigen and 25 μL for SEB). The aliquot in use can be stored at −20 °C.
6. Only heparinized whole blood (with sodium or ammonium heparin) is suitable for stimulation because other anticoagulants interfere with signaling needed to induce specific cytokines. Always use fresh blood within 24 h after venipuncture. If you cannot process immediately after venipuncture, keep the blood sample at 4 °C until processing.
7. Be careful and avoid foaming.
8. If you plan to use additional antibodies for staining of other cytokines or cell surface molecules, please note that the epitopes of some antibodies may be destroyed after fixation of the cells. In this case, each new antibody should first be tested for stability of antigen-recognition. If it does not recognize the epitope after fixation, you have to stain the cells before adding the Lysing solution. Therefore, add the respective antibody directly to the EDTA-treated blood and incubate for 30 min in the dark. Then add 3 mL FACS buffer, pellet the cells by centrifugation for 10 min at 320 × *g*, discard the supernatant and proceed with the lysing step.
9. For dissolving the pellet completely, we think it is most effective to pull the tube several times over a metal grid.
10. Be careful as the pellets may come off the bottom of the tube (and may be nearly invisible).
11. At this point, the samples may be stored in the refrigerator (4–8 °C) for up to 3 days.
12. The remaining 200 μL may be used for another staining with other antibodies of interest.
13. To avoid contamination of the antibody stocks, we recommend using a sterile work bench when preparing the staining mix.
14. To reach homogeneous conditions when staining more than one sample, prepare a master mix by multiplying the per-reaction amounts with $1.08 \times n$ (n = the total number of samples to stain).

15. The cell pellet may be nearly invisible, but it is important to remove the whole supernatant to avoid dilution of the antibodies in the staining mix. Tilting of the tube may facilitate the procedure.
16. After addition of the fluorescently labeled antibodies, avoid unnecessary light exposure to keep the fluorescent dye from bleaching.
17. At this point, samples may be stored for up to 24 h. An exception is the use of antibodies that are coupled with tandem conjugates of fluorescent dyes (e.g., APC-H7, PerCP-Cy5). As these conjugates may spontaneously dissociate, we recommend proceeding to the flow-cytometric measurement immediately after staining.

References

1. Sester M, Sester U, Gärtner B, Heine G, Girndt M, Mueller-Lantzsch N et al (2001) Levels of virus-specific CD4 T cells correlate with cytomegalovirus control and predict virus-induced disease after renal transplantation. Transplantation 71(9):1287–1294
2. Sester M, Gärtner BC, Sester U (2008) Monitoring of CMV specific T-cell levels after organ transplantation. J Lab Med 32(3):121–130
3. Schmidt T, Dirks J, Enders M, Gärtner BC, Uhlmann-Schiffler H, Sester U et al (2012) CD4+ T-cell immunity after pandemic influenza vaccination cross-reacts with seasonal antigens and functionally differs from active influenza infection. Eur J Immunol 42(7): 1755–1766

Chapter 20

Functional NK Cell Cytotoxicity Assays Against Virus Infected Cells

Rebecca J. Aicheler and Richard J. Stanton

Abstract

Natural Killer (NK) cells are crucial to the control of many viral infections. They are able to kill infected cells directly through the secretion of cytotoxic granules or through binding to death receptors on target cells. They also secrete cytokines and chemokines and, through interactions with dendritic cells, can shape adaptive immunity. The activity of NK cells can be controlled by a balance of activating and inhibitory signals conveyed through ligands on target cells binding to receptors on the NK cell. As a result viruses have devised mechanisms to modulate the expression of NK ligands on target cells, interfering with NK cell recognition and prolonging the life of infected cells. An understanding of how viruses modulate the NK response can lead to an understanding both of NK cell function, and of virus pathogenesis. Measuring the ability of NK cells to kill target cells infected with different viruses, or expressing different viral proteins, is an invaluable technique to identify the proteins and mechanisms by which viruses modulate the NK response. Here we describe two methods to measure this; one method measures sodium dichromate ^{51}Cr that is released from target cells as they are killed, and the other uses 7-amino-actinomycin D (7-AAD) to measure apoptosis and death of target cells following incubation with NK cells.

Key words Natural killer cells, NK cells, Chromium, Sodium dichromate, 7-Amino-actinomycin D (7-AAD), Cytotoxicity, Viruses, Virus infection, Immune evasion

1 Introduction

The activity of Natural Killer (NK) cells is controlled by a balance of activating and inhibitory signals conveyed through ligands on target cells binding to receptors on the NK cell. The initial step in NK recognition of a target cell involves the formation of a synapse between the effector (NK) cell and target cell, within which ligands and receptors are spatially arranged in order to permit their interaction. Following this, if an NK cell is activated, cytotoxic granules are released which result in the death of the target cell [1]. NK cells can also secrete cytokines and chemokines, and can bind to death receptors on target cells, inducing target cell death via the extrinsic apoptotic pathway [2, 3]. As a result they are crucial to the control of many viral infections [4, 5].

Susanne M. Bailer and Diana Lieber (eds.), *Virus-Host Interactions: Methods and Protocols*, Methods in Molecular Biology, vol. 1064, DOI 10.1007/978-1-62703-601-6_20, © Springer Science+Business Media, LLC 2013

Cell surface MHC-I acts as a major inhibitory ligand for NK cells. Viruses may downregulate MHC-I in order to evade cytotoxic T cells [6]; however, this should make cells vulnerable to NK cell attack. Surprisingly this is not necessarily the case [7]. Work with a number of viruses has shown that viruses are capable of modulating the expression of NK ligands, thereby interfering with NK cell recognition of infected cells [4, 5, 8, 9]. In particular, the mechanisms by which human cytomegalovirus (HCMV) modulates the NK response have revealed that upregulating inhibitory receptors, and downregulating activating receptors, promotes resistance to killing by NK cells. Thus HCMV encodes its own MHC-I homologue, and this is stabilized by a peptide derived from the signal peptide of a second gene that also stabilizes the nonclassical MHC-I molecule HLA-E [10–12]. Together these functions inhibit both LIR-1$^+$ NK cells [13] and CD94/NKG2A$^+$ NK cells [14]. Viral infection can also upregulate the stress proteins MICA, MICB, ULBP1-3, and RAET1L, which are ligands for the ubiquitous NK cell activating receptor NKG2D [15]. HCMV encodes two proteins [16–19] and a microRNA [20] that suppress cell surface expression of NKG2D ligands while another protein binds directly to NKp30, inhibiting NK cell cytotoxicity [21]. The activating ligands CD112 and CD155 may be exposed following breakdown of intercellular interactions during viral infection [22], and are targeted for downregulation by a further HCMV gene, inhibiting killing by NK cells [7, 23]. Finally HCMV can downregulate TRAIL death receptors in order to inhibit killing by the extrinsic apoptotic pathway [3]. The elucidation of these mechanisms by which HCMV evades NK cell killing has provided fundamental information on both NK cell function and virus pathogenesis.

Here we provide methods for two techniques (chromium release and 7-amino-actinomycin D (7-AAD) assays) that can be used to investigate mechanisms of NK cell activation, by measuring the ability of NK cells to kill target cells. In a chromium release assay the target cells are loaded with sodium dichromate ^{51}Cr and then incubated with purified NK cells for 4 h. As cells die, sodium dichromate ^{51}Cr is released into the supernatant and provides a measure of cytolysis [24]. The assay is performed in a 96-well plate, and each sample is normally analyzed in quadruplicate. It provides robust data by analyzing a wide range of ratios between the effector (NK) and target (virus infected) cells. The final plate will be laid out as shown in Table 1.

The number of target cells in each well is kept constant, and each sample is run in quadruplicate (hence there are four columns per sample). The "maximal" and "spontaneous" rows do not have NK cells added. In the maximal row all the target cells are lysed with Triton X-100. This releases all the sodium dichromate ^{51}Cr that has been taken up by the cells (different cell types can capture

Table 1
Layout for a chromium release assay in a 96-well U-bottom plate. Ratios given are ratio of NK cells–targets

	Sample 1	Sample 2	Sample 3
Maximal			
Spontaneous			
40:1			
20:1			
10:1			
5:1			
2.5:1			
1.25:1			

very variable amounts of sodium dichromate ^{51}Cr). The spontaneous row is target cells on their own—this shows how much sodium dichromate ^{51}Cr is released into the supernatant in the absence of NK cells. The remaining rows have NK cells added at decreasing ratios compared to the target cells. By comparing the amount of sodium dichromate ^{51}Cr taken up by the cells (max), and the amount released in the absence of NK cells (spontaneous), the amount of lysis that is specifically due to NK cells can be calculated.

In a 7-AAD assay, target cells are loaded with fluorescent dye such as carboxyfluorescein succinimidyl ester (CFSE) and then incubated with purified NK cells for 4 h. Following this the target cells are trypsinized (if adherent) and stained with 7-AAD. 7-AAD is a fluorescent dye that is excluded from healthy cells. However, in apoptotic cells, 7-AAD is able to cross the membrane and bind to DNA [3, 25]. Cells are analyzed on a flow cytometer—the CFSE allows target cells to be differentiated from effector (NK) cells, while 7-AAD allows the percentage of dead and dying cells to be calculated. Spontaneous cell death is measured by incubating CFSE labelled target cells in media only.

The chromium release assay offers a number of important advantages. It is highly sensitive, requiring as few as 1,000 target cells per sample. This means a large range of effector–target (E:T) ratios can be simultaneously tested, producing very robust data. The major disadvantage is that it requires the use of radioactive isotopes. 7-AAD assays on the other hand require more cells because target cell death is analyzed by flow cytometry, but these assays do not use radioactive material. K562 cells provide a useful positive control—these cells lack MHC-I and so are efficiently killed by NK cells.

2 Materials

2.1 Chromium Release Assay

1. $Na_2{}^{51}CrO_4$ (sodium dichromate—radioactive) (Amersham).
2. Histopaque (Gibco).
3. Phosphate buffered saline (PBS): KH_2PO_4 1.06 mM, NaCl 155.17 mM, $Na_2HPO_4{\cdot}7H_2O$ 2.97 mM (Gibco).
4. Heparin (Monoparin, CP Pharmaceuticals Ltd).
5. EasySep® Human NK Cell Enrichment Kit: Stem Cell Technologies, catalogue number 19055.
6. RPMI/10: RPMI media (Gibco), 10 % fetal calf serum (FCS).
7. Trypsin (Gibco).
8. Interferon-α (IFN-α) (Roferon, Roche).
9. 5 % (v/v) Triton X-100 in water (Fisher).
10. 50 and 15 mL falcons.
11. 96-well U-bottomed plate.
12. Beta plate.
13. Scintillation fluid (Perkin-Elmer).

2.2 7-Amino-Actinomycin D (7-AAD)-Assay

1. 7-amino-actinomycin D (7-AAD) (Molecular Probes, Invitrogen).
2. Carboxyfluorescein succinimidyl ester (CFSE) (Molecular Probes, Invitrogen).
3. Histopaque (Gibco).
4. Phosphate buffered saline (PBS): 1.06 mM KH_2PO_4, 155.17 mM NaCl, 2.97 mM $Na_2HPO_4{\cdot}7H_2O$ (Gibco).
5. Heparin (Monoparin, CP Pharmaceuticals Ltd).
6. EasySep® Human NK Cell Enrichment Kit: Stem Cell Technologies, catalogue number 19055.
7. RPMI/10: RPMI (Gibco), 10 % fetal calf serum (FCS).
8. Trypsin (Gibco).
9. Interferon-α (IFN α) (Roferon, Roche).
10. 50 and 15 mL falcons.
11. 96-well U-bottomed plate.
12. Flow Cytometer.

3 Methods

All steps are carried out at room temperature unless otherwise specified. When working with ^{51}Cr you should adhere to local guidelines regarding use and disposal of radioactive material.

3.1 Prepare Infected Target Cells

1. This step will differ depending on the virus and cell type in use (*see* **Note 1**).
2. For a single well in a chromium release assay you will need 2,000–5,000 cells (*see* **Note 2**). Chromium release samples are normally run in quadruplicate, and there are eight different conditions, i.e., you need between 6.4×10^4 and 1.6×10^5 infected cells for each sample that you want to test.
3. For a 7-AAD-assay you will need between 5×10^4 and 1×10^5 infected cells for each well in the assay, so if each sample is done in triplicate you will need between 1.5×10^5 and 3×10^5 infected cells for each sample (*see* **Note 3**).

3.2 Prepare Peripheral Blood Mononuclear Cells (PBMC)

This step is done 1 day before the assay date.

1. Warm Histopaque to 37 °C.
2. For a chromium release assay with a starting E:T ratio of 40:1 you will need 6.4×10^5 (for 2,000 target cells/well, in quadruplicate) to 1.6×10^6 (for 5,000 target cells/well, in quadruplicate) purified NK cells for each sample to be tested.
3. For a 7-AAD assay you will need between 1×10^5 and 2×10^5 purified NK cells for each well (depending on the number of target cells). A final effector (NK) to target (virus infected cell) ratio of at least 2:1 will be needed, i.e., for an assay using 1×10^5 target cells you will need 2×10^5 NK cells per sample, each sample will be run in triplicate so a total of 6×10^5 NK cells will be required.
4. You can expect $3–4 \times 10^6$ NK cells from 50 mL blood. Take an appropriate amount of blood from a healthy volunteer and add Heparin at 5 IU/mL (*see* **Notes 4** and **5**).
5. Add Histopaque (approx. 2/3 the volume of blood—e.g., 16 mL Histopaque for 25 mL blood) to a 50 mL falcon.
6. Gently layer the blood on top of the Histopaque (*see* **Note 6**).
7. Centrifuge at $836 \times g$ for 20 min with the brake off.
8. The PBMC should form at the interface of the Histopaque and blood, with red blood cells pelleted at the bottom.
9. Place a pipette through the upper layer, down to the PBMC layer and remove the PBMC layer into a fresh 50 mL falcon.
10. Add PBS up to 50 mL. If volume of PBMC was more than 10 mL, split it into two falcons and top both up to 50 mL (*see* **Note 7**).
11. Centrifuge $252 \times g$ for 6 min with the brake on to pellet cells.
12. A pellet should be visible—discard supernatant and resuspend cells in 50 mL PBS.
13. Centrifuge $252 \times g$ for 6 min with the brake on to pellet cells.

3.3 Isolate NK Cells

Multiple commercial methods can be used for isolating untouched NK cells—we use the EasySep® Human NK Cell Enrichment Kit, although as an alternative you may remove $CD3^+$ cells using anti-CD3 antibodies and Dynal beads, thereby enriching for NK cells.

Follow manufacturer's instructions, then resuspend the NK cells at 2×10^6 cells/mL in RPMI/10. Add Interferon-α (1,000 IU/mL) and incubate overnight at 37 °C.

3.4 ^{51}Cr Release Assay

1. Trypsinise virus infected target cells (if adherent).
2. Neutralize trypsin with RPMI/10 and transfer to 15 mL falcon.
3. Pellet cells by centrifuging (470 × *g*, 3 min) and resuspend in 10 mL PBS (*see* **Note 8**).
4. Pellet cells by centrifuging (470 × *g*, 3 min) and pour off PBS. Remove all PBS with a P200 or P20 pipette (*see* **Note 9**).
5. Resuspend cells in 10 μL (~150 μCi) of sodium dichromate ^{51}Cr (*see* **Note 10**).
6. Incubate for 1 h at 37 °C.
7. Add 10 mL RPMI/10 and pellet cells by centrifuging (470 × *g*, 3 min), then discard supernatant.
8. Add another 10 mL RPMI/10 and incubate 30–60 min at 37 °C. This allows any unbound sodium dichromate ^{51}Cr to leach out of cells.
9. Pellet cells by centrifuging (470 × *g*, 3 min) and resuspend in 0.5 mL RPMI/10. Count cells and resuspend such that 100 μL will contain the desired number of target cells/well (*see* **Note 11**).
10. While the targets are being loaded with sodium dichromate ^{51}Cr, prepare the NK cells. Count them and resuspend in RPMI/10 (without IFNα). The concentration they are resuspended at will depend on the number of targets. Take the number of targets, and multiply by 40. This number of cells should be in 100 μL (*see* **Note 12**).
11. Take a 96-well U-bottomed plate and add samples as follows.
12. Leave the first row blank.
13. To row 2 and rows 4–8 add 100 μL RPMI/10.
14. To row 3 add 200 μL of NK cell suspension/well.
15. Using a multichannel pipette, remove 100 μL NK cells from row 3 and mix with the RPMI/10 in row 4.
16. Using fresh tips, remove 100 μL NK cells from row 4 and mix with the RPMI/10 in row 5.
17. Using fresh tips, remove 100 μL NK cells from row 5 and mix with the RPMI/10 in row 6.

Table 2
Example of raw scintillation counts from a chromium release assay

	Sample 1				Sample 2			
Max	5,375	5,797	5,139	5,172	10,215	9,992	9,992	11,148
Spontaneous	175	155	167	187	1,869	1,650	1,758	1,747
40:1	3,587	3,319	2,874	2,842	3,821	4,616	4,459	6,205
20:1	3,589	3,452	2,767	3,230	4,074	4,092	3,891	3,627
10:1	2,852	3,267	2,654	2,677	3,761	3,551	4,043	4,353
5:1	2,214	2,379	2,719	2,327	3,113	3,833	2,778	2,921
2.5:1	1,804	1,823	1,626	1,531	3,121	2,601	2,552	2,832
1.25:1	1,029	1,043	1,037	1,092	2,208	2,592	2,200	2,305

18. Repeat for each row until row 8. After mixing cells in row 8, remove 100 μL and discard.
19. This should leave you with a serial twofold dilution of NK cells from 40× the number of targets (in row 3) down to 1.25× the number of targets in row 8.
20. To every well, add 100 μL targets.
21. To row 1, add 100 μL 5 % Triton X-100. Every well should now have 200 μL in it.
22. Incubate at 37 °C for 4 h.
23. Centrifuge (470 × *g*, 2 min).
24. Add 150 μL scintillation fluid to each well of a beta plate.
25. Remove 20 μL supernatant from the U-bottomed plate and add to the beta plate (*see* **Note 13**).
26. Agitate plate for 5 min, and read the counts per minute (CPM) in a scintillation counter (Table 2).
27. % specific lysis for each sample is calculated.

$$\frac{\text{Number of dead or dying cells}}{\text{Number of dead or dying cells} + \text{number of live cells}} \times 100.$$

28. Data are plotted as E:T ratio (*x* axis) against % specific lysis (*y* axis) (Fig. 1).

An example of the raw data from the scintillation counter can be seen in Table 2. The percentage specific lysis calculated from these values can be seen in Table 3. The graph drawn from these data can be seen in Fig. 1—clearly sample 2 is not killed as efficiently by NK cells.

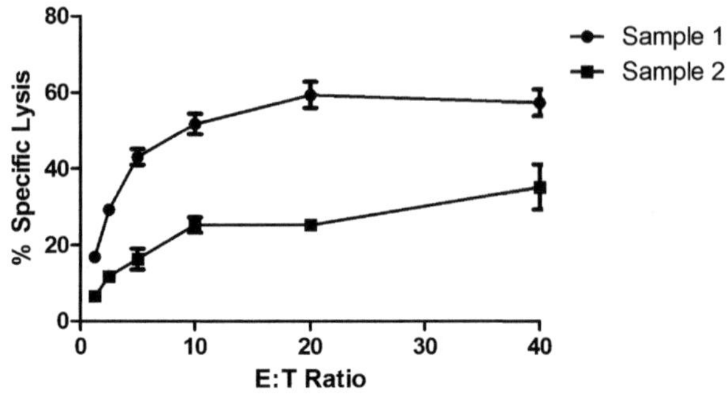

Fig. 1 Results of a chromium release experiment. Data from Table 3 has been plotted. Cells in sample 2 have been killed less efficiently by NK cells

Table 3
Percentage specific lysis calculated from values in Table 2

	Sample 1				Sample 2			
Max (Average)	5,371				10,337			
Spontaneous (Average)	171				1,756			
40:1	65.7	60.5	52.0	51.4	24.1	33.3	31.5	51.8
20:1	65.7	63.1	49.9	58.8	27.0	27.2	24.9	21.8
10:1	51.6	59.5	47.8	48.2	23.4	20.9	26.7	30.3
5:1	39.3	42.5	49.0	41.5	15.8	24.2	11.9	13.6
2.5:1	31.4	31.8	28.0	26.2	15.9	9.8	9.3	12.5
1.25:1	16.5	16.8	16.7	17.7	5.3	9.7	5.2	6.4

3.5 7-AAD Assay

1. Trypsinise virus infected target cells.
2. Neutralize trypsin with RPMI/10 and transfer to 15 mL falcon.
3. Pellet cells by centrifuging (470 × g, 3 min) and resuspend in 10 mL PBS.
4. Pellet cells by centrifuging again (470 × g, 3 min).
5. Resuspend the cells at a concentration of 1 × 10E6 cells/mL in PBS and add CFSE to a final concentration of 1 μM.
6. Incubate the cells at 37 °C for 20 min with gentle agitation every 5 min.
7. After 20 min add an equal volume of warm FCS.
8. Pellet cells by centrifuging (470 × g, 3 min) and resuspend in RPMI/10. Count cells and resuspend such that 100 μL contains the desired concentration—usually 1 × 10E5 cells/100 μL.

9. While the targets are being loaded with CFSE, prepare the NK cells for the assay. Count the NK cells, pellet them by centrifuging (470 × g, 3 min) and resuspending in RPMI/10 (without IFNα) such that 100 μL contains the desired number of cells/well (*see* **Notes 14** and **15**).
10. Take a 96-well U-bottomed plate and add samples as follows.
11. To row 1 add 100 μL of your target cells and 100 μL of RPMI/10, in triplicate (this is your control for background cell death).
12. In the remaining wells add 100 μL of your target cells and 100 μL of your NK cells. Do each well in triplicate. Every well should now have 200 μL in it.
13. Incubate at 37 °C for 4 h.
14. Using a pipette transfer the 200 μL of medium to a fresh 96-well plate.
15. Wash the remaining adhered target cells in PBS before adding 30–50 μL of trypsin and incubating at 37 °C (*see* **Note 16**).
16. When the adherent target cells have been released from the plastic, quench the trypsin by adding back the cells and medium which were removed in step 12.
17. Pellet cells by centrifuging (470 × g, 3 min), discard the supernatant and resuspend in 100 μL RPMI/10 containing 10 μg/mL of 7-AAD.
18. Incubate the cells for 15 min at room temperature prior to running the cells on the flow cytometer.
19. Run the cells on a flow cytometer. To calculate the % of dead and/or dying cells analyze the samples as follows.
20. To identify the target cells from the NK cells, draw a histogram plot with CFSE (FL1) on the x-axis. The target cells were loaded with CFSE so will be CFSE+, draw a gate around the CFSE+ cells (Fig. 2a).
21. Draw a dot-plot of the CFSE+ cells with 7-AAD (FL3) on the y-axis and FSC on the x-axis. Draw gates around the live cells and the dead/dying cells (Fig. 2b). Live cells will be 7-AADlo and will sit to the right on the FSC, dead and dying cells will be 7-AADhi and will be smaller so they will sit to the left on the FSC.
22. Collecting a total of 25,000 cells in the dead/dying and live gates combined is usually sufficient, but you can collect more if you have them (*see* **Note 17**).
23. Use the flow cytometer's analysis software to calculate the number of cells in the live gate and the number of cells in the dead/dying gate.

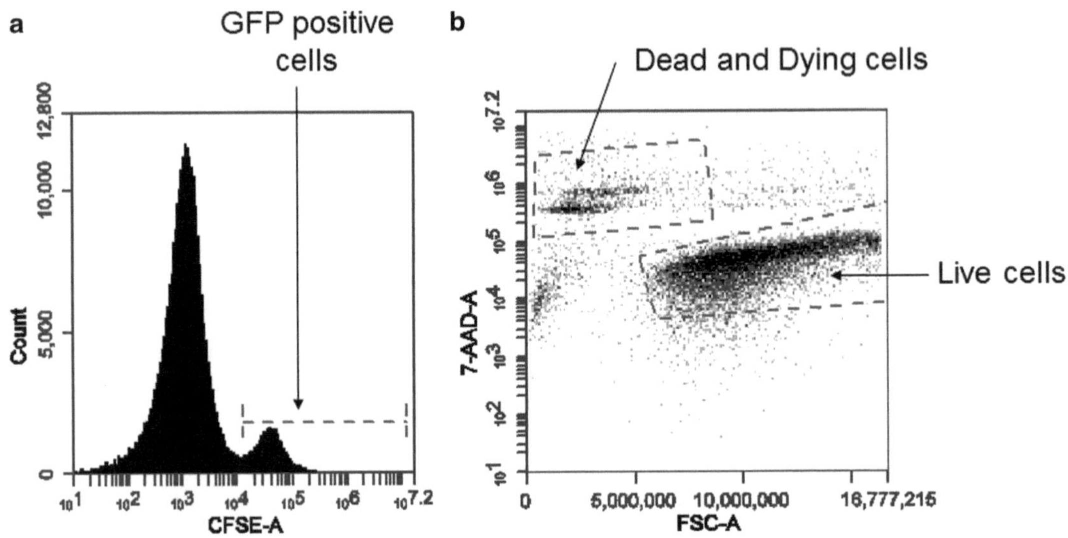

Fig. 2 Analysis of data from a 7-AAD experiment. (**a**) CFSE (FL1) positive target cells are gated. (**b**) CFSE$^+$ cells are plotted on a scatter graph of FSC vs. 7-AAD (FL3), giving two populations. The 7-AAD^{lo} cells are alive, the 7-AAD^{hi} cells are dead/dying

24. The percentage of dead/dying cells is calculated as follows:

$$\frac{\text{Number of dead or dying cells}}{\text{Number of dead or dying cells} + \text{number of live cells}} \times 100.$$

25. Subtract the percentage of dead/dying cells in the background wells from the test wells to obtain your final figure (*see* **Note 18**).

4 Notes

1. The timing for the preparation of the target cells will differ depending on the virus. For example for human cytomegalovirus (HCMV), the cytotoxicity assays are usually done at 48 or 72 h post-infection.
2. For a chromium release assay, the absolute minimum number of target cells required will vary with cell type. Larger cells take up more sodium dichromate ^{51}CR, and so fewer cells are needed. For example fibroblasts are very large, so a minimum of 1,000 cells/well can be used. For smaller cells the minimum will be ~2,000/well. More cells will give a stronger signal at the end, but there is rarely a need to use more than 5,000 cells/well.
3. You will be comparing two or more samples—for example mock infected with infected, or wild-type virus with mutant virus.
4. Buffy coat can also be used, in which case it should be diluted 1:2 in PBS before starting. However, quality can vary and

separation is often not as clean as fresh blood. NK cell lines and NK clones can also be used [26]. However, compared to using cells that have been cultured, using NK cells after just an overnight stimulation with IFN-α offers the advantage that there is little chance for different NK cell populations to outgrow, and thus the NK cells are more representative of the populations found *in vivo*.

5. NK cells isolated from different donors will give very different levels of killing, so you should repeat the assay with different donors. The number of PBMC and NK cells isolated from different donors can also vary widely. If you isolate fewer NK cells than expected, you can reduce the number of targets/well, reduce the number of replicates, reduce the number of samples, or reduce the E:T ratio to match the number of NK cells.
6. Hold the falcon at an angle and gently dribble the blood down the side of the tube.
7. If you do not dilute the Histopaque sufficiently, the PBMC will not pellet.
8. It is important to wash off the RPMI/10 or it will bind the sodium dichromate ^{51}Cr.
9. It is important to remove as much of the PBS as possible in order to maximize ^{51}Cr uptake by cells.
10. Sodium dichromate ^{51}Cr has a half life of 27.7 days. You can calculate the volume to add such that 150 μCi is added, but as a general rule 10 μL fresh solution is sufficient. If it is more than 28 days old, use 20 μL to compensate for decay. Sometimes when the Sodium dichromate ^{51}Cr is older than a month, it becomes slightly toxic. The assay still works, but you will have fewer cells after incubating them with the Sodium dichromate ^{51}Cr than before.
11. As described previously, the concentration will depend on how many cells you wish to use, and how many NK cells you have isolated. The minimum is 1,000 cells/well. 2,000–5,000 cells/well is better.
12. You will use 200 μL of this suspension per sample per replicate. For example for one sample with 3,000 targets/well you will need 2.4×10^5 NK cells per replicate, or 9.6×10^5 total NK cells for a single sample done in quadruplicate.
13. Remove supernatant only, be careful not to aspirate any cells.
14. The concentration they are resuspended at will depend on the number of targets. If you are restricted by the number of NK or target cells, start using a single E:T ratio of 2:1. Take the number of targets, and multiply by 2. For example for one sample with 1×10^5 targets/well you will need 2×10^5 NK cells per well, or 6×10^5 cells in total for triplicate repeats.

15. The E:T ratio needs to be determined empirically and will differ for each donor. Altering the E:T ratio will alter the overall level of killing—if the level of killing is too high, the E:T ratio should be reduced. If it is too low, the E:T ratio can be increased. Ideally the control sample will show a level of killing between 20 and 40 %.
16. Take care when trypsinizing the target cells. Do not over trypsinize them as this could adversely affect cell viability.
17. The cells are not fixed—if you take too long to run them on the FACS machine, they may start to adhere to the plastic (if adherent), or viability may alter. As long as samples are run within 90 min, this should not happen.
18. The cells are trypsinized twice during the protocol—as a result some cell types may give a high level of background cell death, and this can prevent specific killing being visible. In this case you will need to use different cells that are less sensitive to trypsinization and carefully monitor the duration of trypsinization. You may also test whether low-adherence plates avoid this, although in our hands they don't make a difference.

References

1. Bryceson YT, Chiang SC, Darmanin S et al (2011) Molecular mechanisms of natural killer cell activation. J Innate Immun 3:216–226
2. Dickens LS, Powley IR, Hughes MA et al (2012) The 'complexities' of life and death: death receptor signalling platforms. Exp Cell Res 318:1269–1277
3. Smith W, Tomasec P, Aicheler R et al (2013) Human cytomegalovirus glycoprotein UL141 targets the TRAIL death receptors to inhibit host innate defenses. Cell Host Microbe 13:324–335
4. Wilkinson GW, Tomasec P, Stanton RJ et al (2007) Modulation of natural killer cells by human cytomegalovirus. J Clin Virol 41: 206–212
5. Lanier LL (2008) Evolutionary struggles between NK cells and viruses. Nat Rev Immunol 8:259–268
6. Jackson SE, Mason GM, Wills MR (2011) Human cytomegalovirus immunity and immune evasion. Virus Res 157:151–160
7. Tomasec P, Wang EC, Davison AJ et al (2005) Downregulation of natural killer cell-activating ligand CD155 by human cytomegalovirus UL141. Nat Immunol 6:181–188
8. Lodoen MB, Lanier LL (2005) Viral modulation of NK cell immunity. Nat Rev Microbiol 3:59–69
9. McSharry BP, Burgert HG, Owen DP et al (2008) Adenovirus E3/19K promotes evasion of NK cell recognition by intracellular sequestration of the NKG2D ligands MICA and MICB. J Virol 82:4585–4594
10. Prod'homme V, Tomasec P, Cunningham C et al (2012) Human cytomegalovirus UL40 signal peptide regulates cell surface expression of the NK cell ligands HLA-E and gpUL18. J Immunol 188:2794–2804
11. Tomasec P, Braud VM, Rickards C et al (2000) Surface expression of HLA-E, an inhibitor of natural killer cells, enhanced by human cytomegalovirus gpUL40. Science 287:1031
12. Ulbrecht M, Martinozzi S, Grzeschik M et al (2000) Cutting edge: the human cytomegalovirus UL40 gene product contains a ligand for HLA-E and prevents NK cell-mediated lysis. J Immunol 164:5019–5022
13. Prod'homme V, Griffin C, Aicheler RJ et al (2007) The human cytomegalovirus MHC class I homolog UL18 inhibits LIR-1+ but activates LIR-1- NK cells. J Immunol 178:4473–4481
14. Wang EC, McSharry B, Retiere C et al (2002) UL40-mediated NK evasion during productive infection with human cytomegalovirus. Proc Natl Acad Sci USA 99:7570–7575
15. Eagle RA, Trowsdale J (2007) Promiscuity and the single receptor: NKG2D. Nat Rev Immunol 7:737–744
16. Kubin M, Cassiano L, Chalupny J et al (2001) ULBP1, 2, 3: novel MHC class I-related

molecules that bind to human cytomegalovirus glycoprotein UL16, activate NK cells. Eur J Immunol 31:1428–1437
17. Welte SA, Sinzger C, Lutz SZ et al (2003) Selective intracellular retention of virally induced NKG2D ligands by the human cytomegalovirus UL16 glycoprotein. Eur J Immunol 33:194–203
18. Wills MR, Ashiru O, Reeves MB et al (2005) Human cytomegalovirus encodes an MHC class I-like molecule (UL142) that functions to inhibit NK cell lysis. J Immunol 175:7457–7465
19. Chalupny NJ, Rein-Weston A, Dosch S et al (2006) Down-regulation of the NKG2D ligand MICA by the human cytomegalovirus glycoprotein UL142. Biochem Biophys Res Commun 346:175–181
20. Stern-Ginossar N, Elefant N, Zimmermann A et al (2007) Host immune system gene targeting by a viral miRNA. Science 317:376–381
21. Arnon TI, Achdout H, Levi O et al (2005) Inhibition of the NKp30 activating receptor by pp 65 of human cytomegalovirus. Nat Immunol 6:515–523
22. Stanton RJ, McSharry BP, Rickards CR et al (2007) Cytomegalovirus destruction of focal adhesions revealed in a high-throughput Western blot analysis of cellular protein expression. J Virol 81:7860–7872
23. Prod'homme V, Sugrue DM, Stanton RJ et al (2010) Human cytomegalovirus UL141 promotes efficient downregulation of the natural killer cell activating ligand CD112. J Gen Virol 91:2034–2039
24. Brunner KT, Mauel J, Cerottini JC et al (1968) Quantitative assay of the lytic action of immune lymphoid cells on 51-Cr-labelled allogeneic target cells in vitro; inhibition by isoantibody and by drugs. Immunology 14:181–196
25. Philpott NJ, Turner AJ, Scopes J et al (1996) The use of 7-amino actinomycin D in identifying apoptosis: simplicity of use and broad spectrum of application compared with other techniques. Blood 87:2244–2251
26. Morris RJ, Chong LK, Wilkinson GW et al (2005) A high-efficiency system of natural killer cell cloning. J Immunol Methods 307: 24–33

Chapter 21

Assessment of Natural Killer Cell Responses to Human Cytomegalovirus-Infected Macrophages

Zeguang Wu, Giada Frascaroli, and Thomas Mertens

Abstract

Natural killer (NK) cells are major components of the innate immune system and are assumed to play an important role in the defense against cytomegalovirus infection. Human cytomegalovirus (HCMV) is the only virus known that shapes the NK cell receptor repertoire in humans. Assays allowing the simultaneous measurement of multiple cell parameters and the assessment of subpopulations of NK cells are suitable for monitoring the NK cell response to infected cells. Herein, we describe an autologous assay system to assess NK cell responses to HCMV-infected macrophages which is based on flow cytometry to detect CD107a expression and interferon gamma (IFNγ) secretion. Further we established a simple method to handle HCMV-infected macrophages. Our assay provides a practicable approach to investigate NK cell responses to HCMV infection *in vitro*.

Key words Natural killer cells, Human cytomegalovirus, Macrophages, CD107a, Cytometry assay

1 Introduction

NK cell deficiency is associated with severe herpesvirus infections, including HCMV [1, 2]. In a study analyzing HCMV reactivations after bone marrow transplantation, the levels of nonspecific NK cell cytotoxicity correlated with the patient's ability to overcome the infection [3]. A case report also shows that NK cells themselves could control HCMV infection in the absence of T cell help in a $T^{neg}B^{neg}NK^{posi}$SCID patient [4]. It has been also shown that HCMV has evolved several mechanisms to modulate NK cell responses [5, 6].

In our autologous assay, we apply macrophages as target cells for several reasons. First, macrophages can act as antigen presenting cells, and can secrete specific cytokines leading to NK cell activation [7]. Second, there is consensus that myeloid precursors are an important site of HCMV latency and that virus reactivation occurs concomitantly with myeloid cell differentiation [8]. So the NK cell response to macrophages can provide a wealth of information

Susanne M. Bailer and Diana Lieber (eds.), *Virus-Host Interactions: Methods and Protocols*, Methods in Molecular Biology, vol. 1064, DOI 10.1007/978-1-62703-601-6_21, © Springer Science+Business Media, LLC 2013

about the possible mechanisms of latency and reactivation of HCMV. Third, HCMV strain TB40/E replicates efficiently, exhibits a broad cell tropism, and is widely used for infection of endothelial cells and monocyte-derived cells. Furthermore, TB40/E is available as a BAC clone (TB40/E-BAC4) suitable for mutational analysis [7, 9]. Lastly, macrophages can be obtained easily to perform experiments *in vitro*, since monocytes can be isolated from peripheral blood mononuclear cells (PBMCs). Polarized macrophages can attach to tissue culture 96-well plates, which is an advantage for the assessment of infection rates and the possibility to perform coculture assays.

CD107a expression on the NK cells' surface correlates with both cytokine secretion and target cell lysis, and thus can be used as a marker of NK functional activity [10]. Sorted CD107a NK cells exhibited a persistent lytic potential against a wide variety of target cells, including tumor- and virally infected target cells [11]. CD107a can also be used for multiple receptor crosslinking [12]. Combined with IFNγ secretion, flow cytometry-based assays offer the advantage of determining NK cell responses at the single-cell level. However, CD107a detection also has several disadvantages. First, it does not directly measure killing of the target cell. Second, CD107a expression can be influenced by cytokines even in the absence of target cells (unpublished observation). Third, the fact that a substantial proportion of NK cells are lost following culture with target cells needs to be considered when using the CD107a assay to quantify NK cell responses [13].

In summary, we have established and validated an autologous assay system to determine NK cell responses to HCMV-infected macrophages. This system has potential advantages with respect to other systems: First, Macrophages are readily available and easy to handle *in vitro*. Second, Macrophages are efficiently infected by the HCMV stain TB40/E. Third, Flow cytometry-based assays provide the advantage of assessing NK cell responses at the single-cell level. When combined with determination of other markers, the responses of NK cell subsets can be measured.

2 Materials

2.1 Preparation and Cryopreservation of Peripheral Blood Mononuclear Cells

1. Human PBMCs: freshly isolated from buffycoats.
2. RPMI/10 % FBS: RPMI-1640 medium (GIBCO/Invitrogen) supplemented with 10 % fetal bovine serum (FBS), 1 % nonessential amino acid solution (NEAA), 2 mM L-glutamine, 60 mg/mL streptomycin sulfate, 60 U/mL penicillin G.
3. Dulbecco's phosphate buffered saline (Dulbecco's-PBS) without Ca^{2+} and Mg^{2+} (PAA Laboratories GmbH).

4. Lymphocyte Separation Medium LSM 1077 (PAA Laboratories GmbH).
5. Ethylenediaminetetraacetic acid (EDTA), 1 % in PBS without Ca^{2+} and Mg^{2+} (Biochrom AG).
6. PBMCs washing buffer: Dulbecco's-PBS with 2 mM EDTA.
7. Dimethyl sulfoxide (DMSO, Sigma-Aldrich).
8. Cryoprotectant solution I: 100 % FBS.
9. Cryoprotectant solution II: 80 % FBS with 20 % DMSO (*see* **Note 1**).
10. Freezing container ("Mr. Frosty", Nalgene).
11. 30 % Bovine serum albumin, liquid (BSA, Sigma-Aldrich) (*see* **Note 2**).
12. MACS solution: Dulbecco's-PBS containing with 0.5 % BSA and 2 mM EDTA.

2.2 Isolation of Monocytes and Polarization of Macrophages

1. Miltenyi monocyte isolation kit II (Miltenyi Biotec).
2. LS columns and MidiMacs separators (Miltenyi Biotec).
3. Lumox dishes (Sarstedt) (*see* **Note 3**).
4. Recombinant human granulocyte-macrophage colony-stimulating factor (rhGM-CSF) and Recombinant human macrophage colony-stimulating factor (rhM-CSF) (R&D Systems).

2.3 Detachment and Infection of Macrophages

1. HCMV strain TB40/E (kindly provided by Prof. Dr. Christian Sinzger, Institute of Virology, University Medical Center Ulm, Ulm, Germany).
2. Human foreskin fibroblasts (HFFs) (*see* **Note 4**).
3. Acetone (Sigma-Aldrich).
4. Primary antibody: Anti HCMV Immediate Early Antigen (Argene-Biosoft).
5. Secondary antibody: Cy3 conjugated goat anti-mouse IgG (Invitrogen Molecular Probes).
6. 4′,6′-Diamidine-2′-Phenylindole Dihydrochloride (DAPI, Roche).
7. Photoshop CS3 (Adobe).

2.4 Negative Selection of NK Cells from Cryopreserved PBMCs and Coculture of Effector and Target Cells

1. Human PBMCs from liquid nitrogen storage.
2. NK cell isolation kit (Miltenyi Biotec).
3. Fluorochrome-conjugated anti-CD3 mAb and anti-CD56 mAb (BD Biosciences).

2.5 Flow Cytometry-Based Assay for Measuring NK Cell Responses

1. K562 erythroleukemic cells (ATCC, cat. no. CCL-243).
2. Brefeldin A (BFA, Sigma-Aldrich) (*see* **Note 5**).
3. Fluorochrome-conjugated anti-CD107a mAb and anti-CD69 mAb (BD Biosciences), anti-IFNγ mAb (Miltenyi Biotec).
4. BD Cytofix/Cytoperm™ Plus Fixation/Permeabilization Kit (with BD GolgiStop™ protein transport inhibitor containing monensin) (BD Biosciences) (*see* **Note 6**).
5. Sodium azide (Sigma-Aldrich).
6. FACS staining/washing buffer: Dulbecco's-PBS containing with 0.5 % BSA and 0.1 % sodium azide.
7. WinMDI 2.8 Software (The Scripps Institute, Flow Cytometry Core Facility).

3 Methods

3.1 Preparation and Cryopreservation of Peripheral Blood Mononuclear Cells

1. PBMCs are isolated from fresh buffy coats obtained from anonymous healthy blood donors.
2. Fresh buffy coat cells are placed into a 75 cm^2 cell culture flask. Add two volumes of room-temperature PBMCs washing buffer. Mix well.
3. Hold the centrifuge tube at a 45° angle and carefully layer the buffy coat/PBS mixture onto the Ficoll-Hypaque solution. Use 15 mL Ficoll-Hypaque per 35 mL buffy coat/PBS mixture.
4. Centrifuge 30 min at 620 × *g* at room temperature, without brake.
5. Remove the upper layer and transfer the mononuclear cell layer to another centrifuge tube.
6. Wash cells by adding 50 mL PBMCs washing buffer and centrifuging 10 min at 380 × *g*. Repeat the wash step two more times.
7. Preservation of PBMCs (*see* **Note 7**): (a) Adjust cells in 100 % FBS (Cryoprotectant solution I) at a concentration of 50×10^6/mL. (b) Add equal volume 20 % DMSO (Cryoprotectant solution II) over 5 min (1/10 volume per 30 s). (c) Transfer to cryotubes (1 mL/tube). (d) All tubes should be immediately placed into freezing container ("Mr. Frosty", Nalgene) and stored at −80 °C. (e) The next day, transfer the cryovials into vapor phase liquid nitrogen (−135 °C) for long-term storage (*see* **Note 8**).
8. For monocyte isolation: adjust fresh isolated PBMCs in 30 μL ice-cold MACS buffer per 1×10^7 total cells.

3.2 Isolation of Monocytes and Polarization of Macrophages

1. For isolation of monocytes and preparation of rhM-CSF-induced macrophages please refer to Varani and Frascaroli (*see* Chapter 22 of this series). For rhGM-CSF-induced macrophage culture add 100 ng/mL of rhGM-CSF.

2. Refresh cultures by half-volume rhM-CSF/rhGM-CSF containing medium change on day 3.

3.3 Detachment and Infection of Macrophages

1. Carefully calculate the number of macrophages needed. Consider variables like: number of conditions (e.g., mock and infected samples), rhGM-CSF and/or rhM-CSF macrophages and PBMCs and/or purified NK cells as effector cells.
2. Detach macrophages from Lumox dishes by washing with PBS. Wash detached cells twice and resuspend at a concentration of 1×10^6/mL in RPMI/10 % FBS (*see* **Note 9**).
3. Seed 1×10^5 macrophages/well in a 96-well flat-bottom plate. Each condition should be set up in duplicate. An identical parallel plate is needed to determine the rate of infection.
4. Add TB40/E virus stock to macrophages at an MOI = 5 (the virus has previously been prepared and titrated on HFF cells). Adjust the final culture volume to 200 μL with RPMI/10 % FBS. For mock infections, add 100 μL RPMI/10 % FBS.
5. Incubate macrophages 24 h with medium (mock infection) and TB40/E (MOI = 5) at 37 °C (*see* **Note 10**).
6. To determine the infection rate: discard the supernatant and fix macrophages with 80 % acetone for 5 min. Wash three times with PBS.
7. Staining with primary antibody: incubate with 200 μL diluted primary antibody (1:1,000) at room temperature for 90 min. Wash three times with PBS.
8. Staining with secondary antibody: incubate with 200 μL diluted secondary antibody (1:1,000) at room temperature for 45 min. Wash three times with PBS.
9. Add 200 μL diluted DAPI (1:20,000) to each well and incubate at room temperature for 3 min. Wash three times with PBS.
10. The plate is ready now to take pictures or for storage (*see* **Note 11**).
11. Calculate the infection rate: For quantification of the infection efficiency in macrophages, take five frames per well. The number of Cy3 and DAPI signals is determined by an automated counting feature within the Photoshop software and the infection rate is calculated as the ratio of Cy3 positive nuclei to total nuclei (*see* **Note 12**).

3.4 Negative Selection of NK Cells from Cryopreserved PBMCs and Coculture of Effector and Target Cells

1. Thawed PBMCs or purified NK cells will be used as effector cells; they should be prepared around 24 h post macrophage infection.
2. Transfer PBMCs from the liquid nitrogen tank to a 37 °C water bath and thaw by continuous shaking.

3. Slow removal of cryoprotectant (*see* **Note 7**): use ice cold RPMI/10 % FBS, all centrifugations are performed at 320×*g* for 5 min. First washing: 4× 250 μL cold RPMI/10 % FBS, 8× 1 mL cold RPMI/10 % FBS, every 30 s (total 6 min). Second washing: 10× 1 mL cold RPMI/10 % FBS every 20 s (total 3.5 min). Third washing: 10 mL cold RPMI/10 % FBS at once (total 30 s).
4. PBMCs as effector cell: adjust PBMCs in RPMI/10 % FBS at a concentration of 5×10^6/mL.
5. If purified NK cells are to be used: (a) Count and add PBMCs to 40 μL ice-cold MACS buffer reaching 1×10^7 total cells. (b) Isolate NK cells from thawed PBMCs by negative selection using the negative isolation kit from Miltenyi. (c) Count and resuspend the purified NK cells in RPMI/10 % FBS at a concentration of 5×10^5/mL. (d) Verify purity of the separated NK cells by flow cytometry with fluorochrome-conjugated monoclonal antibodies against CD56 and CD3.
6. Perform the coculture: after 24 h of infection of macrophages, discard the supernatants and pipette 0.2 mL indicated effector cells into wells of a 96-well flat-bottom plate (*see* **Note 13**) for 48 h of coculture.

3.5 Flow Cytometry-Based Assay for Measuring NK Cell Responses

1. After 48 h of coculture, calculate and prepare "CD107a/BFA/Monensin ready to use mixture". The final concentration of CD107a is 20 μL/mL, the final concentration of BFA is 5 μg/mL and the final concentration of Monensin is 2.0 μM (*see* **Note 14**).
2. Prepare K562 cells as controls: Maintain K562 cells by twice-weekly subculture at 1×10^4/mL in RPMI/10 % FBS. At the point of coculture, centrifuge the K562 for 5 min at 320×*g*, wash twice in RPMI/10 % FBS and resuspend at a concentration of 1×10^5/50 μL in RPMI/10 % FBS (*see* **Note 15**).
3. Add the indicated volume of CD107a/BFA/Monensin mixture to designated wells of flat bottom 96-well plates. The whole volume of medium is 200 μL. Mix the plate with a vortex mixer. Incubate the plate at 37 °C in a 5 % CO_2 incubator for 5 h (*see* **Note 16**). Figure 1 shows the CD107a expression and IFNγ secretion after PBMCs coculture with K562 cells for 5 h.
4. After 5 h of coculture, wash cells twice with FACS washing buffer. Centrifuge the plate for 5 min at 320×*g*, discarding supernatants after each wash (*see* **Note 17**).
5. Resuspend cells in 50 μL of FACS staining buffer and stain for surface antigen CD3 and CD56, incubate in the dark at 4 °C for 20 min. Wash cells twice with FACS washing buffer. To determine the phenotype of cells, include surface antigens of interest. Figure 2 shows the CD69 expression on NK cells after cocultivation with infected macrophages.

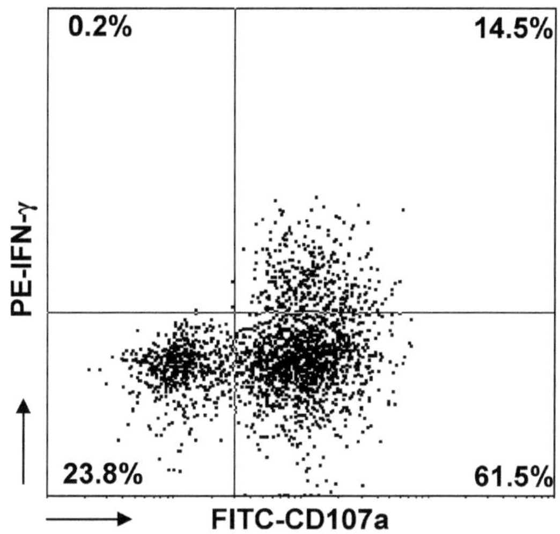

Fig. 1 Detection of CD107a expression and IFNγ secretion of NK cells from unfractionated PBMCs to K562 cell line. An example of the results of one representative donor is presented. Thawed PBMCs after 48 h, then cocultured with target cell line K562 cells for 5 h, as explained in Subheadings 2 and 3, the percentages of CD107a and IFNγ positive NK cells were assessed by flow cytometry

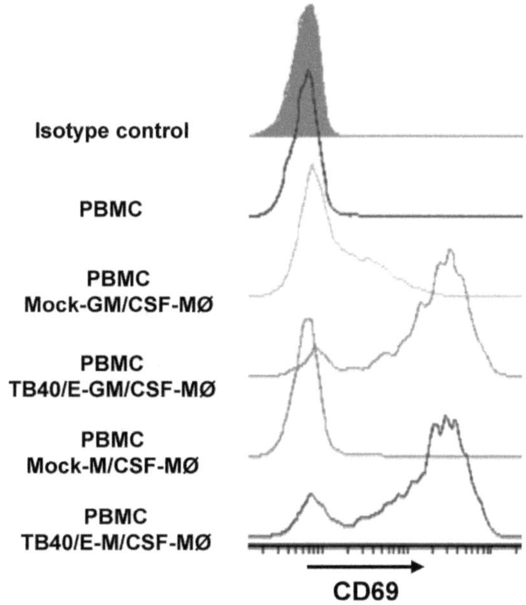

Fig. 2 HCMV-infected macrophages up-regulate CD69 expression on NK cells. PBMCs were cocultured with mock or TB40/E-infected rhGM-CSF/rhM-CSF-induced macrophages for 48 h. CD69 expression on NK cells was assessed by flow cytometry. A representative experiment out of ten is shown

6. Fix and permeabilize cells: Thoroughly resuspend cells and add 200 μL per well of Fixation/Permeabilization solution for 20 min at 4 °C. Wash cells twice in 200 μL 1× Perm/Wash buffer (*see* **Note 18**).

7. Stain for intracellular cytokines: Thoroughly resuspend fixed/permeabilized cells in 50 μL of 1× Perm/Wash buffer containing a pre-determined optimal concentration of a fluorochrome-conjugated anti-IFNγ or appropriate isotype for 20 min at 4 °C (*see* **Note 19**). Wash cells twice in 200 μL 1× Perm/Wash buffer.
8. Resuspend cells in 200 μL FACS washing buffer. Store plates wrapped in foil at 4 °C until run on a flow cytometer (*see* **Notes 20** and **21**).

4 Notes

1. Cryoprotectant solution I and II can be stored at −20 °C for up to 1 year. After thawing, place in a refrigerator at 4 °C for at least 30 min.
2. Sterilized liquid BSA is very convenient to prepare BSA containing solutions.
3. Lumox dishes are used for monocyte polarization, which allows detachment of macrophages simply by PBS washing.
4. HFFs were prepared from newborn foreskin.
5. Brefeldin A stock solution: Dilute to 5 mg/mL in DMSO and store at −20 °C. Working concentration is 5 μg/mL when combined with monensin.
6. Recommended working concentration of Monensin: 2.0 μM (0.7 μL solution per mL).
7. Slow addition of cryoprotectant and slow removal of cryoprotectant helps to preserve the cytotoxic activity of NK cells [14].
8. Avoid liquid phase storage due to safety concerns and to prevent possible problems with label adhesion failure.
9. Use warm PBS. Three rounds of washing and 5 min incubation in 37 °C helps to detach macrophages.
10. 24 h are enough to detect immediate early antigen expression in macrophages.
11. For later use, fill up the plate with PBS and keep the plate at 4 °C (protect from light).
12. Alternative software for counting is available [15].
13. Prepare following plates dependent on the experiment aim. (a) PBMCs with infected and uninfected macrophage cultures. (b) Purified NK cells with infected and uninfected macrophage cultures. (c) PBMCs only (used for K562 coculture).
14. It is time saving and more accurate to first prepare "CD107a/BFA/Monensin ready to use mixture". The "CD107a-isotype/BFA/Monensin" mixture can be prepared at the same time.

15. It is important to use MHC class I-negative K562 cells for each experiment as a positive control.
16. K562 as target cells: discard 50 μL supernatant first, and add 50 μL K562 cell suspension in RPMI/10 % FBS and indicated CD107a/BFA/Monensin mixture.
17. It is time saving to stain the samples in the 96-well plate. When 96-well plates are used for staining, it is important not to disrupt the cell pellet. It is helpful to hold the plates at a 45° angle and use a multi-channel adapter that fits to aspirate the supernatants.
18. Perm/Wash buffer must be maintained in washing steps to keep cells permeabilized.
19. Use 1× Perm/Wash buffer to dilute anti-IFNγ.
20. Plates should be run as soon as possible or within 24 h after staining.
21. It is very helpful to prepare blank, single-stained, and isotype samples for flow cytometer settings, compensation, and analysis. If intracellular staining is performed, the blank, single-stained, and isotype samples all need to undergo the same perm/fix procedures. For single staining and compensation, always use samples expressing high amounts of the protein of interest.

Acknowledgments

The authors thank Ingrid Bennett for secretarial work and colleagues at our institution for discussion and technical help.

References

1. Biron CA, Byron KS, Sullivan JL (1989) Severe herpesvirus infections in an adolescent without natural killer cells. N Engl J Med 320(26): 1731–1735
2. Orange JS (2006) Human natural killer cell deficiencies. Curr Opin Allergy Clin Immunol 6(6):399–409
3. Quinnan GV Jr, Kirmani N, Rook AH et al (1982) Cytotoxic T cells in cytomegalovirus infection: HLA-restricted T-lymphocyte and non-T-lymphocyte cytotoxic responses correlate with recovery from cytomegalovirus infection in bone-marrow-transplant recipients. N Engl J Med 307(1):7–13
4. Kuijpers TW, Baars PA, Dantin C et al (2008) Human NK cells can control CMV infection in the absence of T cells. Blood 112(3): 914–915
5. Wilkinson GW, Tomasec P, Stanton RJ et al (2008) Modulation of natural killer cells by human cytomegalovirus. J Clin Virol 41(3): 206–212
6. Orange JS, Fassett MS, Koopman LA et al (2002) Viral evasion of natural killer cells. Nat Immunol 3(11):1006–1012
7. Romo N, Magri G, Muntasell A et al (2011) Natural killer cell-mediated response to human cytomegalovirus-infected macrophages is modulated by their functional polarization. J Leukoc Biol 90(4):717–726
8. Sinclair J, Sissons P (2006) Latency and reactivation of human cytomegalovirus. J Gen Virol 87(Pt 7):1763–1779
9. Sinzger C, Hahn G, Digel M et al (2008) Cloning and sequencing of a highly productive, endotheliotropic virus strain derived from

human cytomegalovirus TB40/E. J Gen Virol 89(Pt 2):359–368

10. Alter G, Malenfant JM, Altfeld M (2004) CD107a as a functional marker for the identification of natural killer cell activity. J Immunol Methods 294(1–2):15–22
11. Tomescu C, Chehimi J, Maino VC et al (2009) Retention of viability, cytotoxicity, and response to IL-2, IL-15, or IFN-alpha by human NK cells after CD107a degranulation. J Leukoc Biol 85(5):871–876
12. Al-Hubeshy ZB, Coleman A, Nelson M et al (2011) A rapid method for assessment of natural killer cell function after multiple receptor crosslinking. J Immunol Methods 366(1–2):52–59
13. Warren HS (2011) Target-induced natural killer cell loss as a measure of NK cell responses. J Immunol Methods 370(1–2):86–92
14. Duske H, Sputtek A, Binder T et al (2011) Assessment of physiologic natural killer cell cytotoxicity *in vitro*. Hum Immunol 72(11):1007–1012
15. Schuessler A, Laib Sampaio K, Sinzger C (2010) Improvement of a quantitative cell tropism assay for rapid and reliable characterization of human cytomegalovirus mutants. J Virol Methods 167(2):218–222

Chapter 22

Analysis of Cell Migration During Human Cytomegalovirus (HCMV) Infection

Stefania Varani and Giada Frascaroli

Abstract

Previous studies have demonstrated that dendritic cells (DC), macrophages (Mφ), and their precursors monocytes are susceptible to infection by human cytomegalovirus (HCMV) in the natural host as well as *in vitro*. Due to their proficient ability to take up and present antigens to the lymphocytes these cells are also called antigen presenting cells (APC) and represent a crucial component that HCMV needs to disable in order to limit the antiviral immune reaction. It is well known that cell trafficking is an essential property of APC. Monocytes and DC are usually regarded as very motile cells and their trafficking properties through the blood vessels, the peripheral tissues, and the lymphoid organs are intensively studied. On the other hand, although often considered a resident population, Mφ are also motile and can actively migrate into areas of infection, inflammation, and tissue regeneration. The movements of monocytes, DC, and Mφ require a tight control that is mainly assured by chemokines (CK) and their receptors.

While it is quite common to study the expression of chemokine receptors by flow cytometry, methods for the investigation of the chemokine receptor functionality are less widespread. In this chapter, we describe different techniques that can help in the analysis of cell migration in response to CK. *Cell polarization assays* measure the rapid morphological changes that follow the chemokine receptors' engagement by their ligands. *Actin polymerization assays* measure the subsequent conversion of globular units of actin into dynamic filaments. Finally, *chemotaxis assays* quantify the cell movements along a CK gradient.

Key words Chemotaxis, Chemokines (CK), Chemokine receptors, Cell motility, Cell polarization, Cytoskeleton, Antigen presenting cells (APC), Monocytes, Dendritic cells (DC), Macrophages (Mφ), Human cytomegalovirus (HCMV)

1 Introduction

A vast body of literature illustrates the impressive range of strategies adopted by human cytomegalovirus (HCMV), a clinically relevant member of the betaherpesviridae subfamily [1], to promote its own survival and spread [2, 3] in the human host. A well-documented example of HCMV-induced modulation of cellular functions is the sabotage of neutrophil, fibroblast, and smooth muscle cell movements [4–6]. Upon HCMV encounter these cells can transport the intracellular virus to distant areas (thus promoting

Susanne M. Bailer and Diana Lieber (eds.), *Virus-Host Interactions: Methods and Protocols*, Methods in Molecular Biology, vol. 1064, DOI 10.1007/978-1-62703-601-6_22, © Springer Science+Business Media, LLC 2013

viral dissemination) and even recruit new target cells into the area of infection (thus expanding the infected area). Our studies were focused on the migration of antigen presenting cells (APC) because these cells (1) are susceptible to HCMV infection, (2) have a complex and tightly regulated migratory behavior [7], and finally (3) are essential for priming the immune system and eliciting specific antiviral responses [8]. APC includes dendritic cells (DC), macrophages (Mφ), and their precursors monocytes. Under homeostatic conditions, APC are very motile and continuously patrol the human body by recirculating between blood, peripheral tissues, and lymphoid organs [9]. In response to inflammatory chemoattractants APC can infiltrate areas of infection, where they take up virions and infected cellular debris. After antigen uptake and processing, APC become responsive to lymphoid chemoattractants that in turn direct their migration into lymphoid organs where the antigens can be presented to lymphocytes [10–12]. The proper spatial and temporal regulation of APC movements is complex, involving adhesion molecules, selectins, and an entire superfamily of chemoattractant cytokines named chemokines (CK) that signal through seven-transmembrane G protein-coupled receptors (GPCR). CK are small soluble molecules that have been classified into four major families (CC, CXC, C, and CX_3C) according to the configuration of cystein residues near the aminoterminus [13–15]. Binding of the CK to the receptor leads to the dissociation of the G proteins into α and βγ subunits that in turn trigger various biochemical responses, including (1) rapid and transient increases in Ca^{2+} influx, (2) change of cell morphology due to actin polymerization, and (3) cell-oriented migration [16].

The great number of chemokine and chemokine receptor homologues encoded by HCMV (extensively reviewed in [17, 18]) emphasizes the importance of the viral attack to the migratory abilities of host cells. Here we want to describe different techniques that helped us in the analysis of the effects exerted by HCMV on the chemokine-driven migration of human primary APC [19–21].

In this chapter, we will describe (1) how to produce and infect monocytes and Mφ *in vitro* (being DC already well described in the literature), (2) how to analyze the architecture of the cellular cytoskeleton, (3) how to quantify actin polymerization, and (4) how to perform chemotaxis assays.

2 Materials

In order to avoid uncontrolled activation of APC, all solutions and plastic devices should be tested as pyrogen- or endotoxin-free (*see* **Note 1**).

2.1 Cells, Media, and Supplements

1. Buffy-coats and human serum (blood group AB; heat inactivated). All buffy-coats used in the studies were purchased from the Transfusion Center of the Ulm University Hospital (IRB granted to Institut für Klinische Transfusionsmedizin und Immungenetik Ulm GmbH, Ulm, Germany) and were obtained from anonymized healthy blood donors.
2. Human foreskin fibroblasts (HFF).
3. Sterile, single-use plastic tubes and flasks (Falcon, BD Biosciences, Heidelberg, Germany).

   Lumox dishes (Sarstedt, Nümbrecht, Germany).
4. Dulbecco's Phosphate Buffered Saline (DPBS) without Ca^{2+} and Mg^{2+} and Lymphocyte Separation Medium LSM 1077 (PAA Laboratories GmbH, Cölbe, Germany).
5. Monocyte isolation kit II; LS columns and MidiMacs separators (Miltenyi Biotec, Bergisch Gladbach, Germany).
6. Cell culture media: RPMI 1640 and Minimum Essential Media (MEM) (Invitrogen, Life Technologies GmbH, Darmstadt, Germany).
7. Supplements: fetal calf serum (FCS) and lipopolysaccharide (LPS) (Sigma-Aldrich Co., Taufkirchen, Germany); penicillin/streptomycin and L-glutamine (PAA Laboratories); human recombinant macrophage-colony stimulating factor (M-CSF), granulocyte-macrophage-colony stimulating factor (GM-CSF) and interleukin (IL)-4 (R&D Systems, Minneapolis, MN).
8. HFF medium = MEM supplemented with 2 mM L-glutamine, 100 U/mL penicillin and 100 U/mL streptomycin.
9. Monocyte medium = RPMI 1640 supplemented with 2 mM L-glutamine, 100 U/mL penicillin and 100 U/mL streptomycin plus 10 % human serum group AB-heat inactivated.
10. Mφ and DC medium = RPMI 1640 supplemented with 2 mM L-glutamine, 100 U/mL penicillin and 100 U/mL streptomycin plus 10 % FCS.
11. MicoAlert (Cambrex, Rockland, MD).

2.2 Virus and Solution for Virus Stocks

1. The HCMV endotheliotropic strain TB40E was kindly provided by Dr. C. Sinzger, University of Ulm, Germany.
2. Sucrose-phosphate buffer: dissolve 37.31 g sucrose, 0.609 g K_2HPO_4 (dipotassium phosphate) and 0.26 g KH_2PO_4 (monopotassium phosphate; all from Merck KGaA, Darmstadt, Germany) in 500 mL H_2O. Autoclave prior to use.

2.3 Reagents for Confocal Microscopy (Analysis of Cytoskeletal Architecture)

1. Eight wells IBIDI-slides (Integrated Bio Diagnostics ibidi GmbH, Munich, Germany) and 12 wells Diagnostic Slides coated with Teflon (Waldemar Knittel, Braunschweig, Germany).
2. 4 % Paraformaldehyde (PFA, w/v): dissolve 4 g paraformaldehyde (Fluka Chemie AG, Switzerland) in 80 mL of DPBS by heating the solution to 70 °C in a fume hood. Cool to room temperature before adjusting the pH to 7.2. Finally fill volume to 100 mL. Store at 4 °C protected from light in a bottle wrapped with aluminum foil.
3. 0.2 % Triton X-100 (v/v): dilute 0.2 mL of Triton X-100 (Serva, Amstetten, Austria) in 100 mL DPBS.
4. 30 % Bovine serum albumin (BSA) (Sigma Aldrich Co.).
5. Fluorescein (FITC)-labeled phalloidin (P-55282, Sigma-Aldrich Co.): prepare a stock solution 0.5 mg/mL in methanol and store at −20 °C, protected from light.
6. Primary monoclonal antibodies anti-α-tubulin (Molecular Probes, Invitrogen) and anti-vimentin (Oncogene Research Products, La Jolla, CA).
7. Secondary goat-anti mouse immunoglobulins Alexa Fluor 488 conjugated (Molecular Probes, Invitrogen).
8. 4′,6-Diamidine-2′-phenylindole dihydrochloride (DAPI) (Roche, Penzberg, Germany).

2.4 Reagents for Cell Polarization Assays

1. Polypropylene 5 mL Round Bottom Tubes, snap cap (Falcon, BD Biosciences).
2. Medium RPMI 1640 containing 1 % FCS.
3. Chemotactic stimulants such as *N*-Formyl-Met-Leu-Phe (fMLP) and different human recombinant CK (*see* **Note 2**). 10^{-5} M and 10^{-6} M working solutions of fMLP were prepared on the same day of the experiment by diluting the 10^{-3} M stock solution with RPMI 1 % FCS. 10 μg/mL working solutions of CK were prepared the same day of the experiment by thawing the 10 μg/mL stock solutions preserved at −80 °C.
4. Phosphate-buffered formaldehyde 10 % v/v: add 10 mL of 37 % formaldehyde (Fluka Chemie AG) to 27 mL of DPBS and adjust to pH 7.2. Store at 4 °C protected from light and wrapped with aluminum foil.

2.5 Actin Polymerization Reagents

1. Polypropylene 1.5 mL eppendorf tubes (Eppendorf, Wesseling-Berzdorf, Germany) and 5 mL Round Bottom Tubes, snap cap (Falcon, BD Biosciences).
2. Chemotactic stimulants such as fMLP and human recombinant CK (*see* Subheading 2.4).

3. Paraformaldehyde (PFA) 4 % w/v: (*see* Subheading 2.3).
4. 0.1 % Triton X-100 (v/v): dilute 0.1 mL of Triton X-100 (Serva) in 100 mL DPBS.
5. FACS buffer: 2 % FCS, 0.01 % Na azide in DPBS.
6. Fluorescein (FITC)-labeled phalloidin (*see* Subheading 2.3): freshly prepare 1.5 μg/mL working solution by diluting the 0.5 mg/mL stock solution with FACS buffer.

2.6 Chemotaxis (Boyden Chamber or Transwell) Components

1. Medium RPMI 1640 containing 1 % FCS.
2. DPBS.
3. Chemotactic stimulants such as fMLP and human recombinant CK. Working solutions of fMLP (concentration 10^{-7} M and 10^{-8} M) were prepared on the same day of the experiment by diluting the 10^{-3} M stock with RPMI 1 % FCS. 100 ng/mL working CK solutions were prepared by diluting a 10 μL aliquot of the stock (10 μg/mL) into 1 mL RPMI 1640 medium containing 1 % FCS.
4. 6.5 mm Transwell® with 8.0 μm pore polycarbonate membrane insert (Costar, Cambridge, MA).
5. Methanol/acetone 1:1 v/v. Store at −20 °C.
6. Mayer's hematoxylin and eosin (Histolab Products, Gothenburg, Sweden).
7. Cotton swabs.
8. 48-well Boyden chamber and polyvinylpyrrolidone-free polycarbonate filters with 5 μm pores (Neuroprobe, Pleasanton, CA) (*see* **Note 3**).
9. Diff-quick staining kit (Medion Diagnostic, Düdingen, Switzerland).

2.7 Equipments

1. Cell incubator (37 °C in a 5 % CO_2 humidified atmosphere).
2. Optical microscope.
3. Bürker or Neubauer counting chambers.
4. Ultracentrifuge Beckmann with SW28.1 rotor (employed for preparation of virus stocks).
5. Fluorescence microscope Axio Observer.Z1 (Zeiss, Oberkochen, Germany) (employed for confocal microscopy).
6. Cytofluorimeter FACS calibur (BD Biosciences) (employed for actin polymerization assays).
7. Heating block (for actin polymerization assays).
8. Optical microscope with 40× and 100× magnification objectives (employed for chemotaxis assays).

3 Methods

3.1 Cell Purification and Differentiation

Peripheral blood mononuclear cells (PBMC) were isolated from buffy-coats obtained from anonymous blood donors by centrifugation over Lymphocyte Separation Medium according to standard protocols (*see* also Wu et al., Chapter 21 of this series). Monocytes were isolated by negative selection with magnetic microbeads (Miltenyi Monocyte Isolation Kit II) according to the manufacturer's instructions and their purity was >95 % as assessed by flow cytometry. Monocytes were cultured in standard medium (*see* **Note 4**) supplemented with 10 % human AB serum and maintained in polypropylene tubes in order to reduce substrate adhesion (*see* **Note 5**). Mφ were obtained by culturing 3×10^6 monocytes/mL in hydrophobic Lumox dishes (*see* **Note 6**) in standard medium (*see* **Note 4**) supplemented with 10 % FCS and 100 ng/mL of M-CSF. The differentiation of Mφ occurred over a period of 7 days and was evaluated by morphological criteria, phagocytic activity, and by expression of distinct markers employing flow cytometry as shown in [21]. DC were generated as described previously [22]. Briefly, isolated PBMC ($5–10 \times 10^6$/mL) were plated in standard medium (*see* **Note 4**) containing 10 % FCS and allowed to adhere for 2 h at 37 °C. Non-adherent cells were removed by three washes with DPBS. Standard medium (*see* **Note 4**) containing 10 % FCS, 25 ng/mL IL-4 and 100 ng/mL GM-CSF was added, and the cells were cultured for 7 days. During the last 1 or 2 days of culture, 100 ng/mL LPS was added to stimulate DC maturation. Cell differentiation was monitored by light microscopy and flow cytometry as described in [19].

Prior to infection, monocytes, Mφ, and DC were collected (*see* **Note 7**), counted, and resuspended at the concentration of 1×10^6 cells/mL in their respective complete culture media (*see* **Note 8**). Cells were infected in complete medium overnight (*see* **Note 9**) with HCMV by using a multiplicity of infection (MOI) of 5 plaque forming units (PFU) per cell.

3.2 Preparation of Virus Stocks

Cell-free viral stocks of the endotheliotropic HCMV strain TB40E were produced by infecting human foreskin fibroblasts (HFF) that were cultured in minimal essential medium (MEM) with 10 % FCS, 2 mM L-glutamine, 100 U/mL penicillin and 100 U/mL streptomycin.

1. Infected supernatants were recovered when the maximum cytopathic effect was reached and supernatants were cleared of cellular debris by centrifugation ($7,000 \times g$ for 10 min).
2. Virus was then pelleted by ultracentrifugation at $95,000 \times g$ for 60 min, resuspended in sucrose phosphate buffer, frozen at −80 °C, and thawed before single use.

3. Virus stocks were negative for contamination with Mycoplasma, as determined by MicoAlert.
4. Virus titers (PFU/mL) were determined on HFF by plaque assay as previously described [23].

3.3 Analysis of Cytoskeletal Architecture

The cytoskeletal architecture of mock- and HCMV-infected cells (*see* **Note 10**) was visualized at 1 day post infection (dpi) by confocal microscopy. The staining procedure was similar in monocytes [24] and Mφ [21] with the only difference that while monocytes were allowed to adhere to glass microscope slides immediately before the staining (*see* **Note 11**), 0.1×10^6 Mφ per well were seeded in IBIDI slides the day before staining.

1. Remove medium and fix adherent cells with 4 % PFA by 10 min incubation at room temperature (*see* **Note 12**).
2. Wash the cells three times with DPBS.
3. Permeabilize the cells with 0.2 % Triton X-100 by 2 min incubation at room temperature.
4. Wash the cells three times with DPBS.
5. Actin cytoskeleton is visualized by 30 min incubation with 0.1 μg/mL of FITC-labeled phalloidin diluted in DPBS. Vimentin and actin are visualized by 45 min incubation at 37 °C with antibodies anti-vimentin and anti-tubulin (diluted 1:400 and 1:200, respectively in DPBS supplemented with 1 % BSA) followed by 30 min incubation at 37 °C with fluorescently labeled secondary anti-mouse antibodies. Nuclei are stained by 5 min incubation with DAPI. Between primary and secondary antibodies as well as after DAPI counterstaining, cells are washed two to three times with DPBS.

Staining was evaluated using a Zeiss Axioskop2 fluorescence microscope and Fig. 1 shows a representative staining in Mφ.

3.4 Cell Polarization Assays

Cell polarization, an event that occurs within minutes after chemoattractant stimulation, is marked by formation of cell protrusion and pseudopodia that define the direction of locomotion [25]. These protrusions take the form of a broad lamellipodia on the leading front and of a pseudopod on the cellular tail. The effects of HCMV infection on Mφ polarization were assessed by comparing the polarization capacities of uninfected and HCMV-infected Mφ (TB40E MOI of 5 at 1 dpi) [21].

1. Resuspend Mφ in RPMI 1 % FCS at a final concentration of 10^6 cells/mL and dispense at least 180 μL in each tube. Warm the cell suspension by incubating the tubes at 37 °C for 5 min.
2. Prepare the stimulants (i.e., fMLP, C5a, CK) by diluting them in RPMI 1 % FCS to a concentration ten times higher than the final concentration. Use RPMI 1 % FCS as negative control. Allow the solutions to warm up to 37 °C.

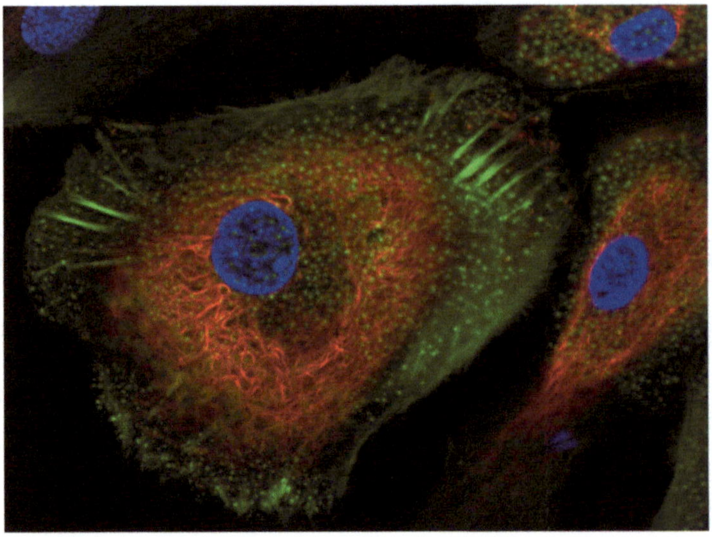

Fig. 1 Cytoskeletal organization of human primary Mφ. Distribution of F-actin (*green*) in discrete dot-like structures called podosomes and supporting microtubules (*red*). Nuclei are counterstained with DAPI (*blue*)

3. Add 20 μL of stimulants (or control medium) to each sample (final volume = 200 μL) and incubate the tubes for 10 min at 37 °C. It is appropriate to prepare each sample in duplicates.
4. Stop the polarization process by adding to each sample an equal volume of ice-cold phosphate-buffered formaldehyde (10 %, pH 7.2) and pipette gently.
5. Transfer 10 μL of cell suspension in a Burker chamber and count at least 200 cells per sample by employing a phase contrast microscopy and differentiating the cells having a bipolar front-tail shape.
6. Calculate the percentage of polarization as follow:

$$\%\ \text{polarized cells} = \frac{\text{Nr bipolar cells} \times 100}{\text{Nr total cells}}$$

3.5 Assessment of Actin Polymerization

The dramatic changes in shapes necessary to direct and propel cell motility are controlled by the rapid assembly and disassembly of actin filaments. The capacity to polymerize actin filaments (F-actin) starting from globular units (G-actin) was measured in uninfected and HCMV-infected Mφ (TB40E MOI of 5 at 1 dpi) [21] (*see* **Note 13**).

1. Resuspend Mφ in RPMI 1640 supplemented with 1 % FCS at a concentration of 2.5×10^6 cells/mL and incubate tubes at 37 °C for 30 min. Plan the experiment calculating how many agonists and time points should be investigated (*see* **Note 14**).
2. Prepare "stimulation tubes" by labeling and adding to each eppendorf tube 10 μL of control medium (RPMI 1640 sup-

plemented with 1 % FCS) or medium with stimulants (i.e., 200 ng/mL RANTES/CCL5, 500 ng/mL VEGF or 10^{-7} M fMLP).

3. Label a corresponding number of eppendorf tubes and add to each of them 300 μL of 4 % PFA. Keep these tubes at room temperature.
4. Transfer the "stimulation tubes" to a heating block and warm up to 37 °C for 2–3 min.
5. Transfer 60 μL cell suspension into the "stimulation tubes" labeled with time point 0, pipette gently once, immediately (or after 15 s, 30 s, 1 min, 5 min, 15 min) transfer the entire volume (10 μL + 60 μL) into the corresponding eppendorf tube containing 4 % PFA and *gently* pipette once.
6. Fix the cells by 10 min incubation at room temperature.
7. Wash the cells by adding 1 mL DPBS. Centrifuge for 15 s at full speed and discard supernatants.
8. Permeabilize the cells by resuspending the pellet in 300 μL of ice-cold 0.1 % Triton X-100 and incubate on ice for 10 min. *Do not incubate in Triton X-100 solution for more than 10 min.* Add 1 mL cold FACS buffer (*see* **Note 15**) and centrifuge for 15 s at full speed. Discard supernatants.
9. Resuspend the pellet in 50 μL FITC-phalloidin staining solution freshly prepared (1.5–2.0 μg/mL) and incubate for 30 min at room temperature in the dark.
10. Add 200 μL FACS buffer and measure fluorescence with a FACScalibur.
11. Measure FL-1 (FITC) mean fluorescence intensity (MFI) values in stimulated and unstimulated Mφ (*see* **Note 16**).
12. Create a graph depicting ΔMFI = (MFI of stimulated cells − MFI of unstimulated cells) on *Y* axis and stimulation time on *X* axis.

3.6 Cell Migration Assays

While the migration of DC was tested using transwell culture chambers [19] the migration of monocytes as well as Mφ was tested using 48-well Boyden chambers [20, 21]. Although the gradient formed into the Boyden chamber is considered by some authors as superior, the two methods are widely used and comparably accepted in the scientific community.

Assessment of DC migration using Transwell plates (*see* **Note 17**):

1. Coat Transwell membrane inserts (polycarbonate filters with 8.0-μm pore size) with 50 μL of gelatin at 0.1 mg/mL on the lower surface and let them dry overnight at room temperature. Wash the coated filters in DPBS and let them dry immediately before use.

2. Resuspend DC in RPMI 1640 at a concentration of 1.0×10^6 cells/mL and add 100 μL cell suspension to the upper compartment of the well.
3. Add 600 μL medium alone (RPMI 1640) or medium containing chemokines to the lower well.
4. Incubate the Transwell plate for 2 h at 37 °C.
5. Wash the filters three times with DPBS and fix the cells with methanol/acetone (1:1) for 5 min at −20 °C. Stain the cells with Mayer's hematoxylin and eosin following manufacturer's instructions.
6. Remove non-migrating DC, corresponding to the cells remaining on the upper surface of the filter, by wiping with cotton swabs.
7. Count migrating DC, corresponding to the cells adherent to the lower surface of the filter, under a light microscope at a magnification of 40×. Randomly select ten fields to be counted in each well. The number of migrated cells is expressed as average of the number of counted cells in ten fields.

Assessment of monocyte and Mφ migration using Boyden chambers (*see* **Note 18**):

1. Resuspend monocytes or Mφ in RPMI 1640 supplemented with 1 % FCS at a concentration of 1.5×10^6 cells/mL.
2. Aliquot 27 μL of an appropriate chemoattractant (e.g., fMLP or chemokines) into the wells of the lower chamber (*see* **Note 19**).
3. Put the polycarbonate filter onto the lower chamber leaving the opaque surface on top. It is important to orient unambiguously the position of the filter. As an example, the right side and upper corner can be marked by cutting the upper right edge.
4. Mount the silicon trimming and then the upper chamber. In order to avoid the mixing of the chemoattractants, press down the upper chamber until completely and tightly screwed.
5. Seed 50 μL of cell suspension in the upper wells by leaning the pipette tip on the border of the well and quickly ejecting the tip content.
6. Incubate the chamber at 37 °C in 5 % CO_2 for 1.5 h.
7. Remove chamber from the incubator. Unscrew and turn the chamber upside down (*see* **Note 20**).
8. Hold the upper chamber tightly and remove the lower chamber. Remove the silicon trimming and hold the filter in front of you.
9. Lift the filter by placing a clamp on each extremity.
10. Wash the opaque surface of the filter, where non-migrated cells remain, by wetting gently this side of the filter with DPBS (*see* **Note 21**).

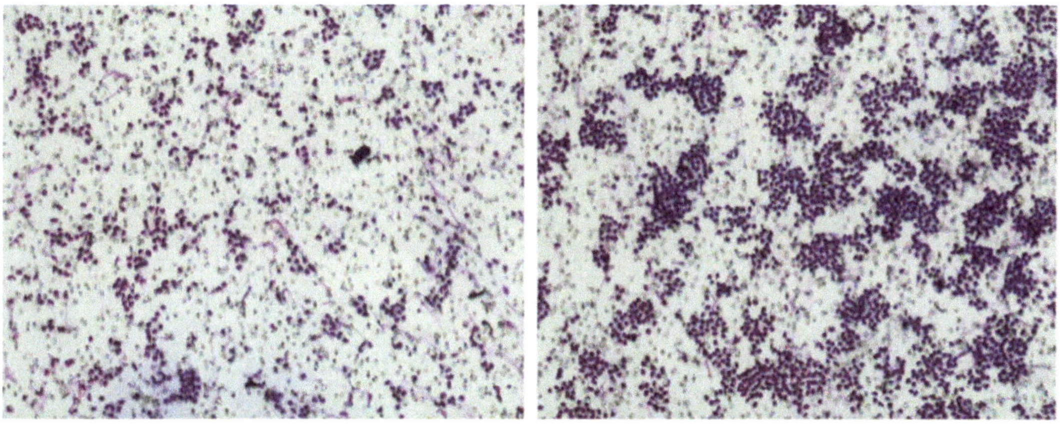

Fig. 2 Monocytes migrated in a Boyden chamber. Monocytes migrated in response to either medium alone (*left panel*, basal migration) or 100 ng/mL RANTES/CCL5 (*right panel*, chemokine-driven migration) were stained with Diff-Quick prior to visualization at the microscope (10× magnification)

11. Remove all non-migrating cells by scraping the opaque surface of the filter against the special rubber policeman.
12. Stain the filter with Diff-Quick stain according to the manufacturer's instruction (1 min each solution).
13. Place the filter on a glass slide and count the migrated cells on the bright surface of the filter. For each well, count five microscopic fields at a magnification of 100×. The number of migrated cells is expressed as average of the cells counted in five fields.

Figure 2 shows monocytes migrated through a polycarbonate filter in a Boyden chamber.

4 Notes

1. Many solutions are certified as endotoxin-free and are available on the market as ready-to-use solutions. Sterile and single-use plastic wares are usually endotoxin-free and can be obtained from various suppliers. Materials and solutions mentioned in this chapter are routinely used in our laboratories.
2. *N*-Formyl-Met-Leu-Phe (fMLP; Sigma-Aldrich Co.) and the human recombinant chemokines MCP-1/CCL2, MIP-1α/CCL3, Rantes/CCL5, MIP-3β/CCL19, Gro-α/CXCL1, IL-8/CXCL8, SDF-1/CXCL12, Fractalkine/CX3CL1 (purchased from R&D Systems and PeproTech, Hamburg, Germany) were solubilised as specified in the product data sheets. fMLP stocks (stored at 4 °C for months or at −20 °C for years) were prepared at the concentration of 10^{-3} M.

Chemokine stocks were prepared at the concentration of 10 μg/mL using DPBS containing 0.1 % BSA as stabilizer. It is important to avoid repeated freeze/thaw cycles.

3. The Neuroprobe web site (www.neuroprobe.com) offers a broad range of pictures and protocols. We ordered 48-wells Boyden chambers and the accompanying accessory pack that contains several items (such as filter wiper, one large and one small filter clamp, and two pairs of curved, stainless steel forceps) to facilitate handling and staining of polycarbonate membranes. The accessory items are used to wipe non-migrated cells off the top of a filter and while fixing and staining the migrated cells.
4. Standard medium = RPMI 1640 supplemented with 2 mM L-glutamine, 100 U/mL penicillin and 100 U/mL streptomycin.
5. Monocytes react to experimental manipulation by changing their activation state. Short time culture, avoidance of substrate adherence as well as replacement of bovine serum with human serum are reasonable attempts to preserve the natural features of these cells. In our hands, monocytes showed unaltered viability and metabolic activity only for the first 24–48 h of *in vitro* culture. Monocytes strongly adhere to glass and plastic surfaces; the employment of polypropylene tubes instead of polystyrene tubes reduces substrate adhesion as well as monocyte activation.
6. Lumox dishes have ultra-thin, gas-permeable film bases. At 25 μm thickness, the Lumox film base provides optical quality, very low auto fluorescence, and high UV transparency for microscopy and imaging. Unlike ordinary dishes with cover glass bases, the Lumox film is gas-permeable for effective gas exchange and homogeneous cell growth. These dishes are essential to avoid excessive adherence of Mφ and to facilitate cell harvesting after 7 days of *in vitro* culture.
7. The detachment and recovery of Mφ out of the Lumox dishes is a simple procedure. At first, the entire volume of a dish is transferred into a new sterile Falcon 15 mL tube. Then, 1 mL of sterile DPBS is added to the Mφ culture, the dish is incubated at 37 °C for 5 min, then, after vigorous pipetting, the content of the dish is transferred into the Falcon tube. This procedure should be repeated three times in total and usually at the third incubation all Mφ detach from the dish bottom. Tubes containing Mφ are then centrifuged at 900 × *g* for 7 min (with brake) and the cell pellet is resuspended in medium.

   Monocyte-derived DC are either floating or loosely adherent to the bottom of the plate, therefore ordinary plastic plates or flasks can be employed for DC differentiation and culture and no specific detachment steps are required to harvest these cells.

8. Monocyte medium = standard medium plus 10 % human serum; Mφ and DC medium = standard medium plus 10 % FCS.
9. Comparable infection rates (measured as percentage of immediate early positive cells at 24 h post infection) were observed when cells were incubated with viral stocks overnight or for 3 h.
10. In stark contrast to the majority of resident cells, migratory cells such as monocytes, DC, and Mφ do not possess stress fibres or large focal adhesions [26]. In fact, while fibroblasts are anchored flat to the underlying matrix via linear bundles of actin filaments called "stress fibres," in DC and Mφ actin filaments form a delicate tracery of bundles that concentrate at a series of discrete foci at the substratum interface. These structures, termed podosomes, are the most prominent part of the actin cytoskeleton of monocyte-derived cells such as DC and Mφ and have important roles in adhesion, migration, and tissue invasion.
11. Monocytes were resuspended in standard medium at the cell concentration of 1×10^6 cells/mL and then 30 μL were distributed in each well of the glass microscope slide. Slides were incubated in a humid box for 30 min at 37 °C to allow monocyte adhesion.
12. In our hands, the fixation with PFA gave the best results while solutions containing methanol and aceton altered the cytoskeletal structures.
13. Change of temperature and vigorous pipetting may cause actin polymerization thus interfering with the experimental conditions. Temperature shift should be avoided by maintaining cells and cell solutions at 37 °C. Always pipette very gently.
14. We stimulated cells for 0 s, 15 s, 30 s, 1 min, 5 min and 15 min. As example of a typical experiment: six eppendorf tubes (one for each point of time) were prepared with 10 μL medium alone, six eppendorf tubes with 10 μL RANTES (200 ng/mL) and six eppendorf tubes with 10 μL VEGF (500 ng/mL). Additionally to the already mentioned 18 "stimulation tubes," 18 eppendorf tubes were labeled accordingly and filled with 300 μL 4 % PFA. For this experiment at least 1.2 mL cell suspension (60 μL per stimulation tube) and 100 μL medium alone or medium plus agonist (10 μL per each time point) are required.
15. FACS buffer: 2 % FCS, 0.01 % Na azide in DPBS.
16. Suggested settings for FACScalibur are FCS (Voltage: E^{-1} A/Gain: 6.0); SSC (Voltage: 435 A/Gain: 1.0); FL1 (Voltage: 290 Compensation: 0 %); FL2 (Voltage: 290 Compensation: 25 %); FL3 (Voltage 289 Compensation: 0 %). 5,000 cells were acquired for each tube.

17. DC and chemoattractants were prepared in serum-free RPMI 1640 medium as described in [27].
18. Monocyte and Mφ cell suspensions as well as chemoattractants were prepared in RPMI 1640 supplemented with 1 % FCS.
19. The 27 μL volume may have some variation (2–3 μL more or less) depending on the microchamber used. It is advisable to check in advance the volume that assures the formation of a small convex surface emerging from the well in order to guarantee a perfect adhesion of the filter and avoidance of air bubbles.
20. Please note that the chamber is now turned upside down: the lower part with chemoattractants faces the researcher while the upper part with the cell suspension faces the table.
21. Do not entirely dip the filter into DPBS, otherwise the cells will be lost.

Acknowledgments

Giada Frascaroli wishes to thank Dr. Silvano Sozzani (University of Brescia, Brescia, Italy) and the passed away Walther Luini for introducing and guiding into this topic.

This study was financially supported by the Deutsche Forschungsgemeinschaft (DFG) project SPP 1175 Me 1740/1, the Bundesministerium für Bildung und Forschung (BMBF) Project C, the Carl Zeiss Stiftung Project "Infektionsbiologie humaner Makrophagen," the Ludovisi-Blanceflor Stiftung, and by RFO2009 from the University of Bologna, Italy.

References

1. Mocarski ES Jr, Shenk T, Pass RF (2007) Cytomegaloviruses. In: Knipe DM, Howley PM, Griffin BE, Lamb RA, Martin MA, Roizman B, Straus SE (eds) Fields virology. Lippincott Williams & Wilkins, Philadelphia, PA, pp 2701–2772
2. Powers C, Defilippis V, Malouli D et al (2008) Cytomegalovirus immune evasion. Curr Top Microbiol Immunol 325:333–359
3. Noriega V, Redmann V, Gardner T et al (2012) Diverse immune evasion strategies by human cytomegalovirus. Immunol Res 54:140–151
4. Penfold ME, Dairaghi DJ, Duke GM et al (1999) Cytomegalovirus encodes a potent alpha chemokine. Proc Natl Acad Sci USA 96:9839–9844
5. Streblow DN, Soderberg-Naucler C, Vieira J et al (1999) The human cytomegalovirus chemokine receptor US28 mediates vascular smooth muscle cell migration. Cell 99: 511–520
6. Boomker JM, Van Luyn MJ, The TH et al (2005) US28 actions in HCMV infection: lessons from a versatile hijacker. Rev Med Virol 15:269–282
7. Notarangelo LD, Badolato R (2009) Leukocyte trafficking in primary immunodeficiencies. J Leukoc Biol 85:335–343
8. Christensen JE, Thomsen AR (2009) Co-ordinating innate and adaptive immunity to viral infection: mobility is the key. APMIS 117:338–355
9. Caux C, Vanbervliet B, Massacrier C et al (2002) Regulation of dendritic cell recruitment by chemokines. Transplantation 73:S7–S11
10. Mellman I, Steinman RM (2001) Dendritic cells: specialized and regulated antigen processing machines. Cell 106:255–258

11. Hume DA (2008) Macrophages as APC and the dendritic cell myth. J Immunol 181: 5829–5835
12. Sallusto F, Lanzavecchia A (2010) Monocytes join the dendritic cell family. Cell 143:339–340
13. Baggiolini M (1998) Chemokines and leukocyte traffic. Nature 392:565–568
14. Murphy PM, Baggiolini M, Charo IF et al (2000) International union of pharmacology. XXII. Nomenclature for chemokine receptors. Pharmacol Rev 52:145–176
15. Zlotnik A, Yoshie O (2012) The chemokine superfamily revisited. Immunity 36:705–716
16. Rossi D, Zlotnik A (2000) The biology of chemokines and their receptors. Annu Rev Immunol 18:217–242
17. Alcami A, Koszinowski UH (2000) Viral mechanisms of immune evasion. Immunol Today 21:447–455
18. Murphy PM (2001) Viral exploitation and subversion of the immune system through chemokine mimicry. Nat Immunol 2:116–122
19. Varani S, Frascaroli G, Homman-Loudiyi M et al (2005) Human cytomegalovirus inhibits the migration of immature dendritic cells by down-regulating cell-surface CCRl and CCR5. J Leukoc Biol 77:219–228
20. Frascaroli G, Varani S, Moepps B et al (2006) Human cytomegalovirus subverts the functions of monocytes, impairing chemokine-mediated migration and leukocyte recruitment. J Virol 80:7578–7589
21. Frascaroli G, Varani S, Blankenhorn N et al (2009) Human cytomegalovirus paralyzes macrophage motility through down-regulation of chemokine receptors, reorganization of the cytoskeleton, and release of macrophage migration inhibitory factor. J Immunol 182: 477–488
22. Sallusto F, Lanzavecchia A (1994) Efficient presentation of soluble antigen by cultured human dendritic cells is maintained by granulocyte/macrophage colony-stimulating factor plus interleukin 4 and downregulated by tumor necrosis factor alpha. J Exp Med 179: 1109–1118
23. Wentworth BB, French L (1970) Plaque assay of cytomegalovirus strains of human origin. Proc Soc Exp Biol Med 135:253–258
24. Straschewski S, Patrone M, Walther P et al (2011) Protein pUL128 of human cytomegalovirus is necessary for monocyte infection and blocking of migration. J Virol 85:5150–5158
25. Ridley AJ, Schwartz MA, Burridge K et al (2003) Cell migration: integrating signals from front to back. Science 302:1704–1709
26. Calle Y, Chou HC, Thrasher AJ et al (2004) Wiskott-Aldrich syndrome protein and the cytoskeletal dynamics of dendritic cells. J Pathol 204:460–469
27. Sato K, Kawasaki H, Nagayama H et al (1999) CC chemokine receptors, CCR-1 and CCR-3, are potentially involved in antigen-presenting cell function of human peripheral blood monocyte-derived dendritic cells. Blood 93:34–42

Chapter 23

Gene Targeting in Mice: A Review

Hicham Bouabe and Klaus Okkenhaug

Abstract

The ability to introduce DNA sequences (e.g., genes) of interest into the germline genome has rendered the mouse a powerful and indispensable experimental model in fundamental and medical research. The DNA sequences can be integrated into the genome randomly or into a specific locus by homologous recombination, in order to: (1) delete or insert mutations into genes of interest to determine their function, (2) introduce human genes into the genome of mice to generate animal models enabling study of human-specific genes and diseases, e.g., mice susceptible to infections by human-specific pathogens of interest, (3) introduce individual genes or genomes of pathogens (such as viruses) in order to examine the contributions of such genes to the pathogenesis of the parent pathogens, (4) and last but not least introduce reporter genes that allow monitoring *in vivo* or *ex vivo* the expression of genes of interest. Furthermore, the use of recombination systems, such as Cre/loxP or FRT/FLP, enables conditional induction or suppression of gene expression of interest in a restricted period of mouse's lifetime, in a particular cell type, or in a specific tissue. In this review, we will give an updated summary of the gene targeting technology and discuss some important considerations in the design of gene-targeted mice.

Key words Gene targeting, Transgenic mice, Knockout mice, Reporter mice, ES cell lines, Targeting vector, Cre/loxP, FRT/FLP, MultiSite Gateway Cloning

1 The Application of Genetically Modified Mice for the Study of Viral Pathogenesis and Antiviral Immunity

The development of mice with germline genetic modifications has advanced our understanding of the mechanisms of viral pathogenesis and antiviral immune responses during virus–host-interactions enormously. For instance, transgenic mouse models that express antigen-specific, major histocompatibility complex (MHC)-restricted T cell receptor (TCR) transgenes have been used extensively to investigate virus-specific T cell responses. Such transgenic mice include for example mice expressing a TCR transgene specific for the influenza virus hemagglutinin (HA) in the context of the MHC class I or II molecules [1, 2], and transgenic mice expressing

Susanne M. Bailer and Diana Lieber (eds.), *Virus-Host Interactions: Methods and Protocols*, Methods in Molecular Biology, vol. 1064, DOI 10.1007/978-1-62703-601-6_23, © Springer Science+Business Media, LLC 2013

MHC-restricted TCR with specificity for a lymphocytic choriomeningitis virus (LCMV) glycoprotein-derived T helper cell epitope [3, 4].

In addition, because mice are not susceptible to many human viruses, such as hepatitis viruses, papillomavirus, poliovirus, human immunodeficiency virus-1 (HIV-1), and measles, the generation of transgenic mice that express human receptors specific for such viruses have rendered those transgenic mice susceptible to infection by human viruses of interest and subsequently enabled to investigate their pathogenesis in *in vivo* models (reviewed in ref. 5).

An alternative to mice expressing human virus receptors is provided by transgenic mice with (conditional) expression of individual genes or genomes of viruses of interest. Such (conditional) expression of viral genes in mice imitates viral infection and thus enables *in vivo* investigation of the pathogenesis and immune responses induced by human viruses of interest [5–7].

Moreover, knockout mice with deletions of specific genes or cell populations of interest have been useful to identify and investigate the cellular and molecular components of the adaptive and innate immune system that play a role in controlling viral infections. Such knockout mice include, for example, TCR-β knockout mice that lack T cells [8]; μMT mice that lack B cells [9]; RAG-1 and RAG-2 knockout mice that lack B and T cells [10, 11]; knockout mice with deletions of immune mediators (such as chemokines and cytokines) including knockout mice for type I and type II interferons or interferon receptors [12–15]; knockout mice for immune receptors, such as Toll like receptors (TLRs), retinoic acid-inducible gene-I (RIG-I)-like receptors (RLRs), nucleotide-binding oligomerization domain (NOD)-like receptors (NLRs) and C-type lectin receptors (CLRs) [16–19]; knockout mice for immunologically relevant transcription factors, such as STAT (Signal Transducer and Activator of Transcription) molecules [20, 21] and interferon regulatory factors (IRFs) [22]; and knockout mice for adaptor molecules (such as MyD88, Rip2, and Trif) that are involved in connecting signals from immune receptors to downstream enzymes and transcription factors [16, 17, 19].

In order to unequivocally identify and investigate the cellular sources of cytokines, major soluble mediators of innate and adaptive immune responses, many cytokine reporter mice have been established that enable to track cytokine production during, e.g., infections [23–25]. A special chapter by Brinkmann and colleagues (*see* Scheibe et al., Chapter 25 of this series) describes in this issue of "Virus-Host-Interactions" the use of a cytokine reporter mouse model, the IFNβ-Luciferase-knockin reporter mice [26], for the analysis of type I IFN induction by mouse cytomegalovirus (MCMV).

2 Techniques to Generate Genetically Modified Mice

While the research of viral pathogenesis and antiviral immunity has taken advantage of genetically modified mice, as described above, it is worth mentioning that, on the other side, it was the use of viruses that opened up the possibility to modify the genome of mice and helped to generate the first transgenic mice in 1976 [27]. To address the question whether exogenous viruses (transmitted horizontally, not hereditarily, from individual to individual) can be converted into endogenous viruses (of which DNA sequences are present in all somatic and germ cells of an individual and passed on to the offspring), Rudolf Jaenisch infected preimplantation mouse embryos (at the 4–8 cell stage) with the Moloney murine leukemia virus (M-MuLV). The mice generated from these infected preimplantation embryos developed M-MuLV-induced leukemia, and the viral DNA was integrated into the germ line of the mice and transmitted to their offspring [27]. Subsequently the direct microinjection of DNA of interest into the pronucleus of fertilized murine eggs was developed as a more commonly used technique to generate transgenic mice [28–32].

However, the generation of genetically modified mice by infection of mouse embryos with retroviruses or microinjection of DNA into fertilized murine eggs results in random integrations of the exogenous DNA into the mouse genome. This in turn can lead to variegated expression of the transgene and inadvertent disruption of genes at the site into which the transgene is inserted. The frequency of phenotypes arising from insertion site mutation by a transgene (almost 10 %) is higher than might be expected from random integration into the genome [33], because introduction of transgenes by pronuclear injection can generate large deletions and complex rearrangements at the site of DNA integration [33]. In contrast, gene targeting by homologous recombination in murine embryonic stem cells (ESCs), a method that was established in the late 1980s, has enabled controlled and specific genetic modification by site-specific integration of exogenous DNA of interest into the genome of mice [34–37].

3 Generation of Mutant Mice from Genetically Modified Embryonic Stem Cells

The realization that genes of interest can be specifically modified in a whole mouse was developed from extensive work to determine whether cultured mammalian cells can mediate homologous recombination between their endogenous DNA and exogenously added DNA molecules [38–40]. Simultaneously, murine embryonic stem cells (ESCs) were successfully cultured *in vitro* without losing their pluripotent potential. Thus, when introduced into a preimplantation embryo they could contribute to the germ line [41–43]. ESCs have subsequently been widely used as vehicles to transfer site-specific genetic modifications of interest to the mouse germline [34–37, 44–46].

ESCs are derived from the pluripotent inner cell mass (ICM) of blastocysts, a structure formed in the early stage of embryogenesis (3.5 days old pre-implantation mouse embryo), and thus ESCs can contribute to all embryonic tissues, including the germ cells, in developing mice (Fig. 1). Isolated ESCs have to be cultured in special culture conditions to maintain their multiplication (self-renewal) capacity without loss of pluripotency. ESCs are typically cultured on a feeder layer of mitotically inactivated (mytomicin C treated or gamma-irradiated) mouse embryonic fibroblasts (MEFs) in order to obtain the necessary factors for self-renewal and pluripotency. Leukemia inhibitory factor (LIF), Wingless/Integrated (Wnt), and ligands of the TGF-β/BMP signaling pathway are among factors supplied by the fibroblasts and were found to influence the state and pluripotency of murine ESCs [47]. Supplementation of ESC culture medium with recombinant LIF helps to increase the maintenance of pluripotency. Recently, it has been shown that the maintenance of pluripotency can also be

Fig. 1 (**a**) Summarized most common steps for isolation of mouse embryonic stem cells (ESCs) and generation of homologous recombinant ESC clones. (1) In the first step 3.5-day-old mouse embryos (blastocysts) are collected from the uterine horn of superovulated (hormone treated) mated female mice with, for example, an agouti coat (strain 129/Sv). (2) ESCs are derived from the inner cell mass of blastocysts and cultured on a feeder layer of mitotically inactivated mouse embryonic fibroblasts (MEFs), in ESC medium (supplemented with leukemia inhibitory factor (LIF)). (3) After electroporation with the targeting vector of interest, (4) successfully transfected ESCs are selected by adding appropriate selection agent to the ESC medium; and (5) ESC clones are picked. (6) Homologous recombinant ESC clones are identified by Southern blot. The genomic DNA isolated from ESC clones should be digested with an appropriate restriction enzyme that produce one cut inside the targeting vector and one cut just outside (upstream or downstream) the targeting vector, in the targeted chromosomal region. The use of an "external" probe outside of the targeting construct will produce a band with a size corresponding to unmodified wild-type allele(s), which is here indicated by *X* kb, and, if homologous recombination occurred, a second band of bigger or smaller size corresponding to the targeted allele, which is here indicated by *X*–*Y* kb. (**b**) Generation of mice with genome modification of interest using homologous recombinant ESC clones. (1) If ESCs are derived from mice with an agouti coat (such as strain 129/Sv), the recipient pre-implantation mouse embryos (blastocysts) should be collected from female mice with black coat (such as strain C57BL/6). (2) The identified and expanded homologous recombinant ESC clones (see Fig. 1a) are injected into recipient pre-implantation mouse embryos (blastocysts) that are collected from female mice with black coat (strain C57BL/6). (3) These injected blastocysts are then surgically transferred to a recipient pseudopregnant foster mother to allow the embryos to develop. Females of CD1 mouse strain make very good mothers, and are thus used by several laboratories as foster mothers. (4) Because ESCs and recipient blastocysts were derived from mouse strains with distinguishable coat-colors, the desired chimeric offspring can be visually recognized by inspection of coat-color chimerism (percentage of black and agouti hair on the mouse black-agouti). (5) Chimeric offspring (usually only the males, because the used ESC lines are usually male) are mated with C57BL/6 mice to produce the F1 generation. (6) The germline transmission is then confirmed by Southern blot analysis or PCR of tail DNA from the agouti (not black) mice of the F1 generation

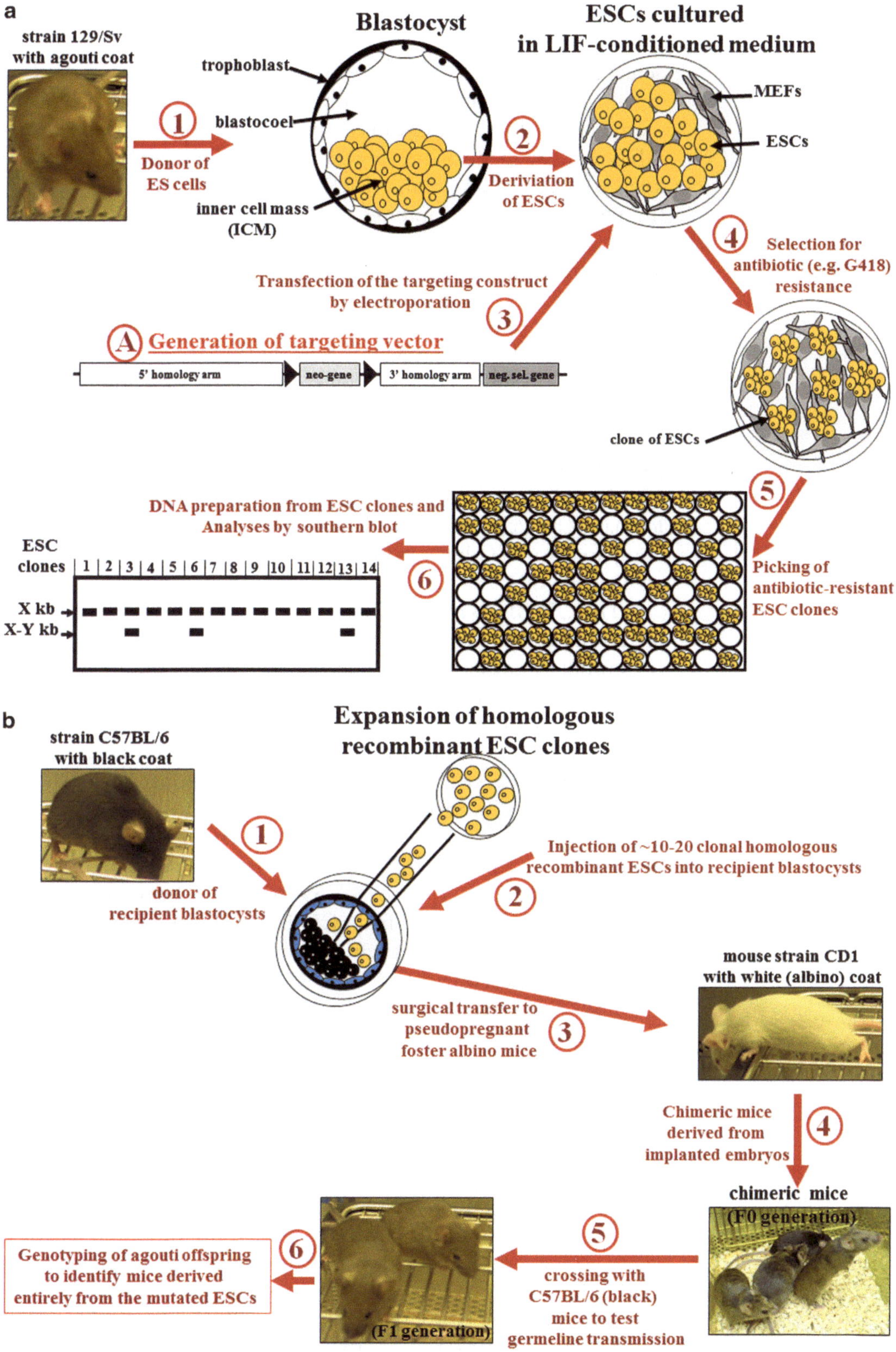
a
strain 129/Sv
with agouti coat
1
Donor of
ES cells
Blastocyst
trophoblast
blastocoel
inner cell mass
(ICM)
2
Deriviation
of ESCs
ESCs cultured
in LIF-conditioned medium
MEFs
ESCs
4
Selection for
antibiotic (e.g. G418)
resistance
Transfection of the targeting construct
by electroporation
3
A Generation of targeting vector
5' homology arm
neo-gene
3' homology arm
neg. sel. gene
clone of ESCs
5
Picking of
antibiotic-resistant
ESC clones
DNA preparation from ESC clones and
Analyses by southern blot
6
ESC
clones
1 2 3 4 5 6 7 8 9 10 11 12 13 14
X kb
X-Y kb
b
strain C57BL/6
with black coat
1
donor of
recipient blastocysts
Expansion of homologous
recombinant ESC clones
Injection of ~10-20 clonal homologous
recombinant ESCs into recipient blastocysts
2
surgical transfer to
pseudopregnant
foster albino mice
3
mouse strain CD1
with white (albino) coat
Chimeric mice
derived from
implanted embryos
4
chimeric mice
(F0 generation)
5
crossing with
C57BL/6 (black)
mice to test
germeline transmission
(F1 generation)
6
Genotyping of agouti offspring
to identify mice derived
entirely from the mutated ESCs

achieved, in MEF-free culture systems, by use of glycogen synthase kinase (GSK)-3-specific inhibitors, such as 6-bromoindirubin-3′-oxime (BIO) [48], or, optimally, by use of a combination of three inhibitors (3i medium): SU5402 (inhibits FGF receptor tyrosine kinases), PD184352 (inhibits ERK signal cascades) and CHIR99021 (a more selective inhibitor of GSK-3) [49]. Interestingly, Domogatskaya et al. [50] demonstrated recently that pluripotency and self-renewal of mouse ESCs can be achieved in the absence of feeder cells, LIF, or differentiation inhibitors by culturing ESCs on plates coated with a recombinant human extracellular matrix protein, the laminin isoform 511 (LN-511) [50].

Several established ESC lines are in common use. Initially, most used ES cells were derived from the 129 mouse strain [51, 52]. Examples include, among others, E14 cell lines [53], D3 cell lines [54], J1 cell line [55], R1 cell line [56], and AB2.1 cell line [57]. 129 ES cell lines are often used because of their more robust performance in cell culture and higher germline transmission rates compared to ESC lines from C57BL/6J and C57BL/6N mouse strains [58–60].

The C57BL/6 mouse strain is however one of the best characterized inbred strains that is widely accepted as the reference strain for immunological, neurobiological, behavioral, and physiological studies in mice, and is the standard reference library for the mouse genome-sequencing program [58–61]. Therefore, mice derived from 129 ESC lines need to be backcrossed for ten or more generations onto C57BL/6 background, which is a time-consuming process taking 2 or more years.

C57BL/6 ESC lines with efficient germline colonization have been generated and these facilitate the direct generation of genetically altered C57BL/6 mice [58, 59, 61–64]. The requirement for backcrossing mice derived from C57BL/6-derived ESCs is considerably less (though not entirely eliminated as the ESCs may harbor mutations acquired *in vitro*).

ESC lines derived from C57BL/6 mice that can be used include: Bruce4 [64], BL/6-III [63], LK1 [59], and JM8 [61]. The JM8 cell lines have been used for the large-scale mouse knockout program to generate mice with targeted mutations in the C57BL/6 genetic background [61, 65]. JM8 cells can be easily propagated using standard ESC culture conditions, in the presence or absence of feeder cells. In addition, to simplify breeding schemes, the dominant agouti coat color gene was restored in JM8 cells by targeted repair of the C57BL/6 nonagouti mutation enabling visual assessment of coat color contribution and germline transmission [61].

Taken together, when selecting an ESC line, several considerations are important: (1) the advantages and drawbacks of C57BL/6 versus 129 ESC lines [60]; (2) the genetic variation and stability among ESC lines from different 129 substrains [52] and

C57BL/6 substrains [66], respectively, which might influence, e.g., gene targeting efficiency and phenotype of mice derived from the respective ESC line; (3) the homology arms of the targeting vector should be from DNA that is isogenic to the ESCs used. Furthermore, the most appropriate ESC line to use is dependent on the cell culture conditions and expertise in the respective laboratories.

In vitro cultured ESCs can be genetically modified and then reinjected into the blastocoel (cavity) of blastocysts (3.5 days old pre-implantation mouse embryo) [67, 68] (Fig. 1), or into morula (2.5 days old pre-implantation mouse embryo) [68]. Usually 10–20 ESCs are injected in a blastocyst [67, 68]. Alternatively, chimeric embryos can be generated by aggregating ESCs with morula [69]. The chimeric embryos (usually five to ten) are then surgically transferred into the uterus of recipient pseudopregnant foster females. The genotype of these females is irrelevant, as long as they are good surrogate mothers. That is, they nurse carefully the newborns to weaning age and they accept pups from another mother. For these reasons, CD1 females or C57BL/6 × BALB/c F1 females are recommended for use [68]. Chimeric mice, in which the injected genetically modified ESCs have contributed to the formation of most or all tissues, will be born at frequencies varying from a few percent to the majority of the pups (Fig. 1). If the genetically modified ESCs have contributed to germ cell formation, the introduced genetic modifications can be passed on to offspring from the chimeric mice.

To facilitate the selection of the desired chimeric offspring, the ESCs and recipient blastocysts are derived from different mouse strains with distinguishable coat-colors (each mouse strain homozygous for the corresponding coat-color allele) (Fig. 1b). For example, the recipient blastocysts may be derived from C57BL/6 mice (black coat-color), and ESCs from a 129 mouse strain (agouti/brown coat-color). The extent of the contribution of the ESCs to the formation of the chimeric mouse can be visually recognized by inspection of coat-color chimerism (percentage of black and agouti hair on the mouse black-agouti) (Fig. 1b).

4 Gene Targeting in ESCs

Using current techniques, there are almost no limitations into the types of modifications that can be introduced, ranging from gene insertion, point mutations, short- and long-range deletions, inversions. Conditional knockouts or knockins are generated by placing loxP or FRT sites flanking selected exons (see also text below).

The introduction of site-specific modifications into the genome of ESCs by homologous recombination, a process called gene

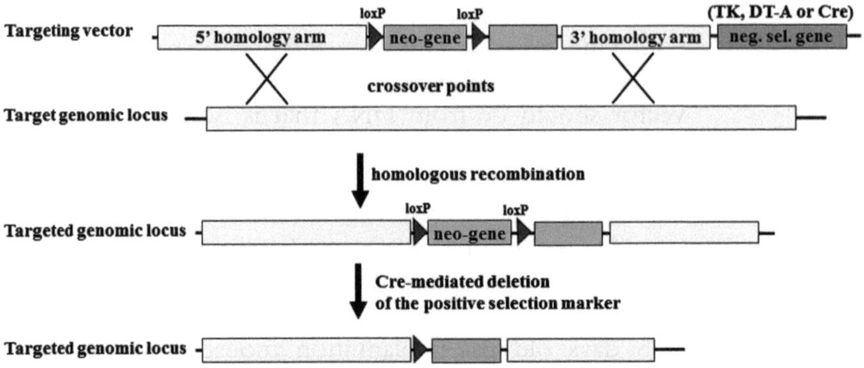

Fig. 2 Typical gene targeting strategy. A targeting vector is typically composed of three basic units: (1) a 5′ homology arm; (2) a gene marker for positive selection (e.g., neomycin resistance gene (neo)); (3) a 3′ homology arm. Furthermore, a negative selection marker (neg. sel. Marker) can be included outside the homology arms, such as thymidine kinase (TK), diphtheria toxin fragment A (DT-A), or, if the positive selection marker is flanked by loxP sites, Cre recombinase gene (Cre). Furthermore, any desired DNA sequence of interest (here *green box*) can be inserted between the homology arms of the targeting vector, in order to introduce it into the target genome by homologous recombination. Homologous recombination between the targeting vector and the target cognate chromosomal region results in the disruption of one genomic copy of the targeted genomic locus and loss of the vector's negative selection marker gene. Crossover points are depicted by "X". The floxed (loxP sites flanked) positive selection marker gene can be removed by expressing Cre recombinase in the recombinant ESCs or by crossing the chimeric mice with Cre-expressing transgenic mice (*see* also Fig. 4a) (Color figure online)

targeting, is achieved through the introduction of a targeting vector into ESCs by electroporation (Figs. 1 and 2).

A targeting vector (DNA construct) is typically composed of three basic units (Fig. 2): (1) a 5′ homology arm, (2) a positive selectable gene marker (such as the neomycin resistance gene (*neo*) or hygromycin (hyg)), (3) a 3′ homology arm. The transfected targeting vector can either insert itself randomly into the genome or be integrated by homologous recombination as determined by the 5′ and 3′ homology arms. Successfully transfected cells are positively selected by culturing ESCs in medium with neomycin (G418) or other appropriate antibiotics, such as hygromycin or puromycin (Fig. 1). If the positive selection marker gene is flanked by loxP or FRT sites, then it can be later removed from targeted loci in ESCs or transgenic mice by expressing Cre recombinase or flippase (FLP) in the recombinant ESCs (through transfection with Cre- or FLP-expressing vector) or by crossing the chimeric mice with Cre- or FLP-expressing transgenic mice (Fig. 2, and *see* also Fig. 4a). This is an important consideration as the introduction of selection marker genes, such as neo, can profoundly affect the expression of endogenous genes neighboring to the targeted gene locus [70–72].

The homologous recombination of a targeting vector into a genomic locus of interest occurs at a very low frequency (at a frequency of 10^{-3} to 10^{-4} relative to nonhomologous recombinants)

[73]. In general, the longer the length of the 5′ and 3′ homology arms the higher the targeting frequency is [74, 75]. Another factor that increases homologous recombination frequencies is whether the homology arms of the targeting vector are isogenic with the ESC DNA [74, 76]. Ideally, the homologous arms should be derived from genomic DNA prepared from the ESCs to be used or at least from the same strain of mice that the ESCs were derived from. Linearization of the targeting construct before its transfection into ESCs also enhances the frequency of homologous recombination [77]. Nevertheless, because homologous recombination is a rare event, the screening of at least 200, and often up to 1,000 clones is required to identify a few clones that have undergone homologous recombination.

A method to enrich the selection for homologous recombinant clones uses negative selection markers, such as thymidine kinase (TK) from herpes simplex virus (HSV) [44], or the diphtheria toxin fragment A (DT-A) from *Corynebacterium diphtheria* [78]. The gene encoding for the negative selection marker is included outside the homology arms of the targeting vector (Fig. 2). During homologous recombination, sequences outside the regions of homology to the target genomic locus are usually lost. By contrast, if the gene targeting vector is integrated randomly in the genome, the negative selection marker is often retained. TK renders the cells sensitive to thymidine analogues, such as 5-iodo-2′-fluoro-2′-deoxy-1-β-D-arabino-furonosyluracil (FIAU) or gancyclovir, that are supplemented in the ESC culture medium to eliminate clones with randomly integrated targeting vector. The TK enzyme activates these thymidine analogues, resulting in their incorporation into replicating DNA, causing premature chain termination and cell death [79]. DT-A exerts toxicity by catalyzing the transfer of ADP-ribose from nicotinamide adenine dinucleotide (NAD) to a modified histidine residue on the elongation factor 2 (eEF2), thereby inhibiting protein synthesis [78].

Recently, a new simple negative selection procedure using an “auto-selecting targeting vector” has been developed (Fig. 2) [80]. As negative selection marker, a cyclization recombination (*cre*) gene—under the control of herpes simplex virus (HSV) promoter—was placed outside the homology region of the targeting vector. Because the positive selection marker, the neomycin resistance gene (*neo*), was flanked by two loxP sites (floxed), random integration of targeting vector into the genome will often result in the maintenance and expression of Cre recombinase which specifically recognizes the recombining loxP sites and subsequently mediates the deletion of the floxed positive selection marker (*neo*). In this case, the ESCs are not able to grow in medium containing G418 (neomycin) [80]. However, after homologous recombination, the Cre gene will in most cases not integrate and the positive selection marker will be retained (though it can be deleted once positive clones have been identified).

5 The Use of Cre/loxP Recombination System in Gene Targeting

Cre recombinase is a 38 kDa protein from the bacteriophage P1 that mediates intramolecular and intermolecular site-specific recombination between two loxP sites (locus of X-over of P1). The loxP sequence is 34 bp long and consists of two 13 bp inverted repeats separated by an 8 bp nonpalindromic (asymmetric) sequence which dictates the orientation of the overall loxP site (Fig. 3). Two loxP sequences in opposite orientation mediate the inversion of the intervening DNA by Cre recombinase rather than excision while two sites in the same orientation mediate excision of the intervening DNA between the sites after which only one loxP site remains. If the loxP sites are located on different chromosomes, Cre recombinase can mediate a chromosomal translocation (Fig. 4) [81–85].

The Cre/loxP system has emerged as a useful tool in genetic manipulations [86]. Generally, any DNA sequence of interest can be deleted by flanking it with loxP sites. Cre/loxP enables, for

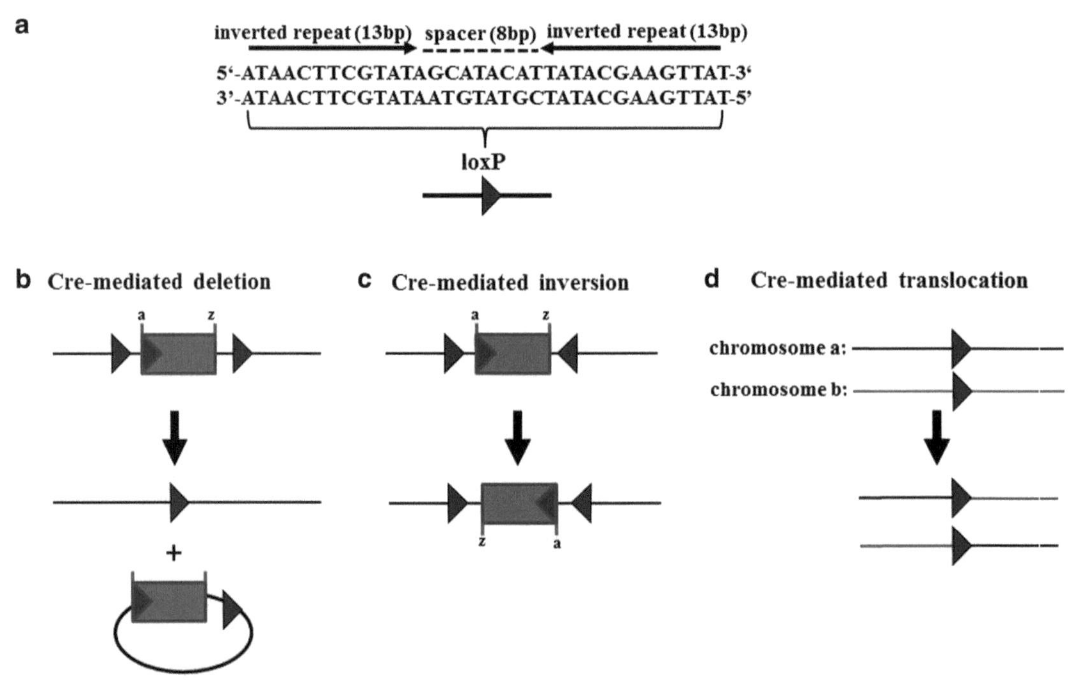

Fig. 3 LoxP structure and Cre recombinase-mediated recombinations. (**a**) Single loxP site that contains two inverted 13 bp repeats, separated by an asymmetric 8 bp long sequence. The type of Cre-mediated recombination is dependent on the orientation and location of the loxP sites: (**b**) Cre excises a circular molecule from between two loxP sites placed in the same orientation; (**c**) Cre inverts the DNA sequence between two loxP sites positioned in opposite orientation; (**d**) Cre-mediated recombination between two different linear DNA molecules (e.g., chromosomes), each containing a loxP site, resulting in the exchange of the DNA regions flanking the loxP sites. Figure was modified from Torres and Kuehn [88]

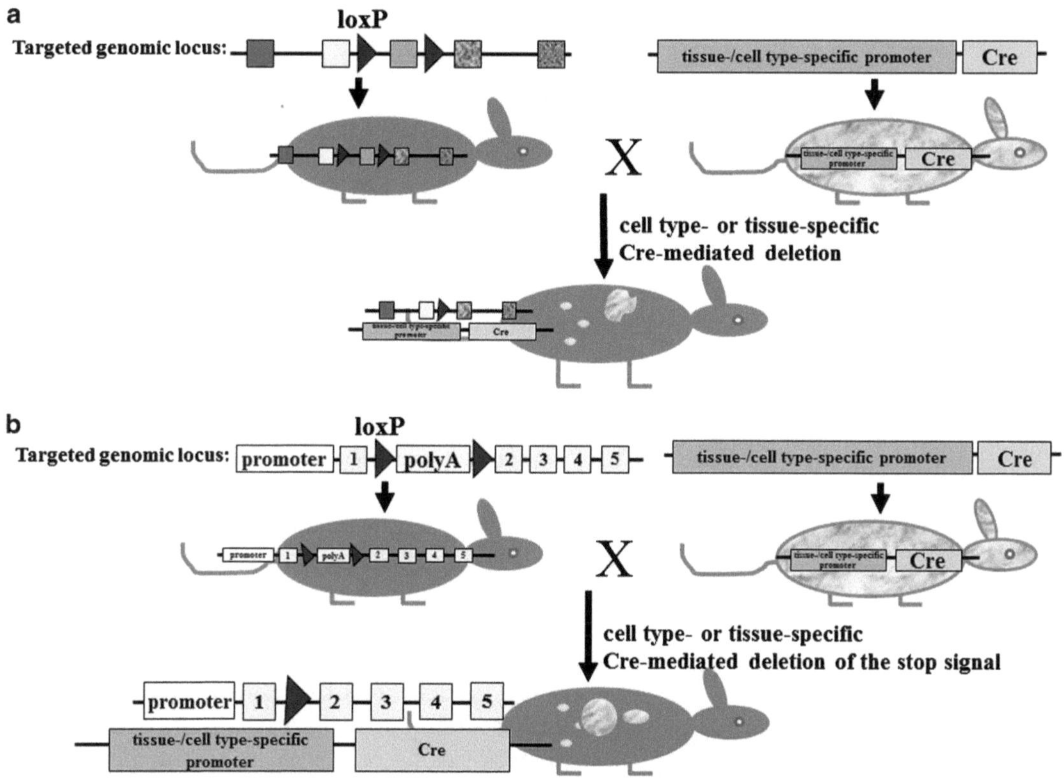

Fig. 4 Conditional gene targeting using the Cre/loxP recombination system. (**a**) Cre-mediated inactivation of a gene of interest. *Left mouse*: introduction of loxP sites into a genomic locus of interest by homologous recombination using ESCs (*see* also Fig. 1). LoxP sites are introduced in a manner that they don't interfere with the function of the targeted gene. *Right mouse*: a transgenic strain that express Cre-recombinase under the control of cell type- or tissue-specific promoter. By crossing the floxed mouse with Cre transgenic mice, Cre mediates the deletion of the floxed genomic sequence, resulting in the inactivation of the targeted gene. Gene deletion is restricted to the "area" (cell types or tissues) where Cre is expressed (*white flecks*). (**b**) Cre-mediated activation of a gene of interest. *Left mouse*: introduction of a floxed intervening sequence (e.g., polyadenylation signal sequences) that prevent the correct transcription of the targeted gene. By crossing the floxed mouse with Cre transgenic mice, Cre mediates the deletion of the floxed intervening sequence, resulting in the reactivation of the targeted gene. Gene activation is restricted to the "area" (cell types or tissues) where Cre is expressed (*white flecks*)

example, the deletion of the selection marker after successful integration of the targeting vector into the genome of ESCs or mice. The loxP sites can be introduced into the genomic locus of interest by homologous recombination as described above. Furthermore, the conditional (timely or spatially controlled) expression of Cre recombinase enables to determine, where (e.g., in which cell type or tissue) and when (at which time of mouse's life or of developmental stage of cells/tissues) the deletion of the floxed DNA sequence should occur (Fig. 4a). Thus, for conditional site-specific genome modification, two mouse lines are usually needed (Fig. 4a):

First, mice that have the DNA sequence of interest flanked by loxP sites (floxed). Second, Cre recombinase transgenic mice, in which Cre is expressed under the control of a promoter that is active in specific cell types or tissues, or Cre is transiently expressed under the control of a promoter that is active at a particular developmental stage of tissues or cells. When crossing the floxed mouse with a Cre transgenic mouse, the floxed DNA sequence is subsequently deleted in the cell types or tissues, where Cre is expressed. Collection databases of several hundreds of Cre transgenic mouse lines expressing the Cre recombinase in specific tissues or cells are available (e.g., http://www.ics-mci.fr/mousecre/; http://nagy.mshri.on.ca/cre_new/index.php; http://www.creportal.org/; http://bioit.fleming.gr/crezoo/; http://creline.org/).

The Cre/loxP system enables not only the deletion (shut off) of genes of interest, but has been proved to be successful also to specifically turn on (activate) the expression of any gene (or transgene) of interest [87, 88]. For this, the gene of interest is rendered quiescent, e.g., by interposing floxed polyadenylation signal sequences mediating premature transcription arrest within a fundamental site of a gene of interest (Fig. 4b). After intercrossing with a Cre-transgenic mouse or delivery of Cre into floxed transgenic mice using, e.g., Cre-expressing adenoviruses, the floxed polyadenylation signal sequences can be removed by Cre-mediated excision, resulting in the activation of the gene expression in a specific cell type or tissue.

Such Cre/loxP-mediated gene activation has been a useful approach to avoid harmful effects of the transgene during mouse embryogenesis, or the induction of immune tolerance against the transgene product, for example in the case of viral genes. For instance, the application of the Cre/loxP technology has enabled to generate transgenic mice that conditionally express human hepatitis C virus transgenes upon intravenous administration of Cre-expressing adenovirus, and thus enabled the investigation of the immune responses to and pathogenesis of HCV infection [89].

Finally, another useful recombination system is the yeast *Fl*i*p*pase (*Flp*)/*F*lp *r*ecognition *t*arget (*FRT*), which is mechanistically identical to that of the Cre/loxP recombination system [90–95], and represents an alternative tool to Cre/loxP. Moreover, the combination of both recombination systems can significantly increase the potentials of conditional gene manipulation in mice.

6 Cloning of the Targeting Vector

The design and construction of the targeting vector is among the most critical steps in generating gene-targeted mice. Many issues should be considered in the design of the targeting vector. These include, among others, the type of the desired genetic modification and the scientific questions of interest to be addressed (e.g., complete or conditional knockout, insertion of a reporter gene to monitor the expression of a gene of interest, insertion of point mutations, and constitutive or conditional expression of a transgene of interest). Furthermore, the design should consider the option to remove the selection marker cassette at a later stage, and to linearize the targeting vector at a unique restriction site outside the homology arms prior to electroporation. It is also important to design restriction sites and probes that will enable the detection of ESC clones that have undergone homologous recombination. If necessary, new restriction sites have to be inserted that enable easy discrimination of homologous recombinant ESC clones by Southern blotting. Alternatively, homologous recombination can also be detected by long-range PCR or by qPCR reactions designed to detect the loss of an endogenous allele. Thus, accurate design of the targeting vector requires knowledge about the sequence and structure of the gene locus to be targeted (e.g., restriction sites, promoter region, 5′UTR, exons, introns, and intron–exon borders, splice donor and acceptor sequences, and 3′UTR). This has been made considerably simpler by access to the data sequences of the whole mouse genome [96] (http://www.informatics.jax.org/; http://www.sanger.ac.uk/resources/mouse/genomes/).

The construction of a gene targeting vector can proceed by conventional restriction enzyme-based cloning strategies, by using MultiSite Gateway Technology, or by using *Recombination-mediated* genetic engi*neering* (*Recombineering*) based protocols. Generation of targeting vectors using MultiSite Gateway Cloning is described in a separate chapter (*see* Bouabe and Okkenhaug, Chapter 24 of this series).

7 Targeting Transgenes into the ROSA26 Locus

The expression of transgenes can be achieved by microinjection of DNA (e.g., BAC) containing the transgene of interest into fertilized murine eggs. However, this results in variegated expression of the transgene and inadvertent disruption of genes at the site into which the transgene might be inserted.

A controlled insertion and expression of a transgene of interest can be achieved by targeting it into the Rosa26 locus. The Rosa26 locus was identified by analyzing pools of embryonic stem cells infected with the retroviral gene trap vector Gen-ROSAβgeo (bifunctional lacZ/neomycin phosphotransferase gene cassette) at low multiplicity [97]. The gene trap vector has integrated into the Rosa26, a chromosomal region that encodes for three transcripts (only the third transcript, originating from the reverse strand, seems to encode for a protein) [98].

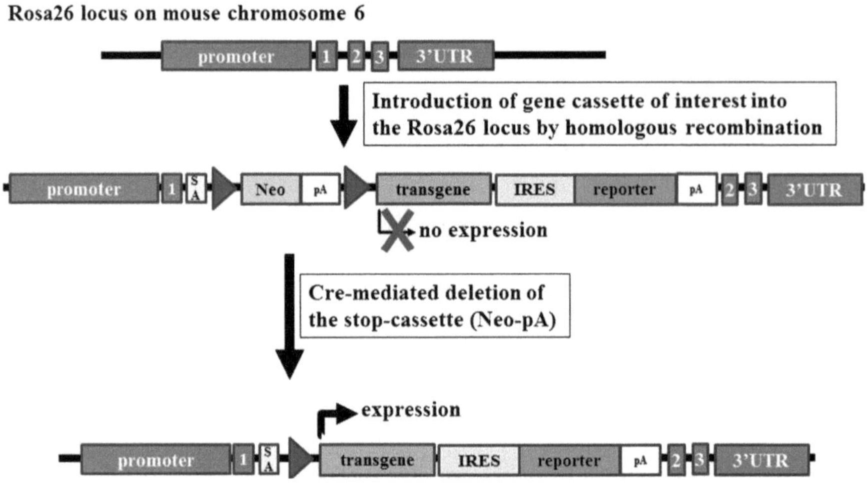

Fig. 5 Targeting transgenes into the ROSA26 locus. Introduction of gene cassettes of interest into Rosa26 locus by homologous recombination in embryonic stem cells. A floxed intervening sequence, neomycin (Neo)-polyadenylation signal sequences (pA), prevents the transcription of the transgene. Cre expression mediates then the deletion of the floxed intervening sequence, resulting in the expression of the transgene and a reporter gene. The reporter gene facilitates to track the expression of the transgene. Numbered (1–3) *blue rectangles*: exons of the Rosa26 locus. *SA* splice acceptor. DNA elements are not drawn to scale (Color figure online)

Rosa26 locus is ubiquitously transcriptionally active, and thus it has become a preferred site for the integration and ubiquitous expression of transgenes and reporter genes. The genes of interest can be introduced into Rosa26 locus by homologous recombination. Conditional expression of Rosa26-targeting genes can be achieved by using Cre/loxP system (Fig. 5 and Subheading 5).

8 Considerations for the Type of Genetic Modification

The simplest way to assess the role of a particular gene is to eliminate its expression, either by germline deletion or by inserting loxP sites flanking functionally fundamental parts of the gene to enable its conditional Cre-mediated deletion. This is usually best achieved by deleting the first exon(s). Care must be taken, however, to rule out expression of the gene from internal transcriptional start sites or by alternative splicing of exons. An alternative strategy is to delete exons that encode for the functionally most important part of the protein such as those that control its catalytic activity or essential interactions with other proteins, DNA, or RNA. However, with this strategy there is a risk of generating peptides or truncated proteins which may not be functionally neutral, i.e., they may have dominant negative function preventing activation of other similar proteins in ways that can be difficult to predict and control. A way around this is to introduce point mutations that inactivate a particular enzymatic function or prevent the protein from interaction with others [99]. Examples of such point mutations in immunologically relevant genes include among others, p110δ^{D910A} (kinase inactive) [100], p110γ^{KD} (kinase inactive) [101], p110γ^{DASA} (loss of interaction with Ras) [102], Vav1^{AA} (lacks guanine exchange activity) [103], Vav1^{R442G} (interrupts lipid binding via the PH domain) [104].

9 Considerations for the Type of Reporter Genes When Generating Reporter Mice

Reporter mice have emerged as important tools that facilitate in/*ex vivo* monitoring of the expression of genes of interest, especially when the gene products are secreted proteins, such as cytokines. In reporter mice the expression of an intracellularly localized reporter protein is linked to that of the endogenous gene of interest by using, for example, an internal ribosome entry site (IRES) or 2A peptides. Alternatively, the reporter gene can be introduced just downstream of the start codon of the endogenous gene of interest, resulting in the expression of the reporter gene instead of the targeted allele [23].

Autofluorescent proteins (AFPs), such as GFP, are the most frequently used reporters because their expression can be detected at single cell level without any invasive treatment of cells and without the need of exogenous substrates. However, a high expression amount of AFP (approximately 10^5 molecules "0.1–1 μM" per cell) is required to detect fluorescence up over background [105]. This is a fundamental limitation when expression of weakly expressed genes, as is the case of many cytokines, should be monitored [24, 106].

A powerful solution to overcome this limitation is to use enzymatic reporters that catalyze strong signal amplification, and consequently a thousand times less catalytic molecules (approximately 10^{-4} μM) than AFP can generate a robust, measurable, reporter signal [107, 108].

An example that emphasizes the importance of considering enzymatic reporters versus AFPs for weakly expressed genes has been revealed recently by studies on IL-10 reporter mice (reviewed in [24]). For instance, in contrast to "regular" or "conventional" IL-10-AFP reporter mice that enabled detection of reporter activity mainly in T cells, because of their small cytosol, low autofluorescence, and high expression level of IL-10 [24], an IL-10 reporter mouse model that is based on the reporter enzyme β-lactamase and the fluorescence resonance energy transfer (FRET) substrate coumarin-cephalosporin-fluorescein 4-AM (CCF4-AM) enabled to easily analyze and quantify IL-10 production at the single-cell level in all myeloid and lymphoid cell types, and thus also in cells exhibiting high autofluorescence and/or low expression level of IL-10 [106].

Table 1
Information was collected from the following references [111–114]

Reporter gene	Substrate	Signal amplification	Cells have to be lysed/permeabilized	Detection by						Endogenous activities similar to the reporter activity
				FACS (at single living cell level)	Fluorescence microscopy	*In vivo* imaging	Fluoro-meter	Lumino-meter		
Autofluorescent proteins (GFP, RFP etc.)	No substrate is needed	No (high detection limit)	No	Yes	Yes	No	Yes	No		No
Firefly luciferase (Luc)	D-luciferin	Yes	Cell lysis	No	No	Yes	No	Yes		No
Secreted alkaline phosphatase (SEAP)	PNPP; FADP; CSPD; PPQ	Yes	Cell lysis	No	No		Yes	Yes		Yes (in some cell types)
Chloramphenicol acetyltransferase (CAT)	^{14}C-labeled chloramphenicol; fluorescent chloramphenicol derivative	Yes	Cell lysis	No	No	No	Yes	Yes		Yes (minimal)
β-galactosidase (LacZ)	ONPG; X-Gal; FDG; DDAOG	Yes	Cell permeabilization	Yes/no	Yes	Yes	Yes	Yes		Yes (high)
β-lactamase (Bla)	CCF2-AM; CCF4-AM; nitrocefin	Yes	No	Yes	Yes	No	Yes	Yes		No

Thus, AFPs and enzymatic reporters along with their corresponding assay systems have specific features, advantages, and limitations that must be carefully considered in choosing a system tailored to the particular questions being studied. Some important properties of commonly used reporter systems are summarized in Table 1.

10 The International Knockout Mouse Consortium

Before embarking on a project to generate a knockout or conditional knockout mouse, it is worth considering the international mouse knockout consortium (IKMC) (http://www.knockoutmouse.org/) whose aim is to generate conditional knockout alleles for all genes in the mouse genome (including also microRNA genes). They can supply targeting vectors, targeted ESCs, or gene-targeted mice [65, 109, 110]. This in turn can result in considerable saving and shortening in time from the design to the execution of experiments involving new gene-targeted strains.

Acknowledgements

Work in our laboratory is supported by grants from the BBSRC and the Wellcome Trust.

References

1. Bot A, Casares S, Bot S, von Boehmer H, Bona C (1998) Cellular mechanisms involved in protection against influenza virus infection in transgenic mice expressing a TCR receptor specific for class II hemagglutinin peptide in CD4+ and CD8+ T cells. J Immunol 160:4500–4507
2. Kirberg J, Baron A, Jakob S, Rolink A, Karjalainen K, von Boehmer H (1994) Thymic selection of CD8+ single positive cells with a class II major histocompatibility complex-restricted receptor. J Exp Med 180:25–34
3. Oxenius A, Bachmann MF, Zinkernagel RM, Hengartner H (1998) Virus-specific MHC-class II-restricted TCR-transgenic mice: effects on humoral and cellular immune responses after viral infection. Eur J Immunol 28:390–400
4. Pircher H, Burki K, Lang R, Hengartner H, Zinkernagel RM (1989) Tolerance induction in double specific T-cell receptor transgenic mice varies with antigen. Nature 342:559–561
5. Rall GF, Lawrence DM, Patterson CE (2000) The application of transgenic and knockout mouse technology for the study of viral pathogenesis. Virology 271:220–226
6. Mador N, Braun E, Haim H, Ariel I, Panet A, Steiner I (2003) Transgenic mouse with the herpes simplex virus type 1 latency-associated gene: expression and function of the transgene. J Virol 77:12421–12429
7. Wang K, Pesnicak L, Guancial E, Krause PR, Straus SE (2001) The 2.2-kilobase latency-associated transcript of herpes simplex virus type 2 does not modulate viral replication, reactivation, or establishment of latency in transgenic mice. J Virol 75:8166–8172
8. Mombaerts P, Clarke AR, Rudnicki MA, Iacomini J, Itohara S, Lafaille JJ, Wang L, Ichikawa Y, Jaenisch R, Hooper ML et al (1992) Mutations in T-cell antigen receptor genes alpha and beta block thymocyte development at different stages. Nature 360:225–231

9. Kitamura D, Roes J, Kuhn R, Rajewsky K (1991) A B cell-deficient mouse by targeted disruption of the membrane exon of the immunoglobulin mu chain gene. Nature 350:423–426
10. Mombaerts P, Iacomini J, Johnson RS, Herrup K, Tonegawa S, Papaioannou VE (1992) RAG-1-deficient mice have no mature B and T lymphocytes. Cell 68:869–877
11. Shinkai Y, Rathbun G, Lam KP, Oltz EM, Stewart V, Mendelsohn M, Charron J, Datta M, Young F, Stall AM et al (1992) RAG-2-deficient mice lack mature lymphocytes owing to inability to initiate V(D)J rearrangement. Cell 68:855–867
12. Dalton DK, Pitts-Meek S, Keshav S, Figari IS, Bradley A, Stewart TA (1993) Multiple defects of immune cell function in mice with disrupted interferon-gamma genes. Science 259:1739–1742
13. Hwang SY, Hertzog PJ, Holland KA, Sumarsono SH, Tymms MJ, Hamilton JA, Whitty G, Bertoncello I, Kola I (1995) A null mutation in the gene encoding a type I interferon receptor component eliminates antiproliferative and antiviral responses to interferons alpha and beta and alters macrophage responses. Proc Natl Acad Sci USA 92:11284–11288
14. Steinhoff U, Muller U, Schertler A, Hengartner H, Aguet M, Zinkernagel RM (1995) Antiviral protection by vesicular stomatitis virus-specific antibodies in alpha/beta interferon receptor-deficient mice. J Virol 69:2153–2158
15. Muller U, Steinhoff U, Reis LF, Hemmi S, Pavlovic J, Zinkernagel RM, Aguet M (1994) Functional role of type I and type II interferons in antiviral defense. Science 264:1918–1921
16. Kawai T, Akira S (2009) The roles of TLRs, RLRs and NLRs in pathogen recognition. Int Immunol 21:317–337
17. Moresco EM, Beutler B (2011) Resisting viral infection: the gene by gene approach. Curr Opin Virol 1:513–518
18. Brennan K, Bowie AG (2010) Activation of host pattern recognition receptors by viruses. Curr Opin Microbiol 13:503–507
19. Hardison SE, Brown GD (2012) C-type lectin receptors orchestrate antifungal immunity. Nat Immunol 13:817–822
20. Meraz MA, White JM, Sheehan KC, Bach EA, Rodig SJ, Dighe AS, Kaplan DH, Riley JK, Greenlund AC, Campbell D et al (1996) Targeted disruption of the Stat1 gene in mice reveals unexpected physiologic specificity in the JAK-STAT signaling pathway. Cell 84:431–442
21. Akira S (1999) Functional roles of STAT family proteins: lessons from knockout mice. Stem Cells 17:138–146
22. Matsuyama T, Kimura T, Kitagawa M, Pfeffer K, Kawakami T, Watanabe N, Kundig TM, Amakawa R, Kishihara K, Wakeham A et al (1993) Targeted disruption of IRF-1 or IRF-2 results in abnormal type I IFN gene induction and aberrant lymphocyte development. Cell 75:83–97
23. Croxford AL, Buch T (2011) Cytokine reporter mice in immunological research: perspectives and lessons learned. Immunology 132:1–8
24. Bouabe H (2012) Cytokine reporter mice: the special case of IL-10. Scand J Immunol 75:553–567
25. Luker KE, Luker GD (2010) Bioluminescence imaging of reporter mice for studies of infection and inflammation. Antiviral Res 86:93–100
26. Lienenklaus S, Cornitescu M, Zietara N, Lyszkiewicz M, Gekara N, Jablonska J, Edenhofer F, Rajewsky K, Bruder D, Hafner M et al (2009) Novel reporter mouse reveals constitutive and inflammatory expression of IFN-beta in vivo. J Immunol 183:3229–3236
27. Jaenisch R (1976) Germ line integration and Mendelian transmission of the exogenous Moloney leukemia virus. Proc Natl Acad Sci USA 73:1260–1264
28. Gordon JW, Scangos GA, Plotkin DJ, Barbosa JA, Ruddle FH (1980) Genetic transformation of mouse embryos by microinjection of purified DNA. Proc Natl Acad Sci USA 77:7380–7384
29. Costantini F, Lacy E (1981) Introduction of a rabbit beta-globin gene into the mouse germ line. Nature 294:92–94
30. Brinster RL, Chen HY, Trumbauer M, Senear AW, Warren R, Palmiter RD (1981) Somatic expression of herpes thymidine kinase in mice following injection of a fusion gene into eggs. Cell 27:223–231
31. Wagner EF, Stewart TA, Mintz B (1981) The human beta-globin gene and a functional viral thymidine kinase gene in developing mice. Proc Natl Acad Sci USA 78:5016–5020
32. Harbers K, Jahner D, Jaenisch R (1981) Microinjection of cloned retroviral genomes into mouse zygotes: integration and expression in the animal. Nature 293:540–542
33. De Deyn PP, Van Dam D (2011) Animal models of dementia. In: Conlon RA (ed) Transgenic and gene targeted models of dementia, vol 48. Springer Science+Business Media, New York, pp 77–90

34. Doetschman T, Gregg RG, Maeda N, Hooper ML, Melton DW, Thompson S, Smithies O (1987) Targetted correction of a mutant HPRT gene in mouse embryonic stem cells. Nature 330:576–578
35. Thomas KR, Capecchi MR (1987) Site-directed mutagenesis by gene targeting in mouse embryo-derived stem cells. Cell 51:503–512
36. Thompson S, Clarke AR, Pow AM, Hooper ML, Melton DW (1989) Germ line transmission and expression of a corrected HPRT gene produced by gene targeting in embryonic stem cells. Cell 56:313–321
37. Zijlstra M, Li E, Sajjadi F, Subramani S, Jaenisch R (1989) Germ-line transmission of a disrupted beta 2-microglobulin gene produced by homologous recombination in embryonic stem cells. Nature 342:435–438
38. Folger KR, Wong EA, Wahl G, Capecchi MR (1982) Patterns of integration of DNA microinjected into cultured mammalian cells: evidence for homologous recombination between injected plasmid DNA molecules. Mol Cell Biol 2:1372–1387
39. Folger K, Thomas K, Capecchi MR (1984) Analysis of homologous recombination in cultured mammalian cells. Cold Spring Harb Symp Quant Biol 49:123–138
40. Smithies O, Gregg RG, Boggs SS, Koralewski MA, Kucherlapati RS (1985) Insertion of DNA sequences into the human chromosomal beta-globin locus by homologous recombination. Nature 317:230–234
41. Evans MJ, Kaufman MH (1981) Establishment in culture of pluripotential cells from mouse embryos. Nature 292:154–156
42. Robertson E, Bradley A, Kuehn M, Evans M (1986) Germ-line transmission of genes introduced into cultured pluripotential cells by retroviral vector. Nature 323:445–448
43. Martin GR (1981) Isolation of a pluripotent cell line from early mouse embryos cultured in medium conditioned by teratocarcinoma stem cells. Proc Natl Acad Sci USA 78:7634–7638
44. Mansour SL, Thomas KR, Capecchi MR (1988) Disruption of the proto-oncogene int-2 in mouse embryo-derived stem cells: a general strategy for targeting mutations to non-selectable genes. Nature 336:348–352
45. Koller BH, Hagemann LJ, Doetschman T, Hagaman JR, Huang S, Williams PJ, First NL, Maeda N, Smithies O (1989) Germ-line transmission of a planned alteration made in a hypoxanthine phosphoribosyltransferase gene by homologous recombination in embryonic stem cells. Proc Natl Acad Sci USA 86:8927–8931
46. Thomas KR, Capecchi MR (1990) Targeted disruption of the murine int-1 proto-oncogene resulting in severe abnormalities in midbrain and cerebellar development. Nature 346:847–850
47. Young RA (2011) Control of the embryonic stem cell state. Cell 144:940–954
48. Sato N, Meijer L, Skaltsounis L, Greengard P, Brivanlou AH (2004) Maintenance of pluripotency in human and mouse embryonic stem cells through activation of Wnt signaling by a pharmacological GSK-3-specific inhibitor. Nat Med 10:55–63
49. Ying QL, Wray J, Nichols J, Batlle-Morera L, Doble B, Woodgett J, Cohen P, Smith A (2008) The ground state of embryonic stem cell self-renewal. Nature 453:519–523
50. Domogatskaya A, Rodin S, Boutaud A, Tryggvason K (2008) Laminin-511 but not -332, -111, or -411 enables mouse embryonic stem cell self-renewal in vitro. Stem Cells 26:2800–2809
51. Downing GJ, Battey JF Jr (2004) Technical assessment of the first 20 years of research using mouse embryonic stem cell lines. Stem Cells 22:1168–1180
52. Simpson EM, Linder CC, Sargent EE, Davisson MT, Mobraaten LE, Sharp JJ (1997) Genetic variation among 129 substrains and its importance for targeted mutagenesis in mice. Nat Genet 16:19–27
53. Hooper M, Hardy K, Handyside A, Hunter S, Monk M (1987) HPRT-deficient (Lesch-Nyhan) mouse embryos derived from germline colonization by cultured cells. Nature 326:292–295
54. Doetschman TC, Eistetter H, Katz M, Schmidt W, Kemler R (1985) The in vitro development of blastocyst-derived embryonic stem cell lines: formation of visceral yolk sac, blood islands and myocardium. J Embryol Exp Morphol 87:27–45
55. Li E, Bestor TH, Jaenisch R (1992) Targeted mutation of the DNA methyltransferase gene results in embryonic lethality. Cell 69:915–926
56. Nagy A, Rossant J, Nagy R, Abramow-Newerly W, Roder JC (1993) Derivation of completely cell culture-derived mice from early-passage embryonic stem cells. Proc Natl Acad Sci USA 90:8424–8428
57. Soriano P, Montgomery C, Geske R, Bradley A (1991) Targeted disruption of the c-src proto-oncogene leads to osteopetrosis in mice. Cell 64:693–702
58. Hansen GM, Markesich DC, Burnett MB, Zhu Q, Dionne KM, Richter LJ, Finnell RH, Sands AT, Zambrowicz BP, Abuin A (2008) Large-scale gene trapping in C57BL/6N mouse embryonic stem cells. Genome Res 18:1670–1679

59. Keskintepe L, Norris K, Pacholczyk G, Dederscheck SM, Eroglu A (2007) Derivation and comparison of C57BL/6 embryonic stem cells to a widely used 129 embryonic stem cell line. Transgenic Res 16:751–758
60. Seong E, Saunders TL, Stewart CL, Burmeister M (2004) To knockout in 129 or in C57BL/6: that is the question. Trends Genet 20:59–62
61. Pettitt SJ, Liang Q, Rairdan XY, Moran JL, Prosser HM, Beier DR, Lloyd KC, Bradley A, Skarnes WC (2009) Agouti C57BL/6N embryonic stem cells for mouse genetic resources. Nat Methods 6:493–495
62. Tanimoto Y, Iijima S, Hasegawa Y, Suzuki Y, Daitoku Y, Mizuno S, Ishige T, Kudo T, Takahashi S, Kunita S et al (2008) Embryonic stem cells derived from C57BL/6J and C57BL/6N mice. Comp Med 58:347–352
63. Ledermann B, Burki K (1991) Establishment of a germ-line competent C57BL/6 embryonic stem cell line. Exp Cell Res 197:254–258
64. Kontgen F, Suss G, Stewart C, Steinmetz M, Bluethmann H (1993) Targeted disruption of the MHC class II Aa gene in C57BL/6 mice. Int Immunol 5:957–964
65. Skarnes WC, Rosen B, West AP, Koutsourakis M, Bushell W, Iyer V, Mujica AO, Thomas M, Harrow J, Cox T et al (2011) A conditional knockout resource for the genome-wide study of mouse gene function. Nature 474:337–342
66. Hughes ED, Qu YY, Genik SJ, Lyons RH, Pacheco CD, Lieberman AP, Samuelson LC, Nasonkin IO, Camper SA, Van Keuren ML et al (2007) Genetic variation in C57BL/6 ES cell lines and genetic instability in the Bruce4 C57BL/6 ES cell line. Mamm Genome 18:549–558
67. Longenecker G, Kulkarni AB (2009) Generation of gene knockout mice by ES cell microinjection. Curr Protoc Cell Biol Chapter 19:Unit 19.14 19.14.11–36
68. Pluck A, Klasen C (2009) Generation of chimeras by microinjection. Methods Mol Biol 561:199–217
69. Pluck A, Klasen C (2009) Generation of chimeras by morula aggregation. Methods Mol Biol 561:219–229
70. Meier ID, Bernreuther C, Tilling T, Neidhardt J, Wong YW, Schulze C, Streichert T, Schachner M (2010) Short DNA sequences inserted for gene targeting can accidentally interfere with off-target gene expression. FASEB J 24:1714–1724
71. Pham CT, MacIvor DM, Hug BA, Heusel JW, Ley TJ (1996) Long-range disruption of gene expression by a selectable marker cassette. Proc Natl Acad Sci USA 93:13090–13095
72. Revell PA, Grossman WJ, Thomas DA, Cao X, Behl R, Ratner JA, Lu ZH, Ley TJ (2005) Granzyme B and the downstream granzymes C and/or F are important for cytotoxic lymphocyte functions. J Immunol 174:2124–2131
73. Jasin M, Berg P (1988) Homologous integration in mammalian cells without target gene selection. Genes Dev 2:1353–1363
74. Deng C, Capecchi MR (1992) Reexamination of gene targeting frequency as a function of the extent of homology between the targeting vector and the target locus. Mol Cell Biol 12:3365–3371
75. Hasty P, Rivera-Perez J, Bradley A (1991) The length of homology required for gene targeting in embryonic stem cells. Mol Cell Biol 11:5586–5591
76. te Riele H, Maandag ER, Berns A (1992) Highly efficient gene targeting in embryonic stem cells through homologous recombination with isogenic DNA constructs. Proc Natl Acad Sci USA 89:5128–5132
77. Thomas KR, Folger KR, Capecchi MR (1986) High frequency targeting of genes to specific sites in the mammalian genome. Cell 44:419–428
78. Yagi T, Ikawa Y, Yoshida K, Shigetani Y, Takeda N, Mabuchi I, Yamamoto T, Aizawa S (1990) Homologous recombination at c-fyn locus of mouse embryonic stem cells with use of diphtheria toxin A-fragment gene in negative selection. Proc Natl Acad Sci USA 87:9918–9922
79. Mortensen R (2006) Overview of gene targeting by homologous recombination. Curr Protoc Mol Biol Chapter 23:Unit 23.21
80. Bouabe H, Moser M, Heesemann J (2011) Enhanced selection for homologous-recombinant embryonic stem cell clones by Cre recombinase-mediated deletion of the positive selection marker. Transgenic Res. doi:10.1007/s11248-011-9522-x
81. Sternberg N, Hamilton D (1981) Bacteriophage P1 site-specific recombination. I. Recombination between loxP sites. J Mol Biol 150:467–486
82. Hoess RH, Ziese M, Sternberg N (1982) P1 site-specific recombination: nucleotide sequence of the recombining sites. Proc Natl Acad Sci USA 79:3398–3402
83. Abremski K, Wierzbicki A, Frommer B, Hoess RH (1986) Bacteriophage P1 Cre-loxP site-specific recombination. Site-specific DNA topoisomerase activity of the Cre recombination protein. J Biol Chem 261:391–396

84. Orban PC, Chui D, Marth JD (1992) Tissue- and site-specific DNA recombination in transgenic mice. Proc Natl Acad Sci USA 89:6861–6865
85. Gu H, Marth JD, Orban PC, Mossmann H, Rajewsky K (1994) Deletion of a DNA polymerase beta gene segment in T cells using cell type-specific gene targeting. Science 265:103–106
86. Albanese C, Hulit J, Sakamaki T, Pestell RG (2002) Recent advances in inducible expression in transgenic mice. Semin Cell Dev Biol 13:129–141
87. Lakso M, Sauer B, Mosinger B Jr, Lee EJ, Manning RW, Yu SH, Mulder KL, Westphal H (1992) Targeted oncogene activation by site-specific recombination in transgenic mice. Proc Natl Acad Sci USA 89:6232–6236
88. Torres RM, Kühn R (2003) Laboratory protocols for conditional gene targeting. Oxford Uninversity Press, New York
89. Wakita T, Taya C, Katsume A, Kato J, Yonekawa H, Kanegae Y, Saito I, Hayashi Y, Koike M, Kohara M (1998) Efficient conditional transgene expression in hepatitis C virus cDNA transgenic mice mediated by the Cre/loxP system. J Biol Chem 273:9001–9006
90. Andrews BJ, Proteau GA, Beatty LG, Sadowski PD (1985) The FLP recombinase of the 2 micron circle DNA of yeast: interaction with its target sequences. Cell 40:795–803
91. Dymecki SM (1996) Flp recombinase promotes site-specific DNA recombination in embryonic stem cells and transgenic mice. Proc Natl Acad Sci USA 93:6191–6196
92. Senecoff JF, Bruckner RC, Cox MM (1985) The FLP recombinase of the yeast 2-micron plasmid: characterization of its recombination site. Proc Natl Acad Sci USA 82:7270–7274
93. Broach JR, Guarascio VR, Jayaram M (1982) Recombination within the yeast plasmid 2mu circle is site-specific. Cell 29:227–234
94. Buchholz F, Angrand PO, Stewart AF (1998) Improved properties of FLP recombinase evolved by cycling mutagenesis. Nat Biotechnol 16:657–662
95. Buchholz F, Ringrose L, Angrand PO, Rossi F, Stewart AF (1996) Different thermostabilities of FLP and Cre recombinases: implications for applied site-specific recombination. Nucleic Acids Res 24:4256–4262
96. Bult CJ, Eppig JT, Blake JA, Kadin JA, Richardson JE (2013) The mouse genome database: genotypes, phenotypes, and models of human disease. Nucleic Acids Res 41:D885–D891
97. Friedrich G, Soriano P (1991) Promoter traps in embryonic stem cells: a genetic screen to identify and mutate developmental genes in mice. Genes Dev 5:1513–1523
98. Zambrowicz BP, Imamoto A, Fiering S, Herzenberg LA, Kerr WG, Soriano P (1997) Disruption of overlapping transcripts in the ROSA beta geo 26 gene trap strain leads to widespread expression of beta-galactosidase in mouse embryos and hematopoietic cells. Proc Natl Acad Sci USA 94:3789–3794
99. Saveliev A, Tybulewicz VL (2009) Lymphocyte signaling: beyond knockouts. Nat Immunol 10:361–364
100. Okkenhaug K, Bilancio A, Farjot G, Priddle H, Sancho S, Peskett E, Pearce W, Meek SE, Salpekar A, Waterfield MD et al (2002) Impaired B and T cell antigen receptor signaling in p110delta PI 3-kinase mutant mice. Science 297:1031–1034
101. Patrucco E, Notte A, Barberis L, Selvetella G, Maffei A, Brancaccio M, Marengo S, Russo G, Azzolino O, Rybalkin SD et al (2004) PI3Kgamma modulates the cardiac response to chronic pressure overload by distinct kinase-dependent and -independent effects. Cell 118:375–387
102. Suire S, Condliffe AM, Ferguson GJ, Ellson CD, Guillou H, Davidson K, Welch H, Coadwell J, Turner M, Chilvers ER et al (2006) Gbetagammas and the Ras binding domain of p110gamma are both important regulators of PI(3)Kgamma signalling in neutrophils. Nat Cell Biol 8:1303–1309
103. Saveliev A, Vanes L, Ksionda O, Rapley J, Smerdon SJ, Rittinger K, Tybulewicz VL (2009) Function of the nucleotide exchange activity of vav1 in T cell development and activation. Sci Signal 2:ra83
104. Prisco A, Vanes L, Ruf S, Trigueros C, Tybulewicz VL (2005) Lineage-specific requirement for the PH domain of Vav1 in the activation of CD4+ but not CD8+ T cells. Immunity 23:263–274
105. Tsien RY (1998) The green fluorescent protein. Annu Rev Biochem 67:509–544
106. Bouabe H, Liu Y, Moser M, Bosl MR, Heesemann J (2011) Novel highly sensitive IL-10-{beta}-lactamase reporter mouse reveals cells of the innate immune system as a substantial source of IL-10 in vivo. J Immunol 187:3165–3176
107. Bronstein I, Martin CS, Fortin JJ, Olesen CE, Voyta JC (1996) Chemiluminescence: sensitive detection technology for reporter gene assays. Clin Chem 42:1542–1546
108. Campbell RE (2004) Realization of beta-lactamase as a versatile fluorogenic reporter. Trends Biotechnol 22:208–211

109. Bradley A, Anastassiadis K, Ayadi A, Battey JF, Bell C, Birling MC, Bottomley J, Brown SD, Burger A, Bult CJ et al (2012) The mammalian gene function resource: the international knockout mouse consortium. Mamm Genome 23(9–10):580–586
110. Prosser HM, Koike-Yusa H, Cooper JD, Law FC, Bradley A (2011) A resource of vectors and ES cells for targeted deletion of microRNAs in mice. Nat Biotechnol 29:840–845
111. Kain SR, Ganguly S (2001) Overview of genetic reporter systems. Curr Protoc Mol Biol Chapter 9:Unit9.6
112. Jiang T, Xing B, Rao J (2008) Recent developments of biological reporter technology for detecting gene expression. Biotechnol Genet Eng Rev 25:41–75
113. Olesen CE, Voyta JC, Bronstein I (1997) Chemiluminescent immunoassay for the detection of chloramphenicol acetyltransferase and human growth hormone reporter proteins. Methods Mol Biol 63:71–76
114. Qureshi SA (2007) Beta-lactamase: an ideal reporter system for monitoring gene expression in live eukaryotic cells. Biotechniques 42:91–96

Chapter 24

A Protocol for Construction of Gene Targeting Vectors and Generation of Homologous Recombinant Embryonic Stem Cells

Hicham Bouabe and Klaus Okkenhaug

Abstract

The completion of human and mouse genome sequencing has confronted us with huge amount of data sequences that certainly need decades and many generations of scientists to be reasonably interpreted and assigned to physiological functions, and subsequently fruitfully translated into medical application. A means to assess the function of genes provides gene targeting in mouse embryonic stem cells (ESCs) that enables to introduce site-specific modifications in the mouse genome, and analyze their physiological consequences. Gene targeting enables almost any type of genetic modifications of interest, ranging from gene insertion (e.g., insertion of human-specific genes or reporter genes), gene disruption, point mutations, and short- and long-range deletions, inversions. Site-specific modification into the genome of ESCs can be reached by homologous recombination using targeting vectors. Here, we describe a protocol to generate targeting constructs and homologous recombinant ESCs.

Key words Embryonic stem cells, Targeting vector, MultiSite Gateway Cloning

1 Introduction

The development of mice with site-specific genome modification has become possible because of the establishment of fundamental techniques that enable to culture embryonic stem cells (ESCs) *in vitro*, without altering their pluripotent potential, and to mediate homologous recombination between specific sites in the genome of ESCs and exogenously added DNA molecules.

In the following section, we describe a protocol to generate gene targeting constructs, to culture ESCs and to generate homologous recombinant ESC clones. For a detailed background overview about gene targeting in mice, we refer to the review "Gene Targeting in Mice: a Review" (*see* Bouabe and Okkenhaug, Chapter 23 of this series).

Susanne M. Bailer and Diana Lieber (eds.), *Virus-Host Interactions: Methods and Protocols*, Methods in Molecular Biology, vol. 1064, DOI 10.1007/978-1-62703-601-6_24, © Springer Science+Business Media, LLC 2013

2 Materials

2.1 Cloning of the Targeting Vector

1. MultiSite Gateway Three-Fragment Vector Construction Kit (Invitrogen).
2. Plasmids containing floxed Neomycin gene (loxP-neo-loxP) (or other resistance genes of preference, such as hygromycin or puromycin etc.). If appropriate, the resistance gene can also be flanked by FRT instead of loxP.
3. If required, vectors containing a reporter gene of interest and negative selection cassette (e.g., thymidine kinase (TK), diphtheria toxin fragment A (DT-A) or Cre recombinase), respectively.
4. Cosmid, bacteria artificial chromosome (BAC) or appropriate vector containing the mouse genomic sequence (gene) that should be targeted, and from which the homology arms for the targeting vector will be derived. Alternatively, the homology arms can be amplified directly from whole genomic DNA, isolated from the ESC line that will be used for gene targeting (this ensures isogeny).
5. CcdB Survival *E. coli* bacteria. This CcdB-resistant strain can be used to propagate and maintain vectors containing the ccdB gene, such as the Gateway Donor vector.
6. Restriction enzymes.
7. T4 DNA ligase.
8. Shrimp alkaline phosphatase.
9. PCR reagents: PfuUltra High-fidelity DNA polymerase (buffer provided with the enzyme) (Stratagene) or Q5 Hot Start High-Fidelity DNA Polymerase (New England Biolabs), dNTPs, primers.
10. Highly competent *E. coli* with highest efficiency cloning of large plasmids (*see* **Note 1**).
11. PCR purification kit.
12. Plasmid Mini- and Midi-preparation kits.
13. Gel DNA extraction kit (kit enabling damage-free extraction of big DNA fragments).
14. Tris-equilibrated phenol.
15. Chloroform:isoamyl alcohol (24:1) mixture.
16. Absolute ethanol.
17. 3 M sodium-acetate at pH 5.2.
18. 70 % ethanol.
19. 0.5 M EDTA.
20. TE-Puffer: 10 mM Tris–HCl (pH 8.0), 1 mM EDTA (pH 8.0).

2.2 Media for Cell Culture

1. MEF medium: Dulbecco's modified Eagle medium (DMEM), containing 10 % fetal bovine serum (FBS), 100 μM β-mercaptoethanol, 100 U/mL Penicillin, 100 μg/mL Streptomycin.
2. ESC medium: DMEM with 4.5 g/L glucose and 1 mM Na-pyruvate (Invitrogen), supplemented with 15 % FBS (tested for ESC culture), 2 mM L-glutamine, 100 μM β-mercaptoethanol, 1× nonessential amino acids of 100× stock solution (Invitrogen), 1,000 U/mL leukemia inhibitory factor (LIF), and if preferred 100 U/mL Penicillin, and 100 μg/mL Streptomycin. The use of Penicillin/Streptomycin can often mask low level contamination of cell culture with microorganisms such as mycoplasma.
3. ESC-G418 medium: ESC medium supplemented with 250–400 μg/mL G418 (Geneticin) (the concentration of G418 depends e.g., on the used ESC line).

2.3 Isolation, Culture and Mitotic Inactivation of Mouse Embryonic Fibroblasts (MEFs)

1. Transgenic mice harboring the same resistance gene (e.g., neomycin or hygromycin) that is used for the positive selection of transfected ESCs (can be purchased from the Jackson Laboratories).
2. Sterile phosphate-buffered saline (PBS).
3. Trypsin-EDTA solution.
4. MEF medium (from Subheading 2.2).
5. 70 % ethanol.
6. Sterile dissecting instruments (e.g., dissection scissors, forceps etc.).
7. Cesium source γ irradiator.
8. Freeze medium: DMEM with 10 % DMSO and 20 % FBS.

2.4 Culture of ESCs

1. ESCs: ESC lines can be purchased from commercial sources, e.g., http://www.lgcstandards-atcc.org/.
2. ESC medium (from Subheading 2.2).
3. ESC-G418 medium (from Subheading 2.2).

2.5 Electroporation of ESCs with the Targeting Vector

1. Linearized targeting vector.
2. PBS.
3. ESCs.
4. ESC medium.
5. Trypsin-EDTA.
6. Mitotically inactive MEFs.
7. Electroporation cuvette for eukaryotic cells.
8. Gene pulser.

2.6 Selection of Positively Transfected ESC Clones

1. ESC medium (from Subheading 2.2).
2. ESC-G418 medium (from Subheading 2.2).
3. MEF medium (from Subheading 2.2).
4. Mitotically inactive MEFs (from Subheading 2.3).
5. 96- and 24-well plates.
6. Trypsin-EDTA solution.
7. Laminar flow cabinet.
8. Gloves.
9. Face mask.

2.7 Freezing of and DNA Preparation from ESC Clones

1. Trypsin-EDTA.
2. Freeze medium: DMEM with 10 % DMSO and 20 % FBS.
3. ESC medium (from Subheading 2.2).
4. Lysis buffer: 100 mM Tris–HCl (pH 8.5), 5 mM EDTA, 200 mM NaCl, 0.2 % SDS and every time freshly added 100 μg/mL Proteinase K.
5. Isopropanol.
6. TE-Puffer: 10 mM Tris–HCl (pH 8.0), 1 mM EDTA (pH 8.0).

2.8 Screening by Southern Blot Analysis for ESC Clones with Homologous Recombination of the Targeting Vector

1. Restriction enzymes.
2. Proteinase K.
3. TBE buffer: 1.1 M Tris-base (54 g), 900 mM Borate (27.5 g), 25 mM EDTA, adjust to pH 8, and bring the final volume to 1 L with deionized water.
4. Depurination solution: 0.2 M HCl (8 mL 37 % HCl in 500 mL H_2O).
5. Denaturing buffer: 0.5 M NaOH, 1.5 M NaCl.
6. Neutralization buffer: 0.5 M Tris–HCl, 1.5 M NaCl, pH 7.2.
7. Positive charged Nylon membrane.
8. 20× SSC: 3 M NaCl, 0.3 M tri-sodium citrate 2-hydrate, pH 7.2.
9. UV lamp for cross-linking of DNA to the membrane.
10. Hybridization solution: 50 % formamide, 5× SSC 10 mM Tris–HCl (pH 7.5), 1 % SDS, 5× denhardts (2 % Ficoll 400, 2 % polyvinylpyrrolidone K30, 2 % BSA), 10 % dextransulfate, 100 μg/mL salmon sperm DNA.
11. Radioactive labelled dATP or dCTP (α-^{32}P-dATP or α-3^{2}P-dCTP) (can be purchased from e.g., Hartmann Analytic or Amersham). Labeling of DNA probe has to be performed in room designated for radioactive work with precautions against radioactive contamination.

12. DNA Labeling Kit (e.g., the HexaLabel&trade DNA Labeling Kit from Fermentas).
13. Hybridization bottles.
14. Hybridization oven.
15. Wash-buffer 1: 2× SSC, 0.1 % SDS.
16. Wash buffer 2: 0.5× SSC, 0.4 % SDS.
17. Autoradiography film.
18. Phosphor imaging plates, Scanner for detection and analyzer of radioactive signals (e.g., Fuji Film Scanner FLA-3000 and Aida Image Analyser v.4.00 software).

2.9 The Use of Cre/loxP Recombination System in Gene Targeting

1. Vectors harboring cre recombinase gene and loxP sequences, respectively.
2. If appropriate, vectors harboring a gene encoding for FLP recombinase and FRT sequences, respectively.
3. Mice harboring a cre gene that can be expressed constitutively in the whole body and all cells or conditionally in a specific tissue or cell type. Collection databases of several hundreds of Cre transgenic mouse lines expressing Cre recombinase in specific tissues or cells are available: e.g., http://www.ics-mci.fr/mousecre/; http://nagy.mshri.on.ca/cre_new/index.php; http://www.creportal.org/; http://bioit.fleming.gr/crezoo/; http://creline.org/.

3 Methods

3.1 Cloning of the Targeting Vector

The construction of a gene targeting vector can proceed by conventional restriction enzyme-based cloning strategies. The 5′ and 3′homology arms can be amplified by PCR from genomic DNA prepared from the ESCs to be used or from a mouse of the same strain. Alternatively, bacterial artificial chromosome (BAC) clones can be ordered that contain the gene of interest. BAC libraries from different mouse strains, including C57BL/6J, 129/Ola, and 129Sv mouse strains are available [1–4] (BAC clones can be supplied e.g., by: http://bacpac.chori.org/; http://dna.brc.riken.jp/en/NBRPB6Nbacen.html; http://www.lifesciences.sourcebioscience.com/clone-products/genomic-dna-clones/mouse-genomic-bac-library---rpci-23/mouse-genomic-bac-library-.aspx).

A targeting vector is typically composed of at least three basic units: a 5′homology arm, a positive selectable gene marker, and a 3′homology arm (*see* **Note 2**). Furthermore, in order to enrich for homologous recombinant ESC clones, a negative selection marker can be included in the targeting vector, outside the homology arms (*see* **Note 2**). However, the assembly of those units (fragments)

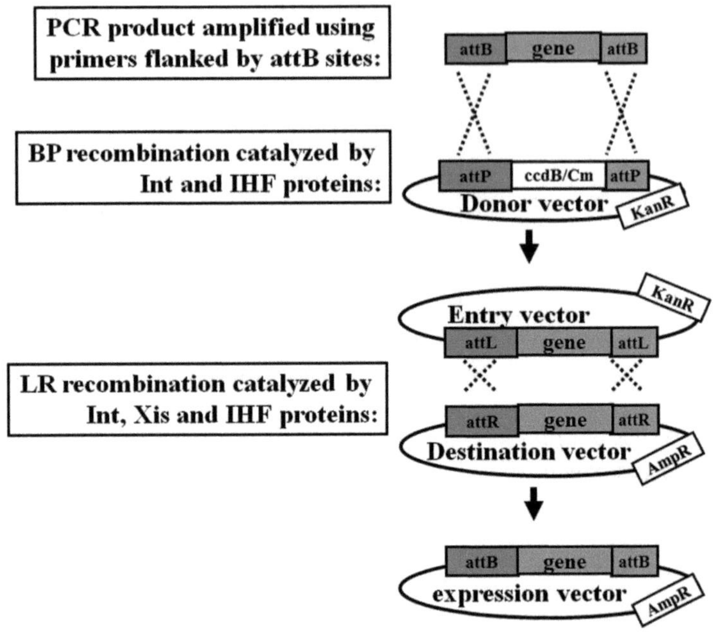

Fig. 1 Gateway cloning technology. (*See* description in Subheading 3.1)

together to build the targeting vector can be a highly time-consuming and complicated process. A cloning method that may help to simplify and speed the construction of targeting vectors is based on the MultiSite Gateway Technology (supplied by Invitrogen, http://tools.invitrogen.com/content/sfs/manuals/multisite_gateway_man.pdf) (Fig. 1) [5, 6]. Indeed, Gateway-mediated cloning has been exploited by the European Conditional Mouse Mutagenesis (EUCOMM) and the National Institutes of Health Knockout Mouse (KOMP) programs in their quest to generate conditional knockout alleles for all protein-coding genes in the mouse genome [7]. The following illustrates how this technique can be used to assemble a basic gene targeting vector (*see* **Note 3** for other alternative methods to generate targeting vectors).

The Gateway Cloning Technology is based on the bacteriophage lambda site-specific recombination system which facilitates the integration of lambda DNA into the chromosome of *E. coli*, and the switch between the lytic (excision of lambda DNA from the bacterial chromosome) and lysogenic pathways (integration of lambda DNA into specific sites of the bacterial chromosome) [8–10]. The phage lambda inserts its DNA into an *E. coli* chromosome via specific recombination sequences designated as *att* sites (*att*achment sites). The att site in the DNA of phage lambda, called *attP* (*att*achment site of *p*hage lambda), recombines with the *att*achment site in the DNA of the *b*acteria (designated *attB*).

The recombination is mediated by two pairs of strand exchanges and ligation at the attB and attP sites, catalyzed by phage lambda Integrase (Int) and *E. coli* Integration Host Factor (IHF) proteins (supplied by Invitrogen as BP Clonase™ II enzyme mix) (Fig. 1) [8–10].

Because att sites are not palindromic, the recombination results in novel sequences designated as attL and attR (Fig. 1). The attL sites always recombine with attR resulting in the excision of lambda DNA (lytic cycle) from the bacterial chromosome (Fig. 1). This reaction is catalyzed by the Integrase (Int) and Excisionase (Xis) proteins of the phage lambda, and the *E. coli* Integration Host Factor (IHF) protein (Fig. 1) (supplied by the company Invitrogen as LR Clonase™ II Plus enzyme mix).

The Gateway cloning system enables efficient transfer of DNA-fragments of interest between plasmids (Fig. 1). The first step in Gateway cloning is the generation of a Gateway Entry vector. Entry vectors are usually generated in two steps: First, amplification of a DNA sequence of interest using specific primers each flanked by an attB site. Second, the attB sites-flanked PCR product is then mixed with a Donor vector containing attP sites flanking a lethal ccdB gene, and BP clonase enzyme mix containing the recombinant proteins phage lambda Integrase (Int) and *E. coli* Integration Host Factor (IHF) (supplied by the company Invitrogen). This enzyme mix catalyzes recombination between attP and attB sites resulting in the insertion of the attB sites-flanked PCR product into attP sites to replace the lethal ccdB gene in the Donor vector (Fig. 1). The resulting vector is called Entry vector, where the inserted PCR product is now flanked by attL sites as a result of recombination between attP and attB sites. The ccdB gene is a lethal gene that serves as negative selection marker for successful BP recombination. CcdB targets DNA gyrase and thus inhibits survival and growth of *E. coli* strains harboring the plasmid containing its coding sequence without the gene coding for its antitoxin, the protein CcdA, which antagonizes the toxic activity of the CcdB protein [11]. Thus, bacteria can survive and propagate only if they contain the "Entry" vector, in which the ccdB gene was removed by BP recombination and replaced by the attB sites-flanked PCR product (Fig. 1).

To propagate and maintain vectors containing the ccdB gene, such as a Gateway Donor vector, CcdB-resistant *E. coli* strains (CcdB Survival strain) should be used.

The attL-flanked DNA sequence in the Entry vector can then be efficiently transferred into any Destination vector of interest that contains attR recombination sites using the enzyme mix, LR Clonase, containing the bacteriophage lambda recombination proteins Int and Excisionase (Xis), and the *E. coli*-encoded protein IHF (supplied by the company Invitrogen) (Fig. 1).

The MultiSite Gateway Cloning has been made possible because of the generation of modified new att sites with very high specificities enabling simultaneous, recombinational cloning of multiple DNA fragments in a single reaction. The modified att sites include among others, attB1, attB2, attB3, attB4, attB1r, of which each reacts (recombines) with the specific corresponding modified attP sites: attP1, attP2, attP3, attP4, attP1r, respectively. Furthermore, the resulting attL sites, such as attL1, attL2, etc., react specifically with the corresponding attR sites, attR1, attR2, etc. (http://tools.invitrogen.com/content/sfs/manuals/multisite_gateway_man.pdf).

3.2 Proposed Strategy to Generate Targeting Vectors Using MultiSite Gateway Cloning (in Combination with Restriction Enzymes/Sites)

Gateway Cloning has been used in our laboratory. However, we didn't yet implement the following strategy which we have designed to generate our future targeting vectors.

1. Amplification of the 5′homology arm using specific primers flanked by attB4 site at 5′primer, and attB1r at the 3′primer. In addition, a unique restriction enzyme site (RS) can be included at the 3′primer, just upstream to the AttB1r-site (*see* **Note 4**). Such RS enables a subsequent insertion of any further DNA sequence of interest, such as a reporter gene to monitor the expression of the tagged gene, or a transgene that should be expressed in the mouse (Fig. 2).
2. Amplification of the 3′homology arm using specific primers flanked by attB2r site (at 5′primer) and attB3 site (at 3′primer). In addition, a unique restriction site (RS) can be included in the 3′primer (*see* **Note 4**). Such RS enables a subsequent insertion of any further DNA sequence of interest, such as a negative selection gene (Fig. 2).
3. Amplification of a floxed resistance gene cassette, such as loxP-neo-loxP cassette, using specific primers flanked by attB1 and attB2 sites (Fig. 2).
4. BP recombination of each amplified fragment into the corresponding donor vectors (according to the manufacturer's instructions), containing the counterpart att sites: pDONR P4-P1R (for attB4 and attB1r-flanked PCR product), pDONR P2R-P3 (for attB2r and attB3-flanked PCR product) and pDONR-221 (for attB1 and attB2-flanked PCR product), resulting in the generation of Entry vectors containing the 3 units of the targeting vector, respectively (Fig. 2) (*see* **Notes 4–6**). To propagate and maintain vectors containing the ccdB gene, such as Gateway Donor vector, ccdB Survival *E. coli* should be used.
5. Amplification of a transgene of interest using specific primers flanked by a unique restriction site (RS) identical to that introduced at the 3′end of the 5′homology arm (*see* **Note 4**). After

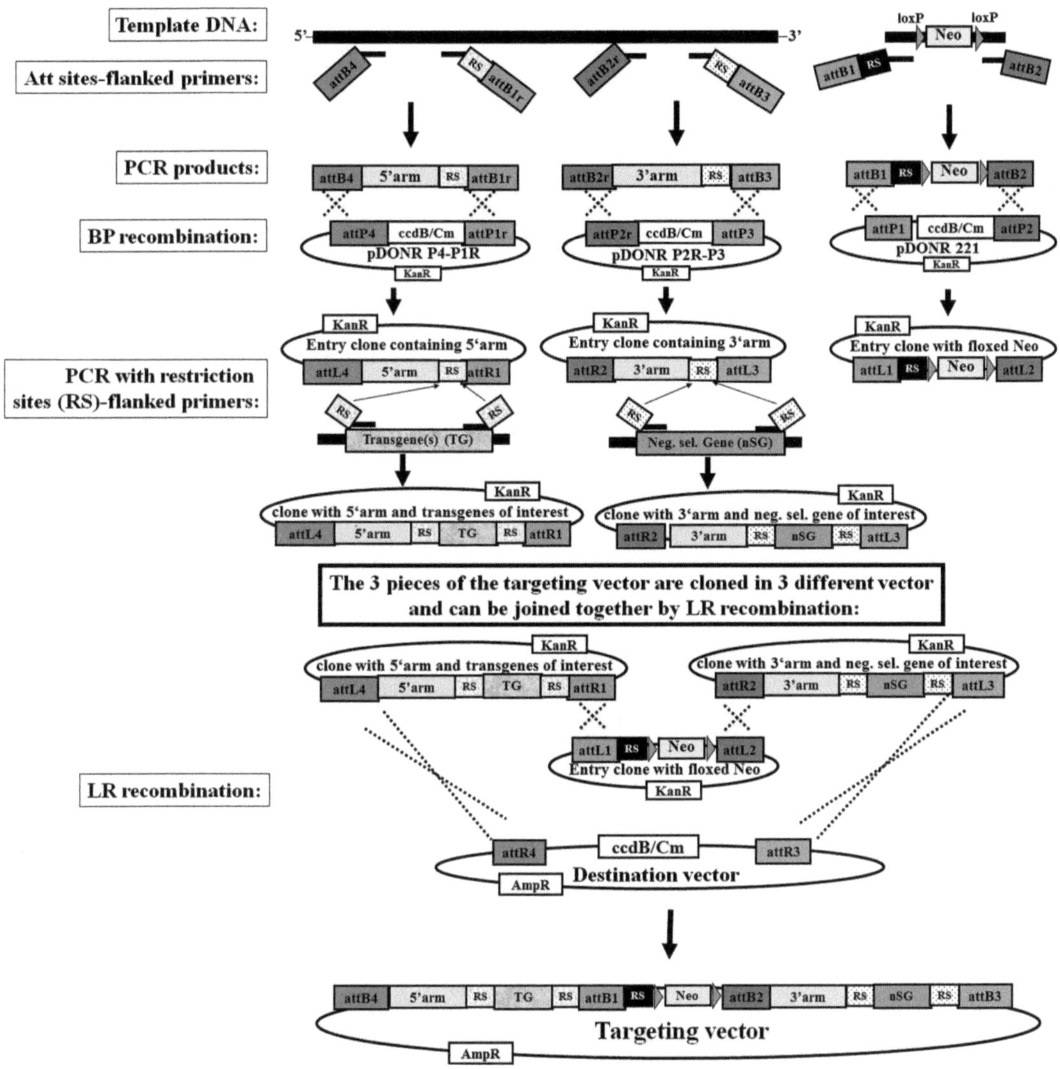

Fig. 2 Generation of targeting vector using MultiSite Gateway Cloning (in combination with restriction enzymes/sites). (*See* description in Subheading 3.2)

its digestion using the appropriate restriction enzyme, the transgene PCR-product can be ligated into the corresponding restriction site (RS) of the Entry vector, downstream of the 5′arm (Fig. 2) (*see* **Notes 2**, **4**, and **5**).

6. Amplification of a gene for negative selection (nSG), such as TK, DT-A, or cre, using specific primers flanked by a unique restriction site (RS) identical to that introduced at the 3′end of the 3′homology arm. After its digestion using the appropriate restriction enzyme, the negative selection gene (nSG) can be ligated into the corresponding restriction site (RS) of the Entry vector, downstream of the 3′arm (Fig. 2) (*see* **Notes 4–6**).

7. LR recombination of the five units, which are subcloned in three different Entry vectors, into a Destination vector containing the appropriate att sites, resulting in the generation of the targeting vector composed of 5′arm, transgene of interest, positive selection cassette, 3′arm, and negative selection gene (Fig. 2) (*see* **Notes 4–6**).

3.3 Isolation, Culture, and Mitotic Inactivation of Mouse Embryonic Fibroblasts (MEFs)

1. Set up breeding pair of mice expressing the same resistance gene (e.g., neomycin or hygromycin) that is used for the positive selection of transfected ESCs. Check plugs and record date of visible plug. Usually, plug is visible 0.5 day after set up breeding.
2. Euthanize pregnant female at 13.5–14.5 days after the appearance of the plug.
3. Rinse dead mouse in 70 % ethanol for ~5 min.
4. Open the abdomen by cutting carefully the abdominal skin using sterile scissor and forceps. Care should be taken not to rip the gut to avoid contamination by gut residing bacteria.
5. Remove both uterine horns containing embryos. Briefly rinse them in 70 % ethanol and place them in petri dish containing sterile PBS on ice.
6. Move to a sterile hood or laminar flow cabinet.
7. Open a small hole through the uterus using sterile scissors and forceps and remove the embryos through the hole.
8. Separate each embryo from its placenta and surrounding membranes.
9. Cut away head of embryos and transfer remnants to a Petri dish containing sterile PBS on ice.
10. Open abdomens of the embryos and remove intestine, liver, kidneys, and spleen.
11. Transfer the rest of the embryos to a small Petri dish containing 2 mL cold PBS.
12. Mince the embryos until they become "pipettable", and add 2 mL of trypsin-EDTA (2 mL per embryo), and incubate at 37 °C for 15 min.
13. Break the remaining tissue pieces by up and down pipetting with a 5 mL then 2 mL pipette, and add 4 mL MEF medium per embryo.
14. Pipette each 2.5 mL of the MEF suspension into a 175 cm^2 tissue culture flask, and add MEF medium to a final volume of ~25–30 mL.
15. Incubate MEFs at 37 °C, 5 % CO_2.

16. Collect MEFs 2–3 days after they reached confluence (confluence is reached after a few days of incubation): wash once carefully with PBS, then add 3 mL trypsin-EDTA per 175 cm^2 flask, and incubate at 37 °C to detach cells from their monolayer.
17. Add 5 mL of MEF medium per cell culture flask to inactivate trypsin, and resuspend the cells thoroughly.
18. Transfer 6 mL of the cell suspension from each flask into a 50 mL tube, and add 25 mL MEF medium to the remaining 2 mL MEF suspension in each 175 cm^2 flask.
19. Pellet the collected MEFs by centrifugation at $400\times g$ for 3 min.
20. Discard supernatants and add 20 mL MEF medium to the MEFs-pellet.
21. γ-irradiate MEFs with 40 Gy (4,000 rad) using a cesium-source irradiator to mitotically inactivate them.
22. Pellet γ-irradiated MEFs by centrifugation at $400\times g$ for 3 min.
23. Discard supernatants and add 3 mL freeze medium to each cell pellet, resuspend the cells thoroughly, and store 1 mL aliquots at −70 °C.
24. The **steps 15–23** should be repeated for the remaining growing MEFs.
25. One vial frozen MEFs can be used to layer a petri dish (10 cm diameter), a 24-well plate, or a 96-well plate.

3.4 Culture of ESCs

ESCs have to be cultured in special culture conditions to maintain their pluripotency (*see* **Note 2**). A combination of mitotically inactivated MEFs and leukemia inhibitory factor (LIF) ensures an optimal culture condition to inhibit differentiation of ESCs. Furthermore, ESCs should always be kept at a low density, and thus have to be frequently split. The present protocol to culture, transfect, select, and pick ESCs was used to successfully generate several knockin mice [12, 13], and is based on the protocol of Talts et al. [14] with some modifications (*see* **Note 7** for alternative protocols using MEFs-free culture systems).

1. Thaw ESCs and MEFs quickly at 37 °C, transfer them into a 15 mL tube and add 5–10 mL ESC medium. Centrifuge for 4 min at $400\times g$.
2. Resuspend cells in appropriate ESC medium and seed ESCs and MEFs together in petri dishes (100 mm × 20 mm) (one vial frozen MEFs per petri dish).
3. Change medium every day.
4. Trypsinize and split the ESCs before the colonies become brownish in their middle (phenotype for ongoing differentiation). Usually, ESCs have to be split every day. For

trypsinization, carefully remove medium, wash one time with sterile PBS, and then add trypsin, and incubate for 5 min at 37 °C. Thereafter add appropriate amount of ESC medium, resuspend the cells by up and down pipetting, and then split them into two dishes (along with new MEFs, one vial frozen MEFs per two petri dish of 100 mm × 20 mm size).

5. 2–4 h before harvesting for electroporation, ESCs should be fed (change of medium).

3.5 Electroporation of ESCs with the Targeting Vector

3.5.1 Linearization of Targeting Vector

1. Linearize at least 150 μg of the targeting construct using a unique restriction site outside the homology arms (use ~2 U enzyme per μg DNA) (*see* **Note 8**).
2. Purify the DNA by two-fold phenol/chloroform extraction. For each extraction add equal volume of phenol/chloroform, mix well by vigorously inverting the tube (don't vortex), and then centrifuge for 3 min at high speed. Transfer the upper DNA-containing aqueous phase to a new vial. Then remove phenol rests by chloroform/isoamylalcohol extraction. For that add equal volume of chloroform/isoamylalcohol to the DNA-containing upper aqueous phase, mix well by vigorously inverting the tube (don't vortex), and then centrifuge for 3 min at high speed. Transfer the upper DNA-containing aqueous phase to a new vial (*see* **Note 9**).
3. Precipitate the DNA by addition of 0.3 M Na-acetate (pH 5.2) to upper aqueous fraction containing the DNA, mixing (don't vortex), and then addition of 2.5 volume of absolute ethanol and mixing by vigorously inverting the tube (don't vortex).
4. Transfer the DNA-precipitate with a pipette tip to a 1.5 mL tube containing 70 % ethanol.
5. The DNA can be stored at this stage at −20 °C until use for electroporation.

3.5.2 Electroporation of ESCs

1. Remove ethanol supernatant and air dry the DNA under sterile hood. Add 500 μL sterile PBS, resuspend by pipetting and incubate at 37 °C to completely dissolve DNA.
2. ESCs (~80 % confluent) should be fed 2–4 h prior to harvesting (change of medium).
3. Collect ESCs by trypsinization, and then wash them twice with PBS (centrifugation steps at 400 × *g* for 4 min).
4. Resuspend 1×10^7 ESCs in 500 μL PBS in 1.5 mL tube.
5. Add the 500 μL PBS/DNA solution, and incubate at RT for 5 min.
6. Split the mixture into electroporation cuvettes. Electroporate at 800 V and 300 μF using a Gene Pulser. The time constant should be ~0.04 ms.

7. Carefully transfer the cells with a sterile pipette into petri dishes (~6–10 dishes) pre-layered with mitotically inactivated MEFs. Add ESC medium to a final volume of 8 mL. Shake the dishes crosswise to distribute the electroporated ESCs and incubate at 37 °C, 5 % CO_2.

3.6 Selection and Picking of Positively Transfected ESC Clones

ESC Clones should be picked on day 6, 7, or 8 after starting selection. The number of ESC clones to be picked depends on the expected ratio of homologous recombination.

1. Begin selection 24 h after electroporation. Exchange ESC medium for G418-ESC medium containing 250–400 μg/mL G418 (or appropriate concentration of alternative antibiotic, depending on the antibiotic resistance gene used). The concentration of the antibiotic depends on the ESC line.
2. Change medium daily.
3. After day 3–4 of selection, cell death begins to be clearly observed. At those days change medium twice per day, in the morning and evening.
4. Several 24-well plates should be layered with mitotically inactivated MEFs in MEF medium (from Subheadings 2.2 and 3.3), 1 day before picking ESC clones (one vial frozen MEFs from Subheading 3.3 can be used to layer one 24-well plate). Just before use, exchange MEF medium for G418-ESC medium (from Subheading 2.2).
5. Depending on the ESC line and the media used, ESC colonies start to be visible ~6–8 days after beginning of selection. The optimal ESC clones to be picked have rounded or oval shape, have tight and bright borders and are closely packed, often with a weak brownish center.
6. Prior to picking of ESC colonies, wear gloves and face mask, and disinfect all surfaces that will be used, including the microscope, with ethanol.
7. Place a petri dish containing transfected ESCs under the microscope at 10× magnification (optimally, the microscope should be in a laminar flow hood, to ensure sterile work).
8. Loosen the ESC colonies from the MEF layer with the tip of a pipette set to 10 μL. Aspirate the colony into the tip, transfer it into one well of a sterile 96-well plate. After filling half of a 96-well plate with 48 clones (takes approximately 60–90 min), add ~100 μL trypsin-EDTA per well, and incubate at 37 °C for 5 min.
9. During the incubation of ESC clones with trypsin, exchange MEF medium of feeder cells (mitotically inactivated MEFs) in the 24-well plates for 500 μL G418-ESC medium (from Subheading 2.2).

10. Add 100 μL G418-ESC medium in each well of the 96-well plate containing the picked ESC clones, mix well by up and down pipetting to break up each colony, avoiding excessive foaming, and then transfer each ESC clone to a well of the prepared 24-well plate (layered with MEFs) (from this Subheading 3.6, **step 4**).
11. Incubate at 37 °C, 5 % CO_2.
12. Change the G418-ESC medium every day until a good coverage of colonies in each well is achieved (this happens ~5–10 days after picking ESC clones).

3.7 Freezing of and DNA Preparation from ESC Clones

1. Remove medium and wash each well with PBS.
2. Add ~100 μL trypsin-EDTA solution to each well and incubate at 37 °C for 5 min.
3. Add 1 mL of ice-cold freeze medium (from Subheading 2.7) to each well.
4. Disperse the cells using a pipette set to 500 μL and transfer 700 μL of the cell suspension from each well to pre-labelled freeze tubes and place them immediately on dry ice.
5. Transfer the tubes to −80 °C freezer.
6. Add 2 mL of G418-ESC medium to the residual cells in each well, and incubate at 37 °C for ~12 h. Then replace medium by 1 mL G418-ESC medium and incubate at 37 °C until the color of the medium turns yellow.
7. Aspirate medium and add 500 μL of fresh lysis buffer (from Subheading 2.7) to each well that turned yellow and incubate at 37 °C for at least overnight.
8. When a well plate is completely lysed, add one volume of isopropanol (0.5 mL) to each well and shake on an orbital shaker for at least 12 h at RT until the DNA becomes visible.
9. Extract the DNA by spooling it on a pipette tip, and transfer it into a pre-labelled tube containing 100–150 μL of TE-buffer (from Subheading 2.7). DNA is visible against a dark background.
10. Incubate for at least 8 h at 55 °C to allow complete dissolving of the DNA.
11. Store the DNA at RT (long storage can be done at 4 °C).

3.8 Screening for ESC Clones with Homologous Recombination of the Targeting Vector by Southern Blot Analysis

Homologous recombinant ESC clones can be identified by Southern blot. The genomic DNA isolated from ESC clones should be digested with an appropriate restriction enzyme that produces one cut inside the targeting vector and one cut just outside (upstream or downstream) the targeting vector, in the targeted chromosomal region. The use of an "external" probe outside of the targeting construct will produce a band with a size corresponding to the unmodified wild-type allele (*see* **Note 2**) and, if homologous recombination occurred, a second band of bigger or smaller size corresponding to the targeted allele (*see* **Note 2**). To analyze whether the targeting vector was also integrated in additional places in the genome of ESCs, an internal probe that hybridizes within the targeting construct (e.g., to the selection marker) should be used.

1. Digest 20–40 μL DNA solution of each ESC clone (from Subheading 2.8) (in a total volume of 50 μL) with 30–60 U of an appropriate restriction enzyme overnight.
2. Fractionate the digested DNA samples on a 0.7 % agarose gel in 1× TBE (from Subheading 2.8).
3. Treat the gel successively with depurination solution (0.2 M HCl), denaturing buffer and neutralization buffer (from Subheading 2.8).
4. Blot the gel overnight to a positively charged Nylon membrane using 20× SSC (from Subheading 2.8).
5. Cross-link the DNA to the membrane by irradiation of the membrane with a UV lamp for 50 s.
6. Incubate the membrane with hybridization solution (from Subheading 2.8) for 1 h.
7. During this incubation time, prepare radioactively labelled probe (*see* **Note 10**).
8. Add radioactively labelled probe to the hybridization solution and incubate overnight at 42 °C.
9. Wash the blot two times with wash-buffer 1 (2× SSC, 0.1 % SDS) for 10 min at RT, then two times with wash buffer 2 (0.5× SSC, 0.4 % SDS) for 20 min at 70 °C.
10. Radioactive signals can then be detected using a phosphor imaging plate, and corresponding Scanner and software (such as Fuji Film Scanner FLA-3000 and Aida Image Analyser v.4.00 software), or by exposing the membrane with an autoradiography film.

3.9 Generation of Chimeric Mice from Genetically Modified ESC Clones

The identified homologous recombinant ESC clones can be used to generate chimeric mice by injecting them into recipient preimplantation mouse embryos (blastocysts) that are collected from female mice with coat color different from that of the mouse strain-parent of the used ESCs (*see* **Note 2**). The injected blastocysts are then surgically transferred to a recipient pseudopregnant foster mother to allow the embryos to develop (*see* **Note 2**). Females of CD1 mouse strain make very good mothers and are thus used by several laboratories as foster mothers. Because ESCs and recipient blastocysts are derived from mouse strains with distinguishable coat-colors, the desired chimeric offspring can be visually recognized by inspection of coat-color chimerism (percentage of black and agouti hair on the mouse black-agouti). Chimeric offspring (usually males) are mated with C57BL/6 mice to produce the F1 generation. The germline transmission is then confirmed by Southern blot analysis or PCR of tail DNA from mice of the F1 generation.

4 Notes

1. It is strongly recommended to use commercially available transformation competent bacteria, because they are highly competent. Transformation competency is among the important factors influencing the speed and success of cloning.
2. For a background overview about gene targeting constructs, ESCs and gene targeting in ESCs, we refer to the review chapter (Chapter 23) in this issue of "Virus-Host-Interactions".
3. A further alternative method to generate targeting vectors uses *Recombi*nation-mediated genetic engi*neering* (*Recombineering*)-based protocols. Particularly, recombineering enables quick BAC-based construction of targeting vectors by introducing any DNA modifications of interest directly into the BAC clone containing the gene of interest, without the need for subcloning steps, restriction enzymes or DNA ligases [15–17]. Recombineering is based on homologous recombination in *E. coli* mediated by bacteriophage recombination proteins, such as RecE, RecT, Redα, Redβ, and RecA [18]. Recombineering can also be used to retrieve homology arms from a BAC clone into a vector.

   In addition, it is becoming increasingly affordable to generate entire or part of the targeting vector (especially complicated parts) by gene-synthesis [19, 20].
4. Be aware that insertion of mutations by PCR site-directed mutagenesis often results in the incorporation of tandem repeats of the used complementary primer pairs at the mutated site. We actually sequence at least five bacteria colonies recovered from each site-directed mutagenesis reaction, and in most cases, at least one clone contained plasmid with the accurate mutation of interest, without primer tandem repeats.
5. According to the manufacturer's instructions, the BP and LR reactions should be incubated at 25 °C for 1 h. However, we realized that, depending on the vectors used, the success of recombination reactions (BP and LR reactions) can be enormously increased when incubated at 25 °C for at least 12 h (overnight).
6. The larger the plasmids ($\gtrsim$10 kb) the more instable they are when used for ligation- or recombination-mediated DNA

transfer. They often undergo unwanted recombinations resulting in "cryptic" plasmids. To reduce the occurrence of such drawbacks and increase the rate of positive colonies, we strictly incubate the transformed bacteria at maximum 27 °C, and we reduce the incubation at the minimum time, just until the bacterial colonies became visible (at 27 °C it takes ~20–24 h to get visible colonies).

Furthermore, we incubate the inoculated fluidic cultures with the recovered bacteria colonies at 27 °C or at 37 °C, and we reduce the incubation at the minimum time, just until the bacterial culture became just slightly cloudy (at 37 °C it takes ~6 h to get a slightly cloudy culture).

7. MEF-free culture systems enabling maintenance of pluripotency use glycogen synthase kinase (GSK)-3-specific inhibitors, such as 6-bromoindirubin-3′-oxime (BIO) [21], or, optimally, a combination of three inhibitors (3i medium): SU5402 (inhibits FGF receptor tyrosine kinases), PD184352 (inhibits ERK signal cascades) and CHIR99021 (a more selective inhibitor of GSK-3) [22].

   In addition, recently a MEF- and inhibitor-free culture system was developed that enables maintenance of pluripotency and self-renewal by culturing ESCs on plates coated with a recombinant human extracellular matrix protein, the laminin isoform 511 (LN-511) [23]. Please see also the following references for detailed protocols on how to culture ESCs without MEFs [21–23].

8. To ensure highly efficient (almost 100 %) linearization of the vector, the digestion reaction should be incubated overnight, and next day, further units of the restriction enzyme (1 U/μg DNA) should be added, and the digestion reaction incubated for further 3–4 h.

9. Do not purify the digested targeting vector by gel extraction. This can be harmful for the DNA, and will also lead to the recovery only of a small amount of the loaded DNA material.

10. The probe can be radioactively labelled using DNA Labeling Kits, such as the HexaLabel&trade DNA Labeling Kit (Fermentas), according to manufacturer's instructions. α-^{32}P-dATP or α-3^{2}P-dCTP can be used as radioactive marker for the probe.

Acknowledgements

Work in our laboratory is supported by grants from the BBSRC and the Wellcome Trust.

References

1. Ohtsuka M, Ishii K, Kikuti YY, Warita T, Suzuki D, Sato M, Kimura M, Inoko H (2006) Construction of mouse 129/Ola BAC library for targeting experiments using E14 embryonic stem cells. Genes Genet Syst 81:143–146
2. Adams DJ, Quail MA, Cox T, van der Weyden L, Gorick BD, Su Q, Chan WI, Davies R, Bonfield JK, Law F et al (2005) A genome-wide, end-sequenced 129Sv BAC library resource for targeting vector construction. Genomics 86:753–758
3. Jansa P, Divina P, Forejt J (2005) Construction and characterization of a genomic BAC library for the Mus m. musculus mouse subspecies (PWD/Ph inbred strain). BMC Genomics 6:161
4. Osoegawa K, Tateno M, Woon PY, Frengen E, Mammoser AG, Catanese JJ, Hayashizaki Y, de Jong PJ (2000) Bacterial artificial chromosome libraries for mouse sequencing and functional analysis. Genome Res 10:116–128
5. Iiizumi S, Nomura Y, So S, Uegaki K, Aoki K, Shibahara K, Adachi N, Koyama H (2006) Simple one-week method to construct gene-targeting vectors: application to production of human knockout cell lines. Biotechniques 41:311–316
6. Walhout AJ, Temple GF, Brasch MA, Hartley JL, Lorson MA, van den Heuvel S, Vidal M (2000) GATEWAY recombinational cloning: application to the cloning of large numbers of open reading frames or ORFeomes. Methods Enzymol 328:575–592
7. Skarnes WC, Rosen B, West AP, Koutsourakis M, Bushell W, Iyer V, Mujica AO, Thomas M, Harrow J, Cox T et al (2011) A conditional knockout resource for the genome-wide study of mouse gene function. Nature 474:337–342
8. Landy A (1989) Dynamic, structural, and regulatory aspects of lambda site-specific recombination. Annu Rev Biochem 58:913–949
9. Moitoso de Vargas L, Kim S, Landy A (1989) DNA looping generated by DNA bending protein IHF and the two domains of lambda integrase. Science 244:1457–1461
10. Nunes-Duby SE, Matsumoto L, Landy A (1989) Half-att site substrates reveal the homology independence and minimal protein requirements for productive synapsis in lambda excisive recombination. Cell 59:197–206
11. Maki S, Takiguchi S, Miki T, Horiuchi T (1992) Modulation of DNA supercoiling activity of Escherichia coli DNA gyrase by F plasmid proteins. Antagonistic actions of LetA (CcdA) and LetD (CcdB) proteins. J Biol Chem 267:12244–12251
12. Bouabe H, Moser M, Heesemann J (2011) Enhanced selection for homologous-recombinant embryonic stem cell clones by Cre recombinase-mediated deletion of the positive selection marker. Transgenic Res. doi:10.1007/s11248-011-9522-x
13. Bouabe H, Liu Y, Moser M, Bosl MR, Heesemann J (2011) Novel highly sensitive IL-10-{beta}-lactamase reporter mouse reveals cells of the innate immune system as a substantial source of IL-10 in vivo. J Immunol 187:3165–3176
14. Talts JF, Brakebusch C, Fassler R (1999) Integrin gene targeting. Methods Mol Biol 129:153–187
15. Valenzuela DM, Murphy AJ, Frendewey D, Gale NW, Economides AN, Auerbach W, Poueymirou WT, Adams NC, Rojas J, Yasenchak J et al (2003) High-throughput engineering of the mouse genome coupled with high-resolution expression analysis. Nat Biotechnol 21:652–659
16. Copeland NG, Jenkins NA, Court DL (2001) Recombineering: a powerful new tool for mouse functional genomics. Nat Rev Genet 2:769–779
17. Malureanu LA (2011) Targeting vector construction through recombineering. Methods Mol Biol 693:181–203
18. Court DL, Sawitzke JA, Thomason LC (2002) Genetic engineering using homologous recombination. Annu Rev Genet 36:361–388
19. Hughes RA, Miklos AE, Ellington AD (2011) Gene synthesis: methods and applications. Methods Enzymol 498:277–309
20. Matzas M, Stahler PF, Kefer N, Siebelt N, Boisguerin V, Leonard JT, Keller A, Stahler CF, Haberle P, Gharizadeh B et al (2010) High-fidelity gene synthesis by retrieval of sequence-verified DNA identified using high-throughput pyrosequencing. Nat Biotechnol 28:1291–1294
21. Sato N, Meijer L, Skaltsounis L, Greengard P, Brivanlou AH (2004) Maintenance of pluripotency in human and mouse embryonic stem cells through activation of Wnt signaling by a pharmacological GSK-3-specific inhibitor. Nat Med 10:55–63
22. Ying QL, Wray J, Nichols J, Batlle-Morera L, Doble B, Woodgett J, Cohen P, Smith A (2008) The ground state of embryonic stem cell self-renewal. Nature 453:519–523
23. Domogatskaya A, Rodin S, Boutaud A, Tryggvason K (2008) Laminin-511 but not -332, -111, or -411 enables mouse embryonic stem cell self-renewal in vitro. Stem Cells 26:2800–2809

Chapter 25

Measurement of Mouse Cytomegalovirus-Induced Interferon-β with Immortalized Luciferase Reporter Cells

Evgenia Scheibe, Stefan Lienenklaus, Tobias May, Vladimir Gonçalves Magalhães, Siegfried Weiss, and Melanie M. Brinkmann

Abstract

The production of cytokines is a crucial element of the host response to viral and bacterial infections. To follow these events *in vivo*, transgenic mice have become a valuable tool to study cytokine production through induction of reporter genes. We describe here the generation and immortalization of cells derived from transgenic reporter mice for development of a high-throughput assay system for virus- or bacteria-induced cytokine induction. As an example we describe mouse cytomegalovirus (MCMV) infection of immortalized fibroblasts derived from mice expressing the firefly luciferase reporter downstream of the IFN-β promoter. Common methods to determine IFN-β production, including ELISA, quantitative real-time PCR (qPCR), and transient reporter assays using plasmid-based reporter constructs, have disadvantages and limitations. Transient transfections influence type I IFN responses in most cell types, and IFN-β ELISA as well as qPCR are both laborious and expensive. The method presented here is highly sensitive as well as cost-effective, and allows monitoring of a robust and dose-dependent induction of IFN-β upon virus infection in cell lysates as well as living cells.

Key words Type I interferon (IFN), IFN-β, Mouse cytomegalovirus (MCMV), Luciferase reporter assay, Ear fibroblast, Luminometer, Immortalization, Lentiviral transduction

1 Introduction

Reporter gene systems enable quantitative determination of the induction and expression of genes of interest. Technologies to detect these reporters have rapidly improved and assays to monitor luciferase activity or fluorescent protein expression are now standard methods in many laboratories. In accordance, recombinant mice equipped with reporter functions have become available in recent years which allow monitoring of gene induction *in vivo* or in *ex vivo* assays. Isolation and immortalization of cells obtained from such mice provide an easily expandable pool of reporter cells that can be used in *in vitro* systems to address a multitude of scientific questions.

Susanne M. Bailer and Diana Lieber (eds.), *Virus-Host Interactions: Methods and Protocols*, Methods in Molecular Biology, vol. 1064, DOI 10.1007/978-1-62703-601-6_25, © Springer Science+Business Media, LLC 2013

We previously generated reporter mice to study analysis of IFN-β promoter induction upon infection *in vivo* [1]. A myc-tagged firefly luciferase was integrated into the mouse genome by targeted mutagenesis to replace the coding sequence of the *ifnβ* gene, leaving the upstream region intact. This strategy ensures that reporter activity mimics the induction of IFN-β in an optimal fashion. The *ifnβ* wild type allele that is still present in cells of heterozygous reporter mice provides functional IFN-β and therefore a physiological response to viral infection.

We describe here the generation of conditionally immortalized cells derived from mice heterozygous for the luciferase transgene, and subsequent *in vitro* luciferase reporter assays. Primary ear fibroblasts were conditionally immortalized by lentiviral-mediated expression of the SV40 large T antigen (TAg) [2, 3] and used for downstream analysis of IFN-β induction upon MCMV infection. We compare luciferase activity measured in cell lysates to luciferase activity measured in living cells. Both methods are robust, reproducible, highly sensitive, and cost-effective, and therefore versatile alternatives to methods such as ELISA and qPCR.

2 Materials

2.1 Cell Culture General

Tissue culture-treated dishes (10 cm dishes, 6-well plates, 96-well plates, T25 flasks), 15 mL conical tubes, scalpels, cover slips (square 20×20 m), cryovials, 70 % ethanol, isopropanol, Dulbecco's Modified Eagle Medium (DMEM) with 4,500 mg/mL glucose, cell freezing container, tissue culture centrifuge.

2.2 Preparation of Lentiviruses in HEK 293T Cells

1. HEK (human embryonic kidney) 293T cells (ATCC CRL-11268).
2. DMEM 2+: DMEM, 10 % fetal bovine serum (FBS), 2 mM glutamine.
3. Phosphate buffered saline (PBS).
4. Trypsin/EDTA solution (trypsin).
5. Lentiviral helper plasmids e.g. ViraPower Lentiviral Expression System containing pLP1 (gag/pol), pLP2 (rev), pLP/VSVg (VSVg) (Invitrogen by LifeTechnologies K4975-00).
6. Lentiviral expression plasmid (pLV): uni-TAg or bi-TAg [3] (*see* **Note 1**).
7. For 293T transfection: 2.5 M $CaCl_2$ solution and HEBS solution: 280 mM NaCl, 50 mM HEPES, 1.5 mM Na_2HPO_4, adjusted to pH 7.1 with 1 N NaOH.
8. Luer Lock plastic syringe (10 mL), 0.45 μm syringe filter.

2.3 Lentiviral Transduction, Selection, and Expansion of IFN-β luciferase Reporter Ear Fibroblasts

1. DMEM 5+: DMEM, 10 % FBS, 2 mM glutamine, 0.1 mM non-essential amino acids, 0.1 mM β-mercaptoethanol, 100 U/mL penicillin; 100 μg/mL streptomycin.
2. G418 (100 mg/mL stock) dissolved in PBS, store at −20 °C.
3. Doxycycline hyclate (2 mg/mL stock) dissolved in 100 % ethanol, store at −20 °C.
4. Polybrene (Hexadimethrine bromide, 4 mg/mL stock) dissolved in H_2O, store at −20 °C.

2.4 Maintenance and Freezing of Cells

1. DMEM 5+ Dox: DMEM 5+, 2 μg/mL doxycycline hyclate (for both uni-TAg and bi-TAg), and 0.4 mg/mL G418 (only for bi-TAg).
2. PBS, trypsin.
3. Freezing solution: final 90 % FBS, 10 % dimethyl sulfoxide (DMSO).

2.5 Infection with Murine Cytomegalovirus (MCMV)

Purified MCMV (e.g. strain Smith, ATCC VR-1399), titered (*see* **Note 2**).

2.6 Luciferase Assay of Cell Lysates

1. White well flat bottom tissue culture-treated polystyrene 96-well plates (*see* **Note 3**).
2. Luciferase assay system (e.g. Promega E4530) containing cell lysis buffer, e.g. Reporter Lysis 5× Buffer, and reconstituted luciferase assay reagent.
3. 96-well plate luminometer (e.g. Promega GloMax 96 Microplate Luminometer with Dual Injectors, E6521).

2.7 Luciferase Assay of Living Cells

1. White well flat, transparent bottom tissue culture-treated polystyrene 96-well plates (*see* **Note 3**).
2. 200× D-luciferin Stock Solution: Dissolve 1 g D-luciferin potassium salt in 33.3 mL of PBS and sterile filter (0.2 μm). Aliquots can be stored at −20 °C.
3. Sensitive light detecting device compatible with 96-well plates (e.g. IVIS®, PerkinElmer).

3 Methods

Murine cells can be efficiently immortalized by ectopic expression of the viral oncogene SV40 large T antigen (TAg). TAg inhibits the activity of the tumor suppressors p53 and retinoblastoma protein [4]. Since constant activation of TAg leads to cellular alterations, we present a conditional immortalization method which is based on transcriptional regulation of TAg mediated by the tet-system [2]. TAg expression is activated by the addition of doxycycline, which leads to cell proliferation. Here, we use continuously proliferating immortalized cells that are grown in the presence of doxycycline. Withdrawal of doxycycline from the cell culture medium (~1 week) leads to complete arrest of cell proliferation which might be advantageous in some assay systems. Cells can be kept in culture without doxycycline for at least 2 weeks.

3.1 Isolation of Primary IFN-β Luciferase Reporter Ear Fibroblasts

1. Sacrifice mouse (e.g. IFN-β luciferase reporter mouse, Fig. 1).
2. Cut off ear, rinse in 70 % ethanol, and transfer to one 10 cm dish with 5 mL PBS (*see* **Note 4**).
3. Cut ear in 40–50 small pieces with a scalpel.
4. Transfer ten pieces into each well of a 6-well plate and cover with a cover slip to immobilize the ear pieces.
5. Add 2 mL of DMEM 5+ per well of a 6-well plate and allow outgrowth of the cells (*see* **Note 5**).
6. Renew medium every 4 to 5 days until >5 colonies (with more than 100 cells) have formed, and then carefully remove the coverslip with forceps.
7. Wash colonies twice with 2 mL PBS, detach with 0.5 mL trypsin for 3–5 min at 37 °C, add 2 mL DMEM 5+, and transfer to a new well of a 6-well plate.
8. When cells have reached confluence, transfer to a T25 flask as described in Subheading 3.1, **step 7** (with a final volume of 5 mL).
9. For long-term storage the primary cells can be frozen. One T25 flask gives one cryovial. Detach cells from a confluent T25 flask (wash with PBS, add 1 mL of trypsin for 3–5 min at 37 °C, add 5 mL DMEM 5+), transfer to 15 mL conical tube and centrifuge at 200 × *g* for 5 min. Remove medium and resuspend pellet in 0.5 mL FBS. Slowly and dropwise add 0.5 mL FBS with 20 % DMSO to generate a freezing solution with a final concentration of 90 % FBS and 10 % DMSO. Mix cell suspension gently after addition of each drop. Transfer to cryovials and put vials in cell freezing container. Put container in −70 °C freezer overnight and transfer vials to liquid nitrogen storage the next day.

3.2 Production of SV40 Large T Antigen Expressing Lentivirus in 293T Cells

The TAg expression cassette is incorporated into a third generation lentiviral vector system which has an improved safety profile due to deletion of most of the viral sequences. Lentiviral transduction allows high immortalization efficiency, as well as transduction and immortalization of different cell types (*see* **Note 6**).

1. Plate HEK293T at a density of 40,000 cells/cm^2 (day 0) in DMEM 2+.

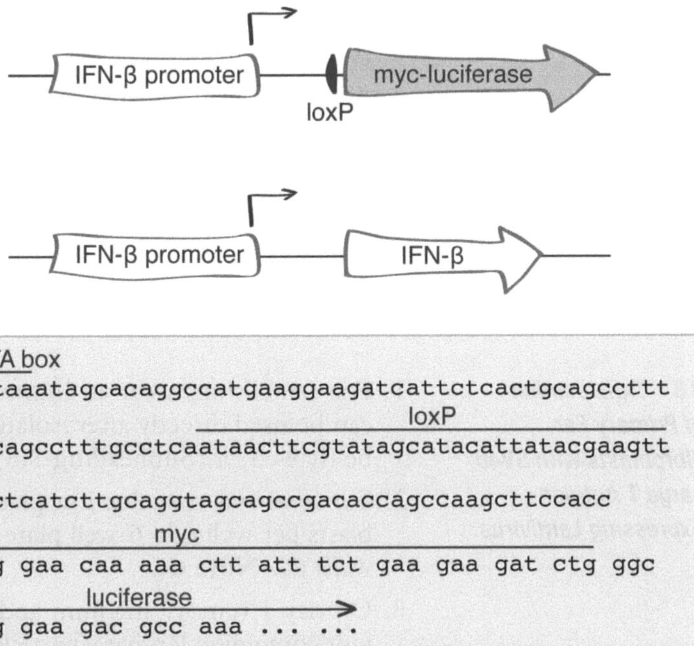

Fig. 1 Scheme of the targeted gene locus of the IFN-β reporter mice. After homologous recombination with the targeting construct, the *ifnβ* locus contains a myc-tagged luciferase gene allowing visualization of virus-induced IFN-β expression [1]. To maintain IFN-β production, one wildtype (wt) allele is preserved in the mice/cells. In the reporter allele the 5′ untranslated region of the IFN-β message is interrupted by a loxP site, which remains from the targeting strategy. Except for this alteration, the sequence preceding the reporter coding sequence is identical to the sequence of the wt locus to avoid interference with regulatory elements driving IFN-β production. The luciferase sequence is derived from pGL3 (Promega), a vector commonly used in reporter gene assays. To facilitate the detection of IFN-β induction in other assays, oligonucleotides coding for a myc tag were inserted in front of the luciferase coding sequence. The myc-luciferase sequence is followed by an SV40 poly(A) site originating from pGL3. The sequence displayed in the lower panel of the figure gives the reporter cassette in the *ifnβ* locus starting from the TATA-box, including the loxP site and a part of the coding sequence for the N-terminally myc-tagged luciferase. The genetic background of the embryonic stem cells as well as the mice used to obtain the reporter cells is C57BL/6

2. On day 1 transfect 293T cells with the three helper plasmids (pLP1, pLP2, pLP/VSVg) and either UniTAg or biTAg (pLV) by calcium phosphate transfection. The following amounts and/or volumes are given for lentiviral production in a 10 cm dish (adjust the amounts accordingly if using other cell culture plates): combine the plasmids (pLV 8 μg, pLP1 5.6 μg, pLP2 1.8 μg, pLP/VSVg 3 μg) with 15 μL 2.5 M $CaCl_2$ and use H_2O to adjust to a final volume of 150 μL. Add the DNA solution dropwise to 150 μL HEBS under vigorous vortexing and incubate at room temperature for 10 min before addition to the cells.

3. On day 2 aspirate medium and add 6 mL of DMEM 2+ (0.1 mL/cm^2).
4. Allow lentivirus production to proceed until day 3.
5. On day 3 collect supernatant containing lentiviral particles. Add fresh medium to cells.
6. Filter harvested supernatant through 0.45 μm pore size filter and store in aliquots at −70 °C until use.
7. On day 4 collect supernatant containing lentiviral particles. Filter and store as described in Subheading 3.2, **step 6**.

3.3 Transduction of Primary Ear Fibroblasts with SV40 Large T Antigen Expressing Lentivirus

1. For establishment of an immortalized cell line, primary cells can be used directly after isolation (*see* Subheading 3.1) or can be thawed (*see* Subheading 3.5).
2. For lentiviral transduction plate 1×10^5 of primary ear fibroblasts per well of a 6-well plate (day 0) in DMEM 5+ (2 mL/well) (*see* **Note 6**).
3. On day 1 remove medium and add 1 mL undiluted supernatant containing lentiviral particles (*see* Subheading 3.2) expressing either uni-TAg or bi-TAg in the presence of 8 μg/mL polybrene and incubate overnight.
4. On day 2 aspirate lentivirus-containing medium and add 2 mL DMEM 5+ Dox.
5. On day 4 or after the cells reach confluence, passage transduced cells with a ratio of 1:3 (wash cells twice with 2 mL PBS, trypsinize with 0.5 mL trypsin for 3–5 min at 37 °C, resuspend cells in DMEM 5+ Dox).
6. Plate cells in desired cell culture plate in DMEM 5+ Dox and either select with 0.4 mg/mL G418 (bi-TAg) or by growth advantage (uni-TAg) (*see* **Note 7**).

3.4 Cell Maintenance

1. For routine maintenance cells are cultivated with DMEM 5+ Dox (*see* **Note 8**).
2. Passage cells when confluent as follows: wash cells once with 0.2 mL/cm^2 PBS and detach with 0.05 mL/cm^2 trypsin. Gently tilt the plate to distribute trypsin equally and incubate at 37 °C, 7.5 % CO_2 for 3–5 min. Add 0.2 mL/cm^2 DMEM 5+ Dox and detach cells with vigorous pipetting or a cell scraper. Pipette up and down to minimize clumps.
3. In the first 4 weeks after infection the cells should be passaged with a ratio of 1:3. At later time points the ratio can be adjusted to 1:5 or 1:10.
4. Cells can be frozen as described in Subheading 3.1.

3.5 Thawing Ear Fibroblasts

1. Remove vial of frozen cells from liquid nitrogen and transfer to a 37 °C water bath.
2. After thawing, transfer the solution into a 15 mL conical tube and add 9 mL pre-warmed medium dropwise (DMEM 2+ for primary fibroblasts and DMEM 5+ Dox for immortalized fibroblasts).
3. Mix gently and centrifuge at 200 × *g* for 5 min at room temperature.
4. Aspirate supernatant, resuspend the pellet in 10 mL of pre-warmed DMEM 2+ or DMEM 5+ Dox, and transfer to a 10 cm tissue culture dish.
5. Check cells the next day and change medium if necessary.
6. When cells are confluent split 1:3 or 1:5. Cells will grow more quickly after complete recovery and can be split 1:10.

3.6 Infection of Immortalized Ear Fibroblasts with MCMV

The luciferase assay is designed for a 96-well plate format to allow high-throughput analysis. The experimental workflow is illustrated in Fig. 2.

1. One day prior to infection, seed 6.25×10^3 cells/well in 75 µL of DMEM 5+ Dox in a 96-well tissue-culture-treated plate. For subsequent detection of luciferase in living cells, use white well transparent bottom 96-well plates (*see* **Note 3**). Cells should be ~80 % confluent the next day for infection.
2. Dilute MCMV in DMEM 5+ Dox to infect with the desired multiplicity of infection (MOI) (*see* **Note 9**). Prepare enough virus suspension to be able to add 75 µL per well.
3. Add 75 µL virus suspension directly to the 75 µL medium on the cells.
4. Spin-infect for 1 h at 800 × *g* at 4 °C (*see* **Note 10**).
5. Carefully remove virus-containing supernatant without touching the cell layer. Add 150 µL of pre-warmed DMEM 5+ Dox.
6. Proceed directly to Subheadings 3.7 or 3.8 (for a 0 h time point measurement) or incubate at 37 °C and 7.5 % CO_2 for the desired time, before proceeding to Subheadings 3.7 or 3.8 (*see* **Note 11**).

3.7 Luciferase Assay with Cell Lysates

1. To make 1× Reporter Lysis Buffer (RLB) add 4 volumes of deionized water to 1 volume of 5× RLB. Equilibrate to RT before use. 50 µL/well of 1× buffer are needed.
2. Carefully remove supernatant and wash cells once with PBS. Then add 50 µL of 1× RLB to each well and incubate at RT for 10 min (*see* **Note 12**). Perform a freeze-thaw cycle to ensure complete lysis (*see* **Note 13**).

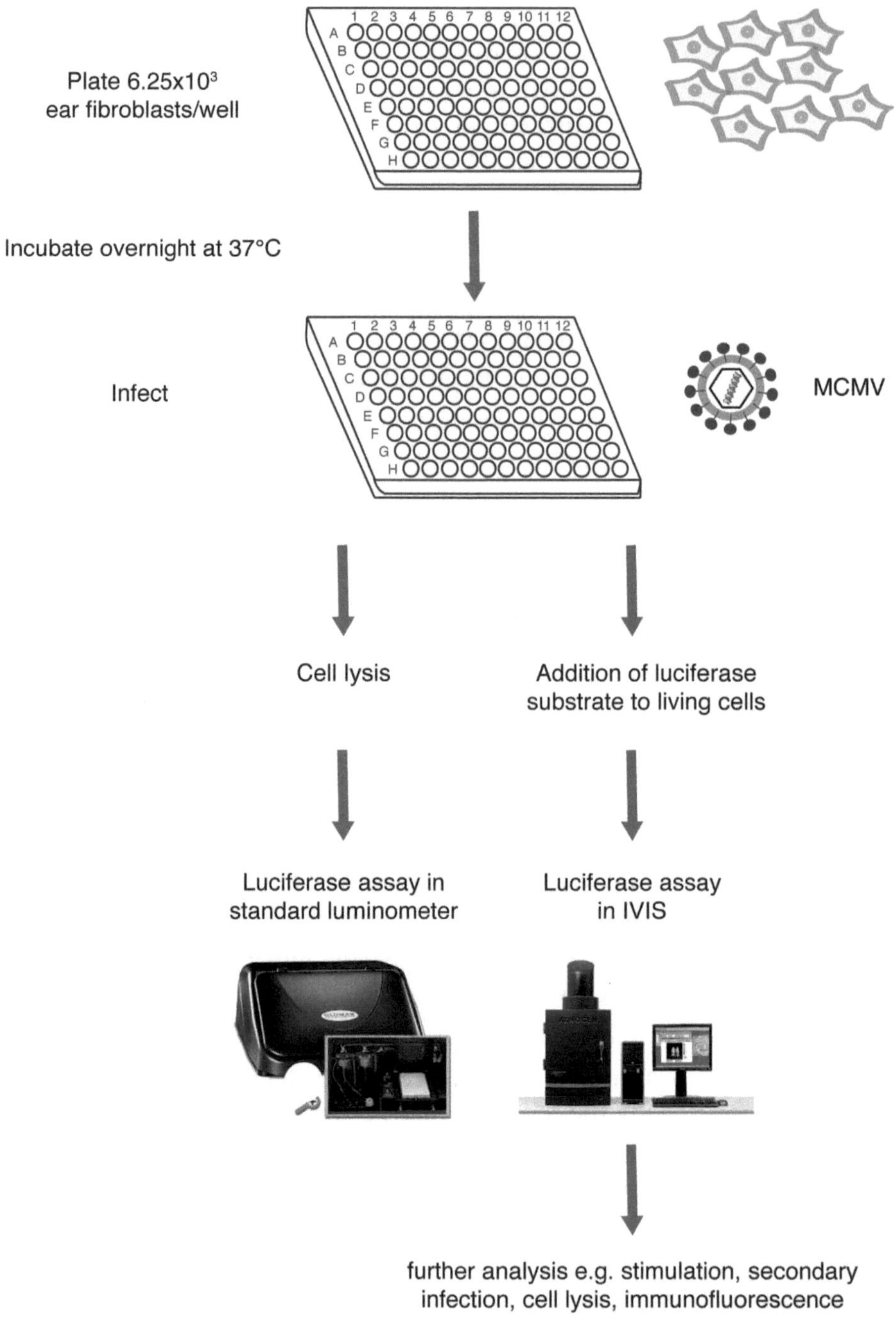

Fig. 2 Luciferase assay experimental setup. 6.25×10^3 cells/well of immortalized ear fibroblasts derived from IFN-β reporter mice are plated the day before infection in DMEM 5+ Dox in a 96-well plate. Cells are incubated overnight at 37 °C and then infected with MCMV at different MOI or left uninfected. For the luciferase assay of cell lysates, cells are lysed in reporter lysis buffer at different time points post-infection and luciferase activity is analyzed in a standard luminometer. To assess luciferase activity in living cells luciferase substrate is directly added to cells without prior cell lysis, and luciferase activity is measured in the IVIS® system

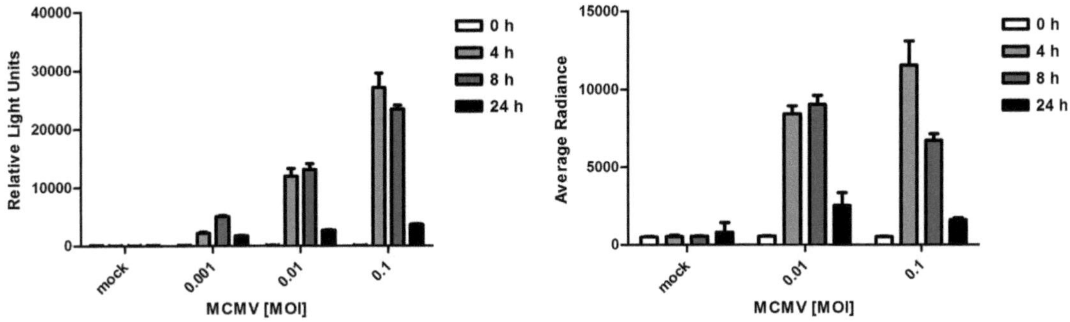

Fig. 3 Kinetics of IFN-β induction in immortalized ear fibroblasts upon MCMV infection. Immortalized ear fibroblasts derived from IFN-β luciferase reporter mice were infected with MCMV at different MOI as indicated in the figures or left uninfected and incubated for 0, 4, 8, and 24 h. *Left panel*: Luciferase activity in cell lysates. After infection, cells were lysed in reporter lysis buffer and luciferase activity was determined in a standard luminometer using a commercial luciferase assay system. The graph shows one representative experiment with samples analyzed in triplicate. *Right panel*: Luciferase activity of living cells. In a parallel experiment, instead of lysing the cells, luciferin was added directly to living cells and light emission was assessed in an IVIS® system. Both readout systems are very well suited to follow MCMV-induced IFN-β expression and show high sensitivity and reproducibility

3. Thaw Luciferase Assay Reagent and equilibrate to room temperature (you will need 100 μL of reagent/sample) (*see* **Note 14**).
4. Transfer 20 μL of each cell lysate to individual wells of a white 96-well plate. Analyze light production according to the manufacturer's instructions. For the Promega Glomax 96 with Dual Injectors, we have utilized automatic injection of 100 μL reconstituted luciferase assay reagent followed by a 2 s measurement delay and a 10 s measurement read (Fig. 3, left panel). The values given are relative light units (RLU).

3.8 Luciferase Assay in Living Cells

This assay variation does not require cell lysis and therefore allows further downstream analysis such as immunofluorescence or secondary stimulations. In contrast to cell lysates, living cells provide necessary cofactors for the luciferase reaction, allowing the use of a cost-effective substrate solution rather than a complex and expensive assay reagent as required for Subheading 3.7.

1. Using cell culture medium, prepare a 4× luciferin working dilution (600 μg/mL) from the 200× (30 mg/mL) D-luciferin stock (*see* Subheading 2.7). Equilibrate to 37 °C.
2. Add 50 μL of the 4× working dilution to each well containing cells (with 150 μL of cell culture medium) to yield a final concentration of 150 μg/mL luciferin.
3. Incubate cells for 5 min at 37 °C.

4. Read the light emission of the plate using high sensitivity settings on the reader (*see* **Note 15**). For the IVIS®, we used the following settings: 4 min integration, binning 16, f/stop 1 (Fig. 3, right panel). Using the unit average radiance (p/s/cm^2/sr) allows comparison of the values to readings that are acquired with other settings.

4 Notes

1. Plasmids uni-TAg and bi-TAg are available upon request from Tobias May. The uni-TAg plasmid harbors a unidirectional Tet-dependent promoter that drives the expression of TAg and a reverse transactivator (tet-on). The bi-TAg plasmid is based on the bidirectional Tet-dependent promoter driving two mRNAs, one encoding for TAg and the other bicistronic mRNA for the reverse transactivator and a fusion gene comprised of eGFP and Neomycin phosphotransferase. The plasmids can be propagated using standard bacterial strains/techniques.
2. MCMV was purified as described in [5] with slight modifications: (a) virus was harvested from cell culture supernatant and not from intact cells and (b) virus was purified through a 10 % Nycodenz cushion instead of a sucrose cushion. Work with MCMV requires biosafety level 1.
3. For detection of luminescent signals white plates are advantageous due to their reflective properties. In addition, scattering of light emitted from one well to neighboring wells is reduced. The use of white well plates with transparent bottoms allows microscopic monitoring of cell growth and viability throughout the experiment. A white piece of paper beneath the plate during measurement increases the signal by reflecting the emitted light to the detector, almost as good as a white bottom.
4. Virus and cell preparation, infections, and stimulations are performed under sterile conditions in a laminar flow hood (*see* **Note 6**). Cells are incubated at 37 °C in a humidified atmosphere with 7.5 % CO_2.
5. To avoid or reduce the risk of bacterial contamination we recommend adding gentamycin to the primary cells for the first 2 weeks in culture (a final concentration of 500 µg/mL is obtained from a 1:200 dilution of a 10 mg/mL stock solution).
6. For the production and use of recombinant lentiviruses (especially for lentiviruses encoding immortalization genes)

biosafety level 2 is required. Lentiviral safety guidelines for the applicable country must be followed.

7. Selection of bi-TAg transduced fibroblasts with G418 usually requires 7–10 days. Selection of uni-TAg transduced fibroblasts by growth advantage requires several subcultivation steps and typically 3–4 weeks to lose untransduced primary cells. Alternatively, cells can be selected by FACS sorting of GFP-positive cells (bi-TAg).
8. The established immortalized fibroblast cell line is cultivated with doxycycline (2 μg/mL) to activate the immortalizing SV40 large T antigen. This activation induces proliferation of the cell line. Without doxycycline the SV40 large T antigen becomes inactivated which leads to growth arrest. The cells stay alive in this state for at least 14 days.
9. The day after seeding cells have approximately doubled and cell numbers are about 1.25×10^4 cells/well. To confirm the cell number at the time of infection, trypsinize one well and count cells. To make a virus suspension of an MOI of 0.1 from a given virus stock with a titer of 10^7 PFU/mL, dilute 0.125 μL of the virus stock in 75 μL medium (MOI × cell number divided by PFU/mL). Prepare enough virus suspension to be able to add 75 μL per well. If less volume is added to cells, they will dry out during the following centrifugation step.
10. Viral attachment to cells is highly enhanced by centrifugation. Centrifugation is performed at 4 °C to avoid virus entry into the cells. Entry will be synchronized upon warming.
11. Prepare separate plates for each time point. This allows freezing of the entire plate to complete cell lysis.
12. Although only 20 μL are needed for analysis, 50 μL of lysis buffer are added to allow an additional luciferase assay or further downstream analysis such as immunoblotting.
13. Reporter lysis buffer requires a single freeze-thaw cycle to fully lyse cells. Cells in 1× RLB are frozen once at −20 °C and thawed either at RT or 37 °C to ensure complete cell lysis.
14. Leftover luciferase assay reagent (LAR) can be frozen at −80 °C and mixed with freshly prepared LAR for subsequent measurements. Repeated freeze-thaw cycles will eventually reduce efficacy.
15. White plates show phosphorescence below 560 nm. To avoid interference of this signal with the luciferase signal, a 620 nm emission filter can be selected on the IVIS® system. Sensitivity can be enhanced by putting a reflective plate (e.g. white paper) underneath the 96-well plate while reading in the IVIS®.

Acknowledgements

This work was supported by the Helmholtz Association through grants VH-NG-637 and the Virtual Institute VISTRIE VH-VI-424 issued to M.M.B. We thank Kendra Ann Bussey for critical reading of the manuscript.

References

1. Lienenklaus S, Cornitescu M, Zięetara N et al (2009) Novel reporter mouse reveals constitutive and inflammatory expression of IFN-β in vivo. The Journal of Immunology 183(5):3229–3236
2. May T, Hauser H, Wirth D (2004) Transcriptional control of SV40 T-antigen expression allows a complete reversion of immortalization. Nucleic Acids Research 32(18):5529–5538
3. May T, Butueva M, Bantner S et al (2010) Synthetic gene regulation circuits for control of cell expansion. Tissue Engineering Part A 16(2):441–452
4. Ahuja D, Saenz-Robles MT, Pipas JM (2005) SV40 large T antigen targets multiple cellular pathways to elicit cellular transformation. Oncogene 24(52):7729–7745
5. Brune W, Hengel H, Koszinowski UH (2001) A mouse model for cytomegalovirus infection. Current Protocols in Immunology 43:19.7.1–19.7.13

ERRATA

Chapter 20

Functional NK Cell Cytotoxicity Assays Against Virus Infected Cells

Rebecca J. Aicheler and Richard J. Stanton

Susanne M. Bailer and Diana Lieber (eds.) *Virus Host Interactions: Methods and Protocols*, Methods in Molecular Biology, vol. 1064, DOI 10.1007/978-1-62703-601-6, pp. 275–288, © Springer Science+Business Media, LLC 2013

DOI 10.1007/978-1-62703-601-6_26

The publisher regrets that in the print and online versions of this book, corrections to chapter 20 was omitted at proof stage. Please see the next page for the corrections.

Page 275:

- In abstract, fourth line from bottom should read: "…one method measures ^{51}Cr..".
- In Keywords, "Sodium dichromate" should read "Sodium chromate".

Page 276:

- Paragraph 2, Line 5 should read: "…the target cells are loaded with ^{51}Cr and..".
- Paragraph 2, Lines 6–7 should read: "..As cells die, ^{51}Cr is released…".
- Paragraph 3, Line 5 should read: "… This releases all the ^{51}Cr …"

Page 277:

- Line 1 should read "…very variable amounts of ^{51}Cr…".
- Lines 2–3 should read: "…how much ^{51}Cr is released..".
- Lines 4–5 should read: "…By comparing the amount of ^{51}Cr taken up by the cells (maximal)…".

The online version of the original chapter can be found at http://dx.doi.org/10.1007/978-1-62703-601-6_20

- Table 1 - NK cells:Target should be a colon missing ^{51}Cr …
- This should be a colon (effector:target)

Page 278:

- Section 2.1, Item 1: "Sodium dichromate" should read "Sodium chromate".
- Section 2.1: Item 14 was omitted, and should read: "Liquid scintillation counter".

Page 280:

- Section 3.4, Item 5: "Sodium dichromate" should read "Sodium chromate".
- Section 3.4, Item 8 should read: "…This allows any unbound ^{51}Cr to"
- Section 3.4, Item 10, Line 1 should read: "While the targets are being loaded with ^{51}Cr,"

Page 281:

- Section 3.4, Item 27, the correct equation is:

$$\frac{(\textit{Test CPM})-(\textit{spontaneous CPM})}{(\textit{max CPM})-(\textit{spontaneous CPM})} \times 100$$

Page 282:

- Table 3, Column 1,: Line 1 should read "Max (Mean)"
- Table 3, Column 2,: Line 2 should read "Spontaneous (Mean)".
- Section 3.5, Item 5 Line 1, should read "Resuspend the cells at a concentration of 1×10^6 cells…".
- Section 3.5, Item 8, final line should read: "…the desired concentration—usually 1×10^5 cells/100 μL".

Page 283:

- Section 3.5, Item 9, Line 3 should read: "(470 × *g*, 3 min) and resuspending…".
- Section 3.5, Item 17, Line 1 should read: "Pellet cells by centrifuging (470 × *g*, 3 min)…"
- Section 3.5, Item 20, Line 1 should read: "To differentiate the target cells….".
- Lines 3-4 should read: "loaded with CFSE so will be CFSE$^+$, draw a gate around the CFSE$^+$ cells..".
- Section 3.5, Item 21, Line 1 should read: "Draw a dot-plot of the CFSE$^+$ cells…".
- Section 3.5, Item 21, Line 3, cells will be 7-AAD^{lo}
- Section 3.5, Item 21, Line 5 be 7-AAD^{hi}

Page 284:

- Figure 2, part A: "GFP" should read "CFSE".
- Section 4, Item 2, should read "more ^{51}Cr…"

Page 285:

- Section 4, Item 8, line 2: should read "^{51}Cr".
- Section 4, Item 10, first line should read "^{51}Cr has a half life…". Line 5 should read "… Sodium chromate ^{51}Cr…". Line 7 should read "…with the Sodium chromate".

INDEX

Susanne M. Bailer and Diana Lieber (eds.), *Virus-Host Interactions: Methods and Protocols*, Methods in Molecular Biology, vol. 1064, DOI 10.1007/978-1-62703-601-6, © Springer Science+Business Media, LLC 2013

W

X

Y

Z

MIX
Papier aus verantwortungsvollen Quellen
Paper from responsible sources
FSC® C105338

If you have any concerns about our products,
you can contact us on
ProductSafety@springernature.com

In case Publisher is established outside the EU,
the EU authorized representative is:
Springer Nature Customer Service Center GmbH
Europaplatz 3, 69115 Heidelberg, Germany

Printed by Libri Plureos GmbH
in Hamburg, Germany

MIX
Papier aus verantwortungsvollen Quellen
Paper from responsible sources
FSC® C105338

If you have any concerns about our products,
you can contact us on
ProductSafety@springernature.com

In case Publisher is established outside the EU,
the EU authorized representative is:
Springer Nature Customer Service Center GmbH
Europaplatz 3, 69115 Heidelberg, Germany

Printed by Libri Plureos GmbH
in Hamburg, Germany